Polyvinyl Chloride-Based Biocomposites and Bionanocomposites

Scrivener Publishing
100 Cummings Center, Suite 541J
Beverly, MA 01915-6106

Publishers at Scrivener
Martin Scrivener (martin@scrivenerpublishing.com)
Phillip Carmical (pcarmical@scrivenerpublishing.com)

Polyvinyl Chloride-Based Biocomposites and Bionanocomposites

Edited by

Shiv Kumari Panda

Department of Chemistry, U.N. Autonomous College of Science & Technology, Adaspur, Cuttack, Odisha, India

and

Visakh P. M

Natural Bioactive Materials Laboratory, Department of Bioengineering, Ege University, Bornova/Izmir, Turkey

Scrivener Publishing

WILEY

This edition first published 2026 by John Wiley & Sons, Inc., 111 River Street, Hoboken, NJ 07030, USA and Scrivener Publishing LLC, 100 Cummings Center, Suite 541J, Beverly, MA 01915, USA

For more information about Scrivener publications please visit www.scrivenerpublishing.com.

Wiley Global Headquarters
111 River Street, Hoboken, NJ 07030, USA

For details of our global editorial offices, customer services, and more information about Wiley products visit us at www.wiley.com.

The manufacturer's authorized representative according to the EU General Product Safety Regulation is Wiley-VCH GmbH, Boschstr. 12, 69469 Weinheim, Germany, e-mail: Product_Safety@wiley.com.

Library of Congress Cataloging-in-Publication Data

ISBN 978-1-394-21495-2

Cover image: Generated with AI using Adobe Firefly
Cover design by Russell Richardson

Set in size of 11pt and Minion Pro by Manila Typesetting Company, Makati, Philippines

Printed in the USA

10 9 8 7 6 5 4 3 2 1

Contents

Preface

This book provides a comprehensive summary of recent research accomplishments in the field of polyvinyl chloride (PVC)-based biocomposites and bionanocomposites. It brings together key developments, innovations, and applications from around the world, reflecting the remarkable growth of this area of materials science.

The book covers a wide range of topics, including:

- PVC/starch-based biocomposites and bionanocomposites
- PVC/polylactic acid-based biocomposites and bionanocomposites
- Flame retardancy in PVC-based biocomposites and bionanocomposites
- Interfacial modification and surface engineering of PVC-based systems
- Sustainable PVC biocomposites and bionanocomposites, charting a path toward green innovation
- PVC/cellulose-based biocomposites and bionanocomposites
- Biodegradation of PVC-based natural fiber bionanocomposites
- Biomedical applications of PVC-based biocomposites and bionanocomposites

This volume serves as a valuable reference for university and college faculty, industry professionals, postdoctoral researchers, senior graduate students, and scientists working in the PVC-based materials field. Each chapter has been contributed by distinguished experts from academia, industry, and government or private research laboratories across the globe. Collectively, these contributions provide a current and detailed record of findings, challenges, and emerging directions in the field.

Chapter one introduces the fundamentals, scope, and state-of-the-art developments in PVC-based biocomposites and bionanocomposites. It also reviews preparation methods, emerging challenges, and future opportunities. The second chapter discusses starch-based materials, emphasizing the synthesis, structure, and properties of PVC/starch biocomposites and bionanocomposites.

Chapter three provides an overview of lactic acid and polylactic acid chemistry, synthesis and processing routes for PVC/PLA composites, and detailed characterization of their properties. The fourth chapter focuses on thermal stability and combustion properties, highlighting key strategies to enhance flame resistance.

Chapter five explores interfacial modification of PVC-based biocomposites and bionanocomposites. It discusses sustainable development of PVC-reinforced materials, enhancement through chemical and physical modifications of biomaterials, and the design of high-performance bionanocomposites. The sixth chapter examines environmentally responsible design approaches, including natural fiber–reinforced systems and hybrid PVC-based materials.

Chapter seven presents a detailed discussion on nanocellulose—its structure, properties, and production—as well as the fabrication, characterization, and applications of PVC–nanocellulose composites. The eighth chapter covers PVC-based natural fiber–reinforced systems, biodegradation mechanisms, and microbial and enzymatic degradation pathways.

Chapter nine highlights the definitions, compositions, and biomedical potential of these materials for use in health-related industries. The tenth chapter explores hybrid structures, their synthesis, and the combination of organic and inorganic phases to achieve multifunctional performance. The book concludes with a chapter on processing techniques and characterization of nano-, micro-, and macro-scale PVC/rubber blends.

We express our deepest gratitude to all the contributors for their dedication, expertise, and collaboration, which made this book possible. We also extend our appreciation to the reviewers for their careful evaluations and constructive feedback on each chapter. Finally, we would like to thank all the reviewers who have taken their valuable time to make critical comments on each chapter. We also thank Martin Scrivener and the team at Scrivener Publishing for recognizing the demand for such a book, realizing the increasing importance of the field, and starting a project which not many other publishers would approach.

Dr. Shiv Kumari Panda
Dr. Visakh P. M

1

Polyvinyl Chloride (PVC)-Based Biocomposites and Bionanocomposites: State-of-the-Art, New Challenges, and Opportunities

Shiv Kumari Panda[1]*, Visakh P. M[2] and Nadir Ayrilmis[3]

[1]*Department of Chemistry, U.N. Autonomous College of Science & Technology, Adaspur, Cuttack, Odisha, India*
[2]*Natural Bioactive Materials Laboratory, Department of Bioengineering, Ege University, Bornova/Izmir, Turkey*
[3]*Department of Wood Mechanics and Technology, Faculty of Forestry, Istanbul University – Cerrahpasa Bahcekoy, Sariyer, Istanbul, Turkey*

Abstract
This chapter provides a short version of all the other chapters. Here we write about the different chapter topics such as polyvinyl chloride/starch-based biocomposites and bionanocomposites; polyvinyl chloride/polylactic acid-based biocomposites and bionanocomposites; flame retardancy of polyvinyl chloride (PVC)-based biocomposites and bionanocomposites; interfacial modification of polyvinyl chloride (PVC)-based biocomposites and bionanocomposites; sustainable PVC biocomposites and bionanocomposites: a path towards green innovation; polyvinyl chloride (PVC)/cellulose-based biocomposites and bionanocomposites; biodegradation of polyvinyl chloride (PVC)-based different fibers bionanocomposites; polyvinyl chloride (PVC)-based biocomposites and bionanocomposites for biomedical applications; polyvinyl chloride (PVC)-based hybrid biocomposites and bionanocomposites; and preparation and PVC/rubber bio nano blends.

***Keywords*:** Polyvinyl chloride, biocomposites, starch, rubber, nano blends, biomedical and bionanocomposites

**Corresponding author*: shivkumaripanda@gmail.com

Shiv Kumari Panda and Visakh P. M (eds.) Polyvinyl Chloride-Based Biocomposites and Bionanocomposites, (1–28) © 2026 Scrivener Publishing LLC

1.1 Polyvinyl Chloride (PVC)/Starch-Based Biocomposites and Bionanocomposites

Starch is one of the most abundant naturally occurring polysaccharides that can be obtained from natural sources like corn, potato, and cassava. It is a renewable, nontoxic, economical, and biodegradable polymer that is second to cellulose in occurrence. It is extensively available in the seeds, roots, and leaves of varieties of plants, and some algae. Starch-based biocomposites and bionanocomposites have been studied widely for their easy availability, inexpensiveness, ease of processibility, and various chemical and physical properties [1, 2]. Starch is a primary energy source. Plants like potato, corn, cassava, wheat, millet, rice, corn, barley, sago, peas, mung beans, arrowroot, banana, yam, sorghum, lentils, and many others are utilized to obtain starch [3–5], which is stored in various parts, including the roots, seeds, tubers, and stems [6]. The nanostarches are of two different types: starch nanocrystals (SNCs) and starch nanoparticles (SNPs). SNCs are crystal-like portions that can result from the disruption of amorphous areas in starch granules, whereas SNPs are produced from amorphous regions of gelatinized starch [7, 8].

Polyvinyl chloride (PVC) is a widely used synthetic thermoplastic polymeric material known for its affordability with various types of characteristics like outstanding insulating capabilities, ease of use, fire resistance, and resistance to chemicals [9, 10]. Also, it has a great utilization in construction [120], electronics [11], automotive [12, 13], packaging [14, 15], and healthcare [16]. But the problem is that it remains chemically and physically inert for many years even after proper disposal due to the evil effect of chlorine, which is the main culprit for environmental pollution. So biodegradable fillers like starch were introduced to produce biocomposites from petroleum-based polymer resins like PVC. According to earlier studies, there are various techniques used for the synthesis of biocomposites through coprecipitation of cross-linked starch xanthate with a PVC latex and dry mixing of starch and PVC and co-concentration of starch and PVC latex [17]. If we consider the characterization of the biocomposite here, then some mechanical studies presented the reduction in tensile strength and elongation at break due to the introduction of starch in PVC [18]. Incorporating nanostarch with PVC can improve the overall biodegradability of the resulting bionanocomposite. In addition to that, samples with higher starch content showed negligible weight loss after prolonged exposure to simulated soil that is due to the encapsulation of starch by the PVC polymer matrix that can hinder the microbial degradation.

PVC is the plastic phase, and starch is the rubbery phase, and due to the combination, the amorphous parts of starch undergo a transition from glassy to rubbery state. Again, thermogravimetric analysis (TGA) indicated that overall maximum degradation value of PVC gets positively influenced by the incorporation of starch. In addition to it, another important conclusion has been made from a study that more concentration of starch can enhance the degree of dechlorination.

1.2 PVC/PLA Composites/Biocomposites

PLA is melt-blended with other elastomeric polymers such as poly(e-caprolactone), poly(butylene succinate), poly(hydroxyl butyrate), and poly butylene adipate-co-terephthalate. PLA is also melt blended with non-biodegradable polymers including polypropylene (PP), polyacrylonitrile (ABS), polystyrene (PS), polyethylene (PET), polymethyl methacrylate (PMMA), polyvinyl alcohol (PVA), polyethylene (PET), polyvinyl chloride (PVC), etc. [19]. PVC goods are widely used because of their low cost and high performance, making it a desirable material for practical applications [20]. This has contributed to the widespread usage of PVC but also has resulted in concern for its non-biodegradable nature. However, PVC/PLA blends can be tailored to have desirable features of both the components in pursuit of sustainable practices [21]. The polycondensation of the two enantiomers of lactic acid leads to the formation of three different forms of PLA: poly(L-lactic acid) (PLLA), poly(D,L-lactic acid) (PDLLA), and poly(D-lactic acid) (PDLA). The crystalline structure of PLLA-α form has been confirmed by employing neutron diffraction and two-dimensional wide-angle X-ray techniques [22]. PLLA exhibits a comparatively elevated melting temperature (T_m). PDLA is classified as a semi-crystalline polymer, whereas PDLLA is categorized as an amorphous polymer. PDLLA is a racemic polymer with an unsystematic sequence of the enantiomers. Several factors, such as the polymer's molecular weight, temperature, processing time, and duration of annealing treatment, can impact the degree of crystallinity [23]. PLA is a polymer derived from sustainable sources and is completely capable of undergoing biodegradation, making it environmentally friendly. Typically, PLA is sourced from plant-based materials like corn, wheat, or rice. PLA requires much less energy to produce compared to petroleum-based polymers. Estimations indicate that in the future, this energy need can be further lowered to less than 10%. Reducing energy consumption also has the potential to make PLA production more cost-effective [24]. Polyvinyl chloride, also known as PVC, is a linear thermoplastic polymer. Typically, PVC has a low number of branches per molecule, usually less than 10. It is unlikely for there to be multiple long chains within a single molecule. When a chlorine atom is present, it enhances the attraction between molecules and contributes to the hardness and stiffness of the polymer [25]. Polyvinyl chloride (PVC) is considered one of the most efficient petroleum-based thermoplastic polymers. It holds a significant share among the six commonly used commercial polymers and is projected to experience a steady growth rate in the coming years. The global production capacity of PVC has reached millions of tons, which is used in a wide range of applications. Sadly, they often wind up in trash heaps after being used as lightweight, single-use products like cooking oil and fruit juice packaging, and they can create major environmental damage when not disposed of appropriately [26].

PVC/PLA blend is a highly promising biodegradable polymer blend that has garnered significant attention from researchers in recent times. Literature reports the development and characterization of novel biomaterials (bioblends) made from poly(vinyl chloride) (PVC) and poly(lactic acid) (PLA) [27]. Hachemi, Belhaneche-Bensemra, and Massardier used melt mixing to obtain PVC/PLA bioblends with and without the addition of MAH as a compatibilizer. Dicumyl peroxide (DCP) was used as an initiator during *in situ* compatibilization with MAH. The incorporation of MAH improved the processability of the bioblends by reducing the mixing temperature [28]. Polymethyl methacrylate (PMMA) was utilized by Singh *et al.* as a compatibilizer to enhance the interfacial activity of PVC/PLA blends to further obtain a biodegradable film. A condensation method was used to synthesize PLA while solution casting technique was used to develop PVC/PLA blend films with and without the addition of PMMA as a compatibilizer [29]. A torque rheometer is generally utilized to melt-mix of PVC/ PLA in addition to employing various analytical techniques to study the phase morphology and compatibility of the blends. Generally, the compatibility of the two polymers is determined by studying the steady-state and dynamic rheological responses [30]. Systematic evaluation of the PVC/PLA blends' static and dynamic rheological properties offers valuable insights into their compatibility. The rheological analyses reveal important information regarding the viscoelastic behavior, melt viscosity, and other rheological properties of the polymer blends [31]. Santana *et al.* examined the biodegradation of PVC/PLA blend under soil burial (up to 180 days) through the use of a range of analytical techniques, including FTIR, TGA, and SEM, to closely monitor the physical and chemical changes occurring in the PLA/PVC blend during the biodegradation process. The degree of biodegradation was evaluated through measurements of mass loss, changes in mechanical properties, and the identification of specific functional group alterations. There were distinct changes in the chemical structure of the blend, indicating the gradual decomposition of the PLA portion. The mechanical properties of the blend, such as tensile strength and elongation at break, decreased over the burial period, further corroborating the ongoing biodegradation process [32].

1.3 Flame Retardancy of Polyvinyl Chloride (PVC)-Based Biocomposites and Bionanocomposites

The most commonly used plastics in manufacturing WPCs are polyethylene (PE) [33, 34], polypropylene (PP) [35, 36], polystyrene (PS) [37], and polyvinylchloride (PVC) [38]. The synergistic effect of ZnO and FRW on veneered WF-PVC also exhibited excellent flame retardancy [39]. The zinc borate was added to WF-PVC composite; the results showed that zinc borate had some flame retardancy on the composite, especially exhibiting an excellent smoke suppression effect [40]. The flame-retardant mechanism of zinc salts in PVC is mainly the Lewis acid catalytic mechanism in condensed phase [41], that is, zinc compounds react with HCl which is removed from PVC to give $ZnCl_2$. In the research of PVC flame retardant and smoke suppression theory, people mainly try to reveal its flame retardant and smoke suppression mechanism by studying the changes in pyrolysis products. Iida [42] studied the changes and pyrolysis mechanism of aromatic compounds in PVC pyrolysis products using Py GC.

Research has found that SnO_2 and ZnO affect the pyrolysis pathway of PVC, leading to the formation of products that favor aliphatic compounds. The production of benzene and its aromatic compounds decreases, while the effects of TiO_2 and Al_2O_3 are not significant. Montaudo [43] found that the pyrolysis process of PVC is divided into two stages, and different stages generate two different types of aromatic compounds: substituted aromatic compounds and condensed aromatic compounds. In the presence of ZnO, there is a significant difference in the generation rate of these two types of aromatic compounds. The addition of ZnO and Fe_2O_3 advances the temperature corresponding to the maximum thermal weight loss in the first stage of WF-PVC material and significantly increases the peak thermal weight loss rate. The peak mass loss rate of WF-PVC material in the first stage is 8.4, while the peak mass loss rate of zinc oxide composite material reaches 21.3, and the peak mass loss rate of Fe_2O_3 composite material is 13.6. Li Bin [44] found that CuO promotes cross-linking reaction in the early stage of PVC, reduces the amount of volatile pyrolysis products, and promotes carbonization more effectively than Fe_2O_3 in PVC when studying the pyrolysis effect of CuO, Fe_2O_3, and MoO_3 on hard PVC. Pihong [45] also studied the effect of ZnO, CuO, and Fe_2O_3 on the changes in the condensed structure and found that CuO had the best effect on promoting the carbon chain of PVC to crosslink structure and improving the gel content, while ZnO had the worst performance. The loss modulus gradually increases from WF/PVC to WF/PVCMZ-6, which is attributed to the MMT@ZnO possibly hindering the movement of PVC molecular chains and requiring more energy during relaxation [46]. The glass transition temperature (Tg) can be regarded as the damping temperature [41]. After adding MMT@ZnO nanosheet, the Tg is prolonged to a relatively higher value due to the strong interaction between MMT@ZnO and WF/PVC.

1.4 Interfacial Modification of Polyvinyl Chloride (PVC)-Based Biocomposites and Bionanocomposites

Modifying PVC with various fillers is crucial for tailoring its properties to meet a wide range of application needs [47]. Mechanical reinforcements using fibers, both natural and synthetic, are commonly employed in polymer composites. While traditional synthetic fibers like aramid, carbon, and glass have been widely used, environmental concerns have driven the shift toward more sustainable natural fibers. Synthetic fibers are known for their superior mechanical properties and thermal stability. Thermal modifiers like organotin, calcium-zinc, and magnesium hydroxide are essential for improving PVC's thermal stability and fire resistance, which are critical for safety in electrical installations and building materials [48–50]. Electrical and magnetic enhancers, including conductive elements like carbon nanotubes, graphene, and various metal nanoparticles, significantly boost PVC's electrical and magnetic properties, making it ideal for advanced electronic applications and effective as shielding materials [51–56]. A variety of fillers have been incorporated into PVC composites to enhance mechanical and thermal properties; extensive research continues to explore the potential of biomaterials in developing high-performance engineering products. Biocomposites, which are not only cost-effective and biodegradable but also characterized by their low density and availability, offer significant adaptability for various applications [57–60]. Augier *et al.* demonstrated that PVC–vegetable fiber composites not only improve the stability and mechanical properties of PVC but also allow the composite to be recycled up to five times without substantial degradation in performance [61]. This recycling capability highlights the potential for internal waste to be effectively reused within the production cycle, reinforcing the sustainability of the composite from an industrial perspective. PVC/rice straw composites have been studied extensively to assess the effects of various factors such as pretreatment, PVC content, and temperature on their properties [81]. Results show that rice straw treated with 5% and 10% NaOH at 80°C resulted in composites with higher tensile strength, while those treated with water at 130°C exhibited the lowest strength, the effect of treatment on the swellability percentage, highlighting the effectiveness of alkali treatment in enhancing the water resistance of PVC composites compared to the untreated control.

In PVC–rice husk biocomposites, epoxidized natural rubber (ENR) has been employed as a plasticizer and a compatibilizer, demonstrating considerable miscibility with PVC and enhancing its mechanical properties due to the polar nature of both polymers [62]. The study further explores the effects of rice husk content, particle size, and ENR content, identifying an optimal formulation of 90:10:20 wt.% for PVC:ENR:rice husk, which results in superior composite properties. The resulting PVC–OHF biocomposites demonstrated substantially improved thermal and mechanical stability with varied OHF content ratios. Composites using chemically treated OHF showed enhanced tensile strength compared to those with untreated OHF, which suffered from poor dispersion and agglomeration, leading to non-uniform stress and reduced binding forces within the composite [63]. Furthermore, the mechanical, thermal, and morphological properties of PVC–Alfa composites have shown remarkable improvement in thermal stability with the addition of maleic anhydride grafted PVC, illustrating the significant role of compatibilizers in enhancing composite performance [64]. Integrating these nanomaterials into PVC biocomposites has shown substantial enhancements across a wide range of properties; effective dispersion and optimal particle size are crucial for maximizing the performance for various applications [65–67]. Nanomaterials in these composites, characterized by their high surface area-to-volume ratio and surface functionality (especially the abundance of –OH groups), tend to aggregate, which can negatively affect properties. To counter this, surface modification of nanoparticles is essential to block these functionalities, prevent aggregation, and ensure uniform dispersion. Building on this knowledge, further reinforcement of PVC nanocomposite films with ZnO nanoparticles, modified by CA and vitamin C, was found to substantially increase flame retardancy [68]. Moreover, the mechanical properties, especially elongation at maximum stress, were markedly improved. These findings emphasize the effectiveness of ZnO as an additive in enhancing both the safety and durability of PVC-based materials. The mechanical parameters, such as tensile strength, Young's modulus, and elongation at maximum stress, were measured and compared between PVC and PVC/ZnO–chitosan composites. The results showed significant improvements in tensile strength, Young's modulus, or the composite films compared to neat PVC. This enhancement is largely due to the presence of chitosan and its strong bonding with both ZnO and PVC.

1.5 PVC–Nanocellulose Biocomposites and Nanocomposites

Nanocellulose, a bio-derived nanomaterial extracted from cellulose-rich sources like wood pulp, agricultural residues, and microbial cultures, has garnered considerable attention for its exceptional mechanical properties, high aspect ratio, large specific surface area, and modifiable surface chemistry [69, 70]. Its three primary forms, cellulose nanofibrils (CNFs), cellulose nanocrystals (CNCs), and bacterial nanocellulose (BNC), differ in crystallinity, morphology, and processing routes [71, 72]. CNCs, characterized by high crystallinity, are typically produced *via* acid hydrolysis, while CNFs are obtained through high-shear mechanical techniques. PVC–nanocellulose composites leverage the synergistic integration of a widely used thermoplastic polymer with a sustainable, high-performance nanofiller [73]. Incorporating CNCs or CNFs into PVC matrices results in substantial improvements in mechanical behavior (e.g., tensile strength, Young's modulus), primarily due to the high aspect ratio and interfacial bonding capacity of nanocellulose. Despite PVC's inherent non-biodegradability, nanocellulose additions reduce the overall environmental impact of the composite, offering partial biodegradability and enhanced recyclability. Nevertheless, end-of-life disposal concerns persist due to potential toxic emissions associated with PVC degradation [74]. Nanocellulose is sourced from wood pulp, agricultural residues, and bacterial fermentation, with each source influencing its properties and effectiveness. Wood pulp, a traditional source, is processed mechanically or chemically to break down fibers into nanoscale dimensions, yielding high crystallinity and mechanical strength suitable for paper and packaging [69]. Agricultural residues, such as wheat straw and rice husks, provide an eco-friendly alternative, reducing dependence on forest resources while aiding in waste management. These residues offer unique properties due to their lignin and hemicellulose content, which affect their mechanical and thermal behavior [75]. Cellulose nanofibers (CNFs), recognized for their unique properties, are composed of elongated, flexible fibers that encompass both crystalline and amorphous regions. These fibers are produced *via* high-pressure homogenization or grinding techniques, resulting in a high aspect ratio and exceptional flexibility, which render them ideal for reinforcing various materials. Additionally, CNFs demonstrate biocompatibility, biodegradability, and an extensive surface area, thereby enhancing their interactions with other substances.

In composites, it enhances mechanical properties and thermal stability, supporting applications in packaging, automotive, and construction [76, 77]. Advanced analytical tools, such as atomic force microscopy (AFM), X-ray diffraction (XRD), and dynamic light scattering (DLS), are used to study nanocellulose morphology and crystallinity. Its shear-thinning rheological behavior is beneficial for food and pharmaceutical applications [78, 79]. Polyvinyl chloride (PVC) is a high-strength thermoplastic polymer known for its durability and versatility. It is synthesized through the polymerization of vinyl chloride monomers derived from petroleum or natural gas [20, 80]. The molecular structure of polyvinyl chloride (PVC) is defined by a two-carbon backbone, in which one hydrogen atom is substituted by a chlorine atom, resulting in the chemical formula $(C_2H_3Cl)_n$. The incorporation of fillers such as clay, wood flour, and plant-based materials significantly improves its mechanical and thermal performance, thereby expanding its applicability in advanced engineering domains [81, 82]. Despite its widespread utility, PVC poses environmental challenges. Its non-biodegradable nature and long lifespan contribute to significant waste management concerns. The incineration of polyvinyl chloride (PVC) waste generates pollutants, including hydrogen chloride gas and, under suboptimal conditions, dioxins. However, contemporary waste-to-energy incinerators, which are equipped with advanced emission control systems, have significantly mitigated these emissions. Despite this progress, the widespread adoption of such technologies continues to pose a challenge [83, 84]. The incorporation of nanocellulose, a renewable and biodegradable material, into PVC matrices significantly improves mechanical properties and thermal stability and reduces environmental impact. However, the hydrophilic nature of nanocellulose, which is attributed to its abundant hydroxyl groups, presents compatibility challenges when combined with hydrophobic polymers such as PVC. These challenges can result in poor dispersion, agglomeration, diminished mechanical performance, and increased moisture sensitivity [85, 86].

1.6 Biodegradation of Polyvinyl Chloride (PVC)-Based Different Fibers Bionanocomposites

PVC can undergo weathering degradation which causes embrittlement and microcracking and leads to the generation of nanoplastics and microplastics which persist for a longer duration of time in the environment. It was discovered that farmland soil samples in China were consistently polluted with various forms of microplastics, especially in open-field agriculture soils (4308 items/kg) with a 40.9% PVC concentration [87, 88]. Bionanocomposites (BNCs) formed by PVC polymer matrix reinforcement with natural fibers are more environmentally friendly than traditional PVC, which are known to contain high chlorine components and considered as potential hazardous material to the ecosystem [89]. Clay-reinforced polyvinyl chloride (PVC) nanocomposites represent a significant advancement in material science, finding utility across a broad range of applications, including the biomedical sector, automotive industry, water purification, packaging, fire protection, and construction [90]. Key clays utilized in the fabrication of these composites include bentonite, montmorillonite, kaolinite, and illite. Incorporating these nanoclays into the PVC matrix enhances its mechanical strength, thermal stability, and yellowness index [91]. Lignin and its derivatives have found great utilization as functional, environmentally friendly fillers in polyvinyl chloride (PVC). Lignin, a complex natural polymer found in plant cell walls, provides essential structural support, rigidity, and protection against microbial attacks [48]. As the second most abundant organic polymer on Earth, after cellulose, lignin is crucial in forming wood and bark, thereby enhancing the plant's strength and durability. Research has explored the partial substitution of PVC with various types of lignin to develop cost-effective materials. These studies have examined both plasticized (soft) and rigid PVC blends [50]. Klapiszewski *et al.* developed a novel SiO_2-lignin hybrid material as a new filler for PVC [92]. They processed PVC compounds as a dry blend with this silica-lignin hybrid material and separately with the individual components used to create the hybrid, utilizing a Brabender mixing chamber for the blending process. Polysaccharides such as cellulose, chitin, glycogen, and starch have been incorporated into PVC to create various blends, offering opportunities to enhance both the properties and sustainability of PVC materials. Cellulose, for example, when blended with PVC, can significantly improve mechanical strength and thermal stability, making it ideal for applications such as biodegradable films and coatings [93].

Cellulose nanocrystals (CNCs), which are derived from renewable sources like wood pulp, are renowned for their high aspect ratio, mechanical strength, and biodegradability. The biodegradation of PVC-based natural fiber-reinforced bionanocomposites is a complex process involving various mechanisms that break down the composite materials into simpler compounds. The degradation of biocomposites is influenced by the chemical, physical, biochemical, and synergistic interaction the polymer matrix, natural fibers, and nanoparticles [94]. It includes the breakdown of covalent bonds in the polymer's backbone, which causes irreversible changes in the chemical and molecular structure of the material, thus breaking down the composite material into simpler byproducts [95]. Biotic degradation occurs due to microbial action, such as bacteria, fungi, and algae [96, 97], that modifies and consumes the polymer or polymeric monomers, producing molecules of low molar mass (terpenes, aldehydes, and acids) and gases (N_2, CH_4, and CO_2). However, surface topography, crystallinity, hydrophobicity, and molecular size of synthetic polymers play a major factor in the biodegradation of the building blocks of plastics [98]. Chemical treatments like alkali treatment (mercerization) are often used to improve fiber–matrix adhesion in composites. These treatments can also alter the biodegradability of the fibers, making the fibers more susceptible to microbial attack [99]. Some fibers may be coated with hydrophobic materials to improve their compatibility with the PVC matrix. While this improves mechanical properties, it can reduce the fiber's biodegradability by limiting microbial access. These organisms are attracted to the natural fibers, which provide organic material for their growth. The initial colonization can lead to the formation of biofilms, which can further contribute to the deterioration of the composite by creating localized microenvironments that promote chemical and biological reactions [100, 101]. Acidic byproducts from microbial activity can also contribute to the chemical deterioration of both the PVC matrix and the natural fibers.

1.7 Polyvinyl Chloride (PVA)-Based Biocomposites and Bionanocomposites for Biomedical Applications

A common thermoplastic polymer for biomedical applications, polyvinyl chloride (PVC) is inexpensive, simple to produce, and has low toxicity [102]. Cardiovascular devices such as artificial heart pumps, blood arteries, dialysis equipment, and catheters are made of PVC [103]. The medical industry makes extensive use of polyvinyl chloride (PVC) thermoplastic polymers, although their applications are somewhat restricted. Other bioactive thermoplastic polymers or biocompatible and bioactive fillers may be added to PVC thermoplastic polymers to improve their characteristics [104]. More than 35 million tons of PVC are produced annually worldwide, and PVC accounts for the majority of medical market materials (blood bags, medical equipment, etc.) [105]. Since synthetic polyvinyl chloride contains hazardous chemical additions like phthalates, lead, cadmium, and organotins, which can be harmful to the health of the environment as well as humans, restrictions on PVC have grown over the past 20 years. These dangerous substances might evaporate or seep into the atmosphere over time. Numerous studies have examined the technique of mixing PVC with different synthetic or natural polymers and inorganic fillers in order to create enhanced PVC materials, such as gels, films, sponges, beads, fibers, sheets, and interconnected polymer (IPN) forms [106]. PVC-hybrid materials exhibit improved mechanical strength and conductivity when combined with nanofillers, which is critical for energy storage applications [107].

In order to enhance the mechanical strength and the electrical and thermal conductivity of PVC, research has focused on the use of various nanofillers over the past decade, such as carbon nanotubes (CNTs), which have high elastic modulus, strength, thermal stability, electrical conductivity, and thermal conductivity [108]. In therapeutic settings, PVC-based mixes, IPNs, and gels are used to treat patients; however, new studies have shown that DEHP, the main plasticizer in PVC, may pose risks. Micropolymers and other advanced biomaterials utilized in tissue engineering must enhance mechanical characteristics and cellular response while also being suitable with biomedical applications [109]. Green PVC blends are suitable for medical applications due to their mechanical properties, resistance to migration, and thermal stability [110]. Along with their long lifespan, PVC-based products have the significant drawback of requiring immediate waste disposal, which can harm the environment [111–113]. Burning waste does not provide a solution to this issue because it releases pollutants into the atmosphere that are dangerous for both people and the environment [114]. Temperature variations may have an impact on the mechanical characteristics of PVC-based WPCs. Closed-loop recycling of wood flour/PVC composites has been researched as a way to cut waste in the production of wood-plastics. According to research, fiber-chain degradation causes bending strength, modulus of elasticity, and impact strength to decrease after four recycling cycles [115]. PVC is an outstanding instance of such polymers, having a high degree of mechanical and chemical resistance, being inert to biological fluids, and being treatable in a variety of ways [116]. Numerous medical devices, including dialysis tubing, lung bypass sets, endotracheal feeds, cannulas, and catheters, employ PVC. Biomaterial-associated infections are becoming more common due to the increasing use of polymeric materials in medical treatment because bacteria or other organisms adhere to the polymer surface and form a pathogenic biofilm.

Several approaches have been shown to be successful in stopping biofilm production, including azide, silver coating, and impregnation of antibacterial compounds into the polymer matrix [117]. In order to develop a biocomposite with improved antimicrobial properties, PVC and gelatin were combined with montmorillonite utilizing a novel antibacterial thermoplastic plasticizer based on the aliphatic anhydride derivative DSA. The remarkable optical, electrical, magnetic, biological, and mechanical properties of composite nanomaterials, coupled with our expanding understanding of fundamental phenomena at the nanoscale, have led to a remarkable increase in interest in these materials in recent years [118]. Green synthesis has recently attracted a lot of attention since environmentally acceptable substitutes for chemical and physical methods for creating nanoparticles have been produced [119]. PVC nanofibers are becoming more popular because of their high porosity, mechanical strength, huge surface area, water resistance, and non-toxicity. Their diameters range from hundreds of nanometers to several micrometers [120]. The biopolymer polyhydroxybutyrate (PHB), which is made by bacteria and archaea, is essential to the bioplastic industry because of its versatility, biodegradability, and environmental friendliness [121]. Dutta *et al.* used rice husk ash (RHA), thermoplastic starch (TPS), and polyvinyl chloride (PVC) as the synthetic matrix to create a bio-based hybrid nanocomposite [122]. The properties of the composite were significantly improved by bio-based additives such as oxidized soybean oil (ESO), tannic acid, calcium oxide (TA-CaO), and organically modified layered double hydroxide (OLDH). The work highlights the potential for creating environmentally friendly materials by demonstrating how nanofillers and bio-based components can be effectively used to enhance the properties of PVC-based composites. A similar study by Dutta & Maji [123] aimed to enhance the value of rice husk by creating composites with microcrystalline cellulose (MCC) and montmorillonite (MMT) using a melt blending process. According to the study, adding MCC and MMT to rice husk-based composites, like PRMCC2, enhances their mechanical and thermal characteristics, making them appropriate for uses requiring strong thermal stability.

1.8 Polyvinyl Chloride (PVC)-Based Hybrid Biocomposites and Bionanocomposites

PVC is a widely used thermoplastic polymer known for its affordability, excellent chemical resistance, flame retardancy, and versatility. Its broad range of applications spans from construction materials (such as pipes and window frames) to healthcare devices and electrical insulation. However, like many synthetic polymers, traditional PVC exhibits poor biodegradability and poses environmental challenges. From an environmental standpoint, PVC-based hybrid biocomposites offer a partial solution to the global plastic waste problem. The use of bio-fillers not only displaces a portion of the petroleum-based polymer but also enables valorization of agricultural waste and natural resources. Furthermore, these composites can be engineered to be recyclable, biodegradable, or biocompatible, depending on the selection of reinforcements and processing techniques. This aligns with the global shift toward a circular economy, where materials are designed for reuse, recycling, or safe degradation at the end of their life cycle. Hybrid-based bionanocomposites represent a frontier in sustainable materials engineering, offering high performance, multifunctionality, and biodegradability. By intelligently combining nano-sized reinforcements of varied origins, researchers can achieve superior mechanical, thermal, and barrier properties while maintaining environmental compatibility. Although technological and economic challenges exist, the ongoing surge in research and innovation indicates that hybrid bionanocomposites will play a pivotal role in the development of next-generation eco-functional materials for a variety of applications.

Polyvinyl chloride (PVC) is a widely used thermoplastic polymer recognized for its chemical resistance, dimensional stability, flame retardancy, and cost-effectiveness. Despite being non-biodegradable, PVC continues to dominate several industrial sectors such as construction, automotive, packaging, and electronics. However, environmental concerns associated with its disposal and plasticizer leaching have led to renewed interest in modifying PVC with eco-friendly reinforcements. In this context, hybrid-based biocomposites, where PVC is reinforced with a combination of natural fibers and other bio-derived or synthetic fillers, offer a compelling solution. These composites seek to bridge the gap between PVC's performance limitations and the need for more sustainable materials by incorporating bio-fillers, agricultural waste fibers, and hybrid nanoparticles to improve mechanical, thermal, and environmental characteristics. The mechanical performance of PVC/hybrid biocomposites is significantly influenced by filler composition, morphology, orientation, and interfacial adhesion. Common trends observed in such composites include the following: PVC/hybrid-based biocomposites represent a powerful evolution in composite materials technology—blending cost-effective, high-performance PVC with natural and inorganic reinforcements to create systems that are mechanically robust, thermally stable, and environmentally conscious. Through intelligent hybridization, these composites overcome the limitations of conventional PVC while reducing ecological footprints. Their growing role in structural, automotive, and consumer applications demonstrates their viability in a world increasingly focused on sustainable development. Bionanocomposites represent a new frontier in materials science, where bio-based materials and nanotechnology converge to enhance mechanical, thermal, and barrier properties of polymer matrices. These materials incorporate nanoscale reinforcements (such as nanoclay, nano-silica, carbon nanotubes, graphene oxide, or metal oxides) along with biodegradable or bio-derived fillers (like cellulose nanocrystals, lignin, chitosan, starch, etc.) into a conventional polymer such as PVC. When applied to PVC systems, hybrid bionanocomposites exploit the flame retardancy, chemical resistance, and structural stability of PVC and augment them using green nanomaterials. This results in lightweight, high-strength composites that can meet the growing demand for eco-efficient and smart materials across multiple industries.

1.9 Preparation and PVC/Rubber Bio Nano Blends

The mentioned rubbers are recognized to indicate numerous properties such as oil resistance, low gas permeability and high mechanical strength. A wide variety of nano additives are used in the PVC/rubber blend for several purposes, of which the most significant are reduction in material cost and improvement in reinforcement and processing. For example, nitrile rubbers are highly miscible with PVC because of near solubility parameters. Blending with PVC exhibits excellent ozone, oil, and flame resistance. Passador and co-workers improved the toughness of PVC using blending with NBR [124]. Yang and co-workers investigated probability of toughening of PVC loaded with $CaCO_3$ nanoparticles. Here, silicone rubber was used an interfacial modifier in order to reduce the incompatibility between polymer–nanoparticle interfaces [125]. Emulsion polymerization is a type of radical addition polymerization that usually happens in a heterogeneous system containing monomer, surfactant, and water. In this process, monomer droplets emulsified in continuous water phase called oil in water emulation, which is the most common type of emulsion. Also, researchers applied this method for preparation of PVC/rubber-based bio nano blend [126]. The operation of elongation flow is based on circulation flows between two opposite cylindrical chambers that are created by removable static mixing element. In this process, all components are discharged into a mixer and mixed at a controlled speed and pressed for preparation of a dry blend of pellets. The determined speed and presser settings depend on components' viscosity [127]. In general, the interfacial adhesion between polymer matrix and nanoparticles results from the blending of PVC/rubber, which has an effective role in the toughening of polymers. Mechanical properties include hardness, strength, modulus of elasticity and fracture toughness, fatigue behavior, and properties that determine the behavior of materials against the forces involved. The existence of nanoparticles in polymer matrix provides a higher surface for the interaction between nanoparticles and the polymer matrix; as a result, it highlights the reinforcement role of nanoparticle in tensile strength [128]. Gheno and co-workers utilized the AFM technique to evaluate the influence dynamically vulcanized PVC/NBR blends on surface morphology. The dark and bright regions related to the valleys and hills of the surface that helped to identify the phase-separated regions. It was observed that the particle size became higher by increasing the NBR content in the blends that were induced by aggregating the NBR particles. We have published some of the research works related to the petroleum-based polymer-biobased composites, nanocomposites and their applications [129–135]. For the future, we have already started some research work on PVC-based biocomposites and bionanocomposites. We are learning from past work and trying to make more improvements in our research work.

AI disclosure statement: The author confirms that no artificial intelligence (AI)-generated content or AI-assisted writing tools were used in the preparation of this chapter (including summaries and the abstract). All text, analysis, and conclusions are the result of the author's original work.

References

1. Ates, B., Koytepe, S., Ulu, A., Gurses, C., Thakur, V.K., Chemistry, Structures, and Advanced Applications of Nanocomposites from Biorenewable Resources. *Chem. Rev.*, 120, 9304–9362, 2020.

2. Saad, E.M., Elshaarawy, R.F., Mahmoud, S.A., El-Moselhy, K.M., New Ulva Lactuca Algae Based Chitosan Bio-Composites for Bioremediation of Cd(II) Ions. *J. Bioresour. Bioprod.*, 6, 3, 223–242, 2021.
3. Alves, Z., Abreu, B., Ferreira, N.M., Marques, E.F., Nunes, C., Ferreira, P., Enhancing the Dispersibility of Multiwalled Carbon Nanotubes within Starch-Based Films by the Use of Ionic Surfactants. *Carbohydr. Polym.*, 273, 118531, 2021.
4. Basiak, E., Lenart, A., Debeaufort, F., Effect of Starch Type on the Physico-Chemical Properties of Edible Films. *Int. J. Biol. Macromol.*, 98, 348–356, 2017.
5. Chaudhary, A.K. and Vijayakumar, R.P., Synthesis of Polystyrene/Starch/CNT Composite and Study on Its Biodegradability. *J. Polym. Res.*, 27, 187, 2020.
6. Arora, B., Bhatia, R., Attri, P., Bionanocomposites: Green Materials for a Sustainable Future, in: *New Polymer Nanocomposites for Environmental Remediation*, C.M. Hussain and A.K. Mishra (Eds.), pp. 699–712, Elsevier, Wilmington, NC, USA, 2018.
7. Qiu, C., Hu, Y., Jin, Z., McClements, D.J., Qin, Y., Xu, X., Wang, J., A Review of Green Techniques for the Synthesis of Size-Controlled Starch-Based Nanoparticles and Their Applications as Nanodelivery Systems. *Trends Food Sci. Technol.*, 92, 138–151, 2019.
8. Le Corre, D. and Angellier-Coussy, H., Preparation and Application of Starch Nanoparticles for Nanocomposites: A Review. *React. Funct. Polym.*, 85, 97–120, 2014.
9. Ahmad, N., Kausar, A., Muhammad, B., Perspectives on polyvinyl chloride and carbon nanofiller composite: a review. *Polym. Plast. Technol. Eng.*, 55, 10, 1076–1098, 2016.
10. Mazhar, S., Promising PVC/MXene based flexible thin film nanocomposites with excellent dielectric, thermal and mechanical properties. *Ceram. Int.*, 46, 8, 12593–12605, 2020.
11. Mohamed, A.T., Thermal experimental verification on effects of nanoparticles for enhancing electric and dielectric performance of polyvinyl chloride. *Measurement*, 89, 28–33, 2016.
12. Zattini, G., Ballardini, S., Benelli, T., Mazzocchetti, L., Giorgini, L., Safer plasticized polyvinyl chloride synthetic leathers for the automotive industry: evaluation of alternatives to antimony compounds as flame retardants. *Polym. Eng. Sci.*, 59, 12, 2488–2497, 2019.
13. Kim, J., Lee, J., Jo, C., Kang, C., Development of low cost carbon fibers based on chlorinated polyvinyl chloride(CPVC) for automotive applications. *Mater. Des.*, 204, 109682, 2021.
14. Moustafa, H., Morsy, M., Ateia, A.M., Abdel-Haleem, M.F., Ultrafast response humidity sensors based on polyvinyl chloride/graphene oxide nanocomposites for intelligent food packaging. *Sensor Actuator Phys.*, 331, 112918, 2021.

15. Hazer, B., Karahaliloglu, Z., Ashby, R.D., Poly (vinyl chloride) derived food packaging applications with Antioxidative and Anticancer properties. *ACS Food Sci. Technol.*, 3, 4, 761–771, 2023.
16. Alshammari, A.H. and Taha, T.A., Structure, thermal and dielectric insights of PVC/PVP/ZnFe2O4 polymer nanocomposites. *Eur. Phys. J. Plus*, 136, 12, 1201, 2021.
17. Richard, P.W., Felix, H.O., Charles, L.M., Charles, R.R., Starch-Filled Polyvinyl Chloride Plastics-Preparation and Evaluation. *Ind. Eng. Chem. Prod. Res. Develop.*, 13, 2, 1974.
18. Rosa, D.S., Rios, A.R., Viana, H.M., Tavares, M.I.B., Characterization of Poly (vinyl chloride) Compounds Containing Different Levels of Starch. *J. Vinyl Addit. Technol.*, 22, 4, 396–404, 2016.
19. Kühnert, I., Spörer, Y., Brünig, H., Tran, N.H.A., Rudolph, N., *Processing of Poly(Lactic Acid) BT - Industrial Applications of Poly(Lactic Acid)*, M.L. Di Lorenzo and R. Androsch (Eds.), pp. 1–33, Springer International Publishing, Cham, 2018, https://doi.org/10.1007/12_2017_30.
20. Ait-Touchente, Z., Khellaf, M., Raffin, G., Lebaz, N., Elaissari, A., Recent Advances in Polyvinyl Chloride (PVC) Recycling. *Polym. Adv. Technol.*, 35, 1, e6228, 2024. https://doi.org/https://doi.org/10.1002/pat.6228.
21. Kumar, S., Singh, R., Singh, T.P., Batish, A., On Investigations of Thermal Conductivity, Circumferential Compressive Strength, and Surface Characterization of 3D-Printed Hybrid Blended Magnetostrictive PLA Composite. *J. Thermoplast. Compos. Mater.*, 35, 5, 631–650, 2022. https://doi.org/10.1177/0892705720907651.
22. Othman, N., Xu, C., Mehrkhodavandi, P., Hatzikiriakos, S.G., Thermorheological and Mechanical Behavior of Polylactide and Its Enantiomeric Diblock Copolymers and Blends. *Polymer*, 53, 12, 2443–2452, 2012. https://doi.org/https://doi.org/10.1016/j.polymer.2012.03.068.
23. de França, J.O.C., da Silva Valadares, D., Paiva, M.F., Dias, S.C.L., Dias, J.A., Polymers Based on PLA from Synthesis Using D,L-Lactic Acid (or Racemic Lactide) and Some Biomedical Applications: A Short Review. *Polymers*, 14, 12, 2022. https://doi.org/10.3390/polym14122317.
24. Ramakrishna, S., Rangaswamy, V., Jain, D., Jagdambalal, R., Patel, P., Kar, D., Ramachandran, S., Ganeshpure, P., Satpathy, U., Process for the Production of Polylactic Acid (PLA) from Renewable Feedstocks, 20052005. https://patents.google.com/patent/US20060036062A1/en.
25. DE Vries, A.J., Bonnebat, C., Carrega, M., Dilute Solution Properties and Molecular Characterization of Polyvinyl Chloride. *Pure Appl. Chem.*, 26, 2, 209–240, 1971. https://doi.org/10.1351/pac197126020209.
26. Kudzin, M.H., Piwowarska, D., Festinger, N., Chruściel, J.J., Risks Associated with the Presence of Polyvinyl Chloride in the Environment and Methods for Its Disposal and Utilization. *Materials*, 17, 1, 2024. https://doi.org/10.3390/ma17010173.

27. Saikia, A. and Karak, N., Introduction to Biodegradable and Biocompatible Polymer Nanocomposites: Synthesis, Structure, Fundamental Properties, Biocompatibility, and Biodegradability, in: *Biodegradable and Biocompatible Polymer Nanocomposites*, K. Deshmukh, M.B.T.-B. Pandey and B.P.N. (Eds.), pp. 1–34, Elsevier, 2023, https://doi.org/10.1016/B978-0-323-91696-7.00007-6.
28. Hachemi, R., Belhaneche-Bensemra, N., Massardier, V., Elaboration and Characterization of Bioblends Based on PVC/PLA. *J. Appl. Polym. Sci.*, 131, 7, 1–7, 2014. https://doi.org/10.1002/app.40045.
29. Singh, S.K., Chowdhury, A., Thakur, R.R., Mangaraj, S., Sami, R., Aljahani, A.H., Aljuraide, N.I., Helal, M., Development and Characterization of Polyvinyl Chloride/Poly Lactic Acid Blend Based Biodegradable Polymeric Films. *Mater. Express*, 13, 4, 632–643, 2023. https://doi.org/10.1166/mex.2023.2388.
30. Xie, D., Zhao, Y., Li, Y., LaChance, A.M., Lai, J., Sun, L., Chen, J., Rheological, Thermal, and Degradation Properties of PLA/PPG Blends. *Mater. (Basel Switzerland)*, 12, 21, 2019. https://doi.org/10.3390/ma12213519.
31. Ramli, H., Zainal, N.F.A., Hess, M., Chan, C.H., Basic Principle and Good Practices of Rheology for Polymers for Teachers and Beginners. *Chem. Teach. Int.*, 4, 4, 307–326, 2022. https://doi.org/10.1515/cti-2022-0010.
32. Santana, V.T., Gonçalves, S.P.C., Agnelli, J.A.M., Martins-Franchetti, S.M., Biodegradation of a Polylactic Acid/Polyvinyl Chloride Blend in Soil. *J. Appl. Polym. Sci.*, 125, 1, 536–540, 2012. https://doi.org/10.1002/app.35685.
33. Pan, M., Mei, C., Du, J., Li, G., Synergistic effect of nano silicon dioxide and ammonium polyphosphate on flame retardancy of wood fiber–polyethylene composites. *Compos. Part A Appl. Sci. Manufacturing*, 66, 128–134, 2014.
34. Park, Y.-K., SulJung, J., JunghoJae, B.S., Watanabe, A., Young-MinKim, Catalytic fast pyrolysis of wood plastic composite over microporous zeolites. *Chem. Eng. J.*, 377, 119742, 2019.
35. Mashkour, M. and Ranjbar, Y., Superparamagnetic Fe_3O_4@wood flour/polypropylene nanocomposites: Physical and mechanical properties. *Ind. Crop Prod.*, 111, 47–54, 2018.
36. Arbelaiz, A., Fernández, B., Ramos, J.A., Mondragon, I., Thermal and crystallization studies of short flax fibre reinforced polypropylene matrix composites: Effect of treatments. *Thermochim. Acta*, 440, 2, 111–121, 2006.
37. Nair, K.C.M., Thomas, S., Groeninckx, G., Thermal and dynamic mechanical analysis of polystyrene composites reinforced with short sisal fibres. *Compos. Sci. Technol.*, 61, 2519–2529, 2001.
38. Hoang, D., Pham, T., Nguyen, T., An, H., Kim, J., Organo-Phosphorus Flame Retardants for Poly(vinyl chloride)/Wood Flour Composite. *Polym. Compos.*, 39, 961–970, 2016.
39. Zong, G., Hao, J., Hao, X., Fang, Y., Song, Y., Wang, H., Ou, R., Wang, Q., Enhancing the flame retardancy and mechanical properties of veneered wood flour/polyvinyl chloride composites. *Poly. Compos.*, 41, 848–857, 2019.

40. Fang, Y., Wang, Q., Guo, C., Song, Y., Cooper, P.A., Effect of zinc borate and wood flour on thermal degradation and fire retardancy of Polyvinyl chloride (PVC) composites. *J. Anal. Appl. Pyrol.*, 100, 230–236, 2013.
41. Yang, R., Fang, Y., Zhu, J., Fabrication ZnO functionalized nano-montmorillonite for enhancing the fire-safety and mechanical properties of wood flour/poly (vinyl chloride) composites. *Constr. Build. Mater.*, 420, 135522, 2024.
42. Iida, T., Nakanishi, M., Gotō, K., Investigations on poly (vinyl chloride). I. Evolution of aromatics on pyrolysis of poly (vinyl chloride) and its mechanism. *J. Polym. Science Polym. Chem. Edition*, 12, 4, 737–749, 1974.
43. Montaudo, G. and Puglisi, C., Evolution of aromatics in the thermal degradation of poly (vinyl chloride): a mechanistic study. *Polym. Degrad. Stabil.*, 33, 229–262, 1991.
44. Li, B., A study of thermal degradation and decomposition of rigid poly (vinyl chloride) with metal oxides using thermogravimetry and cone calorimetry. *Polym. Degrad. Stab.*, 68, 2, 197–204, 2000.
45. Hong, P., *Study on the structural evolution and flame retardant and smoke suppressant mechanism during the thermal decomposition process of polyvinyl chloride/flame retardant and smoke suppressant system*, Sichuan Universtiy, Chendu, 2003.
46. Hao, X., Xu, J., Zhou, H., Tang, W., Li, W., Wang, Q., Ou, R., Interfacial adhesion mechanisms of ultra-highly filled wood fiber/polyethylene composites using maleic anhydride grafted polyethylene as a compatibilizer. *Mater. Des.*, 212, 110182, 2021.
47. Pandey, J.K., Ahn, S.H., Lee, C.S., Mohanty, A.K., Misra, M., Recent Advances in the Application of Natural Fiber Based Composites. *Macromol. Mater. Eng.*, 295, 975–989, 2010.
48. Tomaszewska, J., Wieczorek, M., Skórczewska, K., Klapiszewska, I., Lewandowski, K., Klapiszewski, L., Preparation, Characterization and Tailoring Properties of Poly(Vinyl Chloride) Composites with the Addition of Functional Halloysite–Lignin Hybrid Materials. *Materials*, 15, 8102, 2022.
49. Bula, K., Klapiszewski, L., Jesionowski, T., A novel functional silica/lignin hybrid material as a potential bio-based polypropylene filler. *Polym. Compos.*, 36, 913–922, 2015.
50. Klapiszewski, L., Tomaszewska, J., Skórczewska, K., Jesionowski, T., Preparation and Characterization of Eco-Friendly Mg(OH)2/Lignin Hybrid Material and Its Use as a Functional Filler for Poly(Vinyl Chloride). *Polymers*, 9, 258, 2017.
51. Haruna, H., Pekdemir, M.E., Tukur, A., Coşkun, M., Characterization, thermal and electrical properties of aminated PVC/oxidized MWCNT composites doped with nanographite. *J. Therm. Anal. Calorim.*, 139, 3887–3895, 2020.
52. Chand, N. and Das Jhod, B., Mechanical, Electrical, and Thermal Properties of Maleic Anhydride Modified Rice Husk Filled Pvc Composites. *Bioresources*, 3, 1228–1242, 2008.

53. Mousavi, S.R., Estaji, S., Paydayesh, A., Arjmand, M., Jafari, S.H., Nouranian, S., Khonakdar, H.A., A review of recent progress in improving the fracture toughness of epoxy-based composites using carbonaceous nanofillers. *Polym. Compos.*, 43, 1871–1886, 2022.
54. Qin, J., Wang, C., Yao, Z., Ma, Z., Gao, Q., Wang, Y., Wang, Q., Wei, H., Growing carbon nanotubes on continuous carbon fibers to produce composites with improved interfacial properties: A step towards commercial production and application. *Compos. Sci. Technol.*, 211, 108870, 2021.
55. Petrény, R. and Mészáros, L., Moisture dependent tensile and creep behaviour of multi-wall carbon nanotube and carbon fibre reinforced, injection moulded polyamide 6 matrix multi-scale composites. *J. Mater. Res. Technol.*, 16, 689–699, 2022.
56. Sayam, A., Rahman, A.N.M.M., Rahman, M.S., Smriti, S.A., Ahmed, F., Rabbi, M.F., Hossain, M., Faruque, M.O., A review on carbon fiber-reinforced hierarchical composites: mechanical performance, manufacturing process, structural applications and allied challenges. *Carbon Lett.*, 32, 1173–1205, 2022.
57. Paul, S.A., Joseph, K., Mathew, G.D.G., Pothen, L.A., Thomas, S., Influence of polarity parameters on the mechanical properties of composites from polypropylene fiber and short banana fiber. *Composites, Part A*, 41, 1380–1387, 2010.
58. Ramesh, M., Kenaf (Hibiscus cannabinus L.) fibre based bio-materials: A review on processing and properties. *Prog. Mater Sci.*, 78–79, 1–92, 2016.
59. Sliwa, F., El Bounia, N.-e., Charrier, F., Marin, G., Malet, F., Mechanical and interfacial properties of wood and bio-based thermoplastic composite. *Compos. Sci. Technol.*, 72, 1733–1740, 2012.
60. Nair, K.C.M. and Thomas, S., Effect of interface modification on the mechanical properties of polystyrene-sisal fiber composites. *Polym. Compos.*, 24, 332–343, 2003.
61. Augier, L., Sperone, G., Vaca-Garcia, C., Borredon, M.-E., Influence of the wood fibre filler on the internal recycling of poly(vinyl chloride)-based composites. *Polym. Degrad. Stab.*, 92, 1169–1176, 2007.
62. Bui, Q., Colin, J., Nguyen, T., Mao, N., Nguyen, T., Perré, P., Mechanical and thermal properties of a biocomposite based on polyvinylchloride/epoxidized natural rubber blend reinforced with rice husk microfiller. *J. Thermoplast. Compos. Mater.*, 34, 1180–1192, 2021.
63. Amar, B., Salem, K., Hocine, D., Chadia, I., Juan, M.J., Study and characterization of composites materials based on polypropylene loaded with olive husk flour. *J. Appl. Polym. Sci.*, 122, 1382–1394, 2011.
64. Hammiche, D., Boukerrou, A., Djidjelli, H., Corre, Y.-M., Grohens, Y., Pillin, I., Hydrothermal ageing of alfa fiber reinforced polyvinylchloride composites. *Constr. Build. Mater.*, 47, 293–300, 2013.

65. Khaleghi, M., Didehban, K., Shabanian, M., Effect of new melamine-terephthaldehyde resin modified graphene oxide on thermal and mechanical properties of PVC. *Polym. Test.*, 63, 382–391, 2017.
66. Alghamdi, M.M., El-Zahhar, A.A., Asiri, B.M., Incorporation of magnetite nanoparticles in poly(vinyl chloride) microfiltration membrane for improving antifouling property and desalination performance. *Desalin. Water Treat.*, 165, 54–62, 2019.
67. El-Zahhar, A.A., Alghamdi, M.M., Asiri, B.M., Poly (vinyl chloride)-MMT composite membranes with enhanced properties and separation performance. *Desalin. Water Treat.*, 155, 381–389, 2019.
68. Mallakpour, S. and Javadpour, M., Comprehensive study on reinforcement of poly(vinyl chloride) nanocomposite films with ZnO nanoparticles modified by citric acid and vitamin C. *Int. J. Polym. Anal. Charact.*, 23, 415–429, 2018.
69. Mishra, R.K., Sabu, A., Tiwari, S.K., Materials chemistry and the futurist eco-friendly applications of nanocellulose: Status and prospect. *J. Saudi Chem. Soc.*, 22, 2018. https://doi.org/10.1016/j.jscs.2018.02.005.
70. Kargarzadeh, H., Mariano, M., Gopakumar, D., Ahmad, I., Thomas, S., Dufresne, A., Huang, J., Lin, N., Advances in cellulose nanomaterials. *Cellulose*, 25, 2151–2189, 2018. https://doi.org/10.1007/s10570-018-1723-5.
71. Zhang, Y., Nypelö, T., Salas, C., Arboleda, J., Hoeger, I.C., Rojas, O.J., Cellulose Nanofibrils. *J. Renew. Mater.*, 1, 195–211, 2013. https://doi.org/10.7569/JRM.2013.634115.
72. Sultana, S., Reja, L.N., Mahbub Hilary, M., Saiful Islam, M., Zamsed Uddin, M., Ajmotgir, W.Md., Preparation and characterization of nanocellulose from Albizia lebbeck sawdust and their application in nanocomposites using poly (vinyl chloride) (PVC), Res Sq, 2021.
73. Khalid, M.Y., Al Rashid, A., Arif, Z.U., Ahmed, W., Arshad, H., Recent advances in nanocellulose-based different biomaterials: types, properties, and emerging applications. *J. Mater. Res. Technol.*, 14, 2601–2623, 2021. https://doi.org/10.1016/j.jmrt.2021.07.128.
74. Brebu, M., Environmental Degradation of Plastic Composites with Natural Fillers—A Review. *Polym. (Basel)*, 12, 166, 2020. https://doi.org/10.3390/polym12010166.
75. Kaur, P., Sharma, N., Munagala, M., Rajkhowa, R., Aallardyce, B., Shastri, Y., Agrawal, R., Nanocellulose: Resources, Physio-Chemical Properties, Current Uses and Future Applications. *Front. Nanotechnol.*, 3, 2021. https://doi.org/10.3389/fnano.2021.747329.
76. Mathew, A.P., Thielemans, W., Dufresne, A., Mechanical properties of nanocomposites from sorbitol plasticized starch and tunicin whiskers. *J. Appl. Polym. Sci.*, 109, 4065–4074, 2008. https://doi.org/10.1002/app.28623.
77. Ladhar, A., Bendahou, A., Arous, M., Dufresne, A., Kaddami, H., Dielectric Spectroscopy: An Efficient Tool to Study the Interfacial Adhesion and Properties of Natural Rubber/Nanocellulose-Based Green Nanocomposites, in: *Handbook of Nanocellulose and Cellulose Nanocomposites*, pp. 627–648,

Wiley-VCH Verlag GmbH & Co. KGaA, Weinheim, Germany, 2017, https://doi.org/10.1002/9783527689972.ch19.

78. Jia, X., Chen, Y., Shi, C., Ye, Y., Abid, M., Jabbar, S., Wang, P., Zeng, X., Wu, T., Rheological properties of an amorphous cellulose suspension. *Food Hydrocoll.*, 39, 27–33, 2014. https://doi.org/10.1016/j.foodhyd.2013.12.026.
79. Li, M.C., Wu, Q., Song, K., Lee, S., Qing, Y., Wu, Y., Cellulose Nanoparticles: Structure-Morphology-Rheology Relationships. *ACS Sustain. Chem. Eng.*, 3, 821–832, 2015. https://doi.org/10.1021/acssuschemeng.5b00144.
80. Asadinezhad, A., Lehocký, M., Sáha, P., Mozetič, M., Recent progress in surface modification of polyvinyl chloride. *Materials*, 5, 2012. https://doi.org/10.3390/ma5122937.
81. Elgharbawy, A.S., Poly Vinyl Chloride Additives and Applications - A Review. *J. Risk Anal. Crisis Response*, 12, 2022. https://doi.org/10.54560/jracr.v12i3.335.
82. Hashmi, S.U.M., Iqbal, M.A., Malik, M., Qamar, M.T., Khan, M., Zahid, A., Islam, Md.R., Al-Bahrani, M., Morsy, K., Lai, W.-C., Synthesis and Characterization of Polyvinyl Chloride Matrix Composites with Modified Scrap Iron for Advanced Electronic. *Photonic Opt. Systems Nanomater.*, 12, 3147, 2022. https://doi.org/10.3390/nano12183147.
83. Tomaszewska, J., Sterzyński, T., Walczak, D., Thermal Stability of Nanosilica-Modified Poly(vinyl chloride). *Polym. (Basel)*, 13, 2057, 2021. https://doi.org/10.3390/polym13132057.
84. Andrew, L.J., Lizundia, E., MacLachlan, M.J., Designing for Degradation: Transient Devices Enabled by (Nano)Cellulose. *Adv. Mater.*, 2024. https://doi.org/10.1002/adma.202401560.
85. Manimaran, M., Norizan, M.N., Kassim, M.H.M., Adam, M.R., Norrrahim, M.N.F., Knight, V.F., Critical assessment of the thermal stability and degradation of chemically functionalized nanocellulose-based polymer nanocomposites. *Nanotechnol. Rev.*, 13, 2024. https://doi.org/10.1515/ntrev-2024-0005.
86. Chu, Y., Sun, Y., Wu, W., Xiao, H., Dispersion Properties of Nanocellulose: A Review. *Carbohydr. Polym.*, 250, 116892, 2020. https://doi.org/10.1016/j.carbpol.2020.116892.
87. Nizzetto, L., Futter, M., Langaas, S., Are Agricultural Soils Dumps for Microplastics of Urban Origin?, 2016. https://doi.org/10.1021/acs.est.6b04140.
88. Tang, K.H.D., Climate Change and Plastic Pollution: A Review of Their Connections. *Trop. Environ. Biol. Technol.*, 1, 110–120, 2023.
89. Braun, D., Poly(vinyl chloride) on the Way from the 19th Century to the 21st Century. *J. Polym. Sci. Part A Polym. Chem.*, 42, 578–586, 2004.
90. Boraei, A.S.B., Bakhshandeh, B., Mohammadzadeh, F., Haghighi, D.M., Mohammadpour, Z., Clay-reinforced PVC composites and nanocomposites. *Heliyon*, 10, e29196, 2024.
91. Adeyemo, A.A., Adeoye, I.O., Bello, O.S., Adsorption of dyes using different types of clay: a review. *Appl. Water Sci.*, 7, 543–568, 2017.

92. Klapiszewski, L., Pawlak, F., Tomaszewska, J., Jesionowski, T., Preparation and characterization of novel pvc/silica-lignin composites. *Polym. (Basel)*, 7, 1767–1788, 2015.
93. Ali, M.I., Perveen, Q., Ahmad, B., Javed, I., Razi-Ul-Hussnain, R., Andleeb, S., Atique, N., Ghumro, P.B., Ahmed, S., Hameed, A., Studies on biodegradation of cellulose blended polyvinyl chloride films. *Int. J. Agric. Biol.*, 11, 577–580, 2009.
94. Shah, A.A., Hasan, F., Hameed, A., Ahmed, S., Biological degradation of plastics: A comprehensive review, 2008. https://doi.org/10.1016/j.biotechadv.2007.12.005.
95. Silva, R.R.A., Marques, C.S., Arruda, T.R., Teixeira, S.C., de Oliveira, T.V., Biodegradation of Polymers: Stages, Measurement, Standards and Prospects. *Macromol.*, 3, 371–399, 2023.
96. Yang, X.G., Wen, P.P., Yang, Y.F., Jia, P.P., Li, W.G., Pei, D.S., Plastic biodegradation by *in vitro* environmental microorganisms and *in vivo* gut microorganisms of insects, 2023. https://doi.org/10.3389/fmicb.2022.1001750.
97. Danso, D., Chow, J., Streita, W.R., Plastics: Environmental and biotechnological perspectives on microbial degradation, 2019. https://doi.org/10.1128/AEM.01095-19.
98. Miri, S., Saini, R., Davoodi, S.M., Pulicharla, R., Brar, S.K., Magdouli, S., *Biodegradation of microplastics: Better late than never*, Elsevier Ltd, 2022, https://doi.org/10.1016/j.chemosphere.2021.131670.
99. Bledzki, A., Gassan, J., Composites reinforced with cellulose.pdf. *Prog. Polym. Sci.*, 24, 221–274, 1999.
100. Zhang, Z., Peng, H., Yang, D., Zhang, G., Zhang, J., Ju, F., Polyvinyl chloride degradation by a bacterium isolated from the gut of insect larvae. *Nat. Commun.*, 13, 5360, 2022.
101. Banerjee, A., Chatterjee, K., Madras, G., Enzymatic degradation of polymers: A brief review, 2014. https://doi.org/10.1179/1743284713Y.0000000503.
102. Singh, J. and Agrawal, K.K., Modification of Poly(Vinyl Chloride) for Biocompatibility Improvement and Biomedical Application-Review. *Polym. Plast. Technol. Eng.*, *31*, 3–4, 203–212, 1992. https://doi.org/10.1080/03602559208017742.
103. Banoriya, D., Purohit, R., Dwivedi, R.K., Advanced Application of Polymer based Biomaterials. *Mater. Today Proc.*, *4*, 2, Part A, 3534–3541, 2017. https://doi.org/10.1016/j.matpr.2017.02.244.
104. Ranjan, N., Chitosan with PVC polymer for biomedical applications: A bibliometric analysis. *Mater. Today Proc.*, *81*, 894–898, 2023. https://doi.org/10.1016/j.matpr.2021.04.274.
105. Wang, K., He, Y., Song, X., Cui, X., Effects of the metakaolin-based geopolymer on high-temperature performances of geopolymer/PVC composite materials. *Appl. Clay Sci.*, *114*, 586–592, 2015. https://doi.org/10.1016/j.clay.2015.07.008.

106. Feldman, D. and Barbalata, A., Synthetic Polymers: Technology, properties, applications By Dorel Feldman (Concordia University) and Alla Barbalata (Université de Moncton). Chapman & Hall: Florence. 1996. Xv + 370 pp. ISBN 0-412-71040-4. *J. Am. Chem. Soc.*, 1996. https://pubs.acs.org/doi/10.1021/ja965754l.
107. Althobiti, R.A., Morsi, M.A., Alzahrani, E., Al-Muntaser, A.A., Enhancing the performance of PVC/PMMA polymer blend through hybrid nanofiller of TiO2 NPs/GNPs for capacitive energy storage applications. *Ceram. Int.*, *50*, 11, Part A, 19039–19047, 2024. https://doi.org/10.1016/j.ceramint.2024.03.001.
108. Moghadam, F.G., Pirayandeh, S., Mohammadi, T., Tofighy, M.A., Nanocomposite of PVC with CNT, in: *Poly(Vinyl Chloride) Based Composites and Nanocomposites*, A. H & T. Sabu (Ed.), pp. 241–260, Springer International Publishing, 2024, https://doi.org/10.1007/978-3-031-45375-5_11.
109. Andleeb, R., Sajid, M.A., Ashraf, A., Arooj, J., 22—Applications of polyvinyl chloride-based blends, IPNs, and gels for biomedical fields, in: *Poly(vinyl chloride)-based Blends, Interpenetrating Polymer Networks (IPNs), and Gels*, S. Thomas and H. Akhina (Eds.), pp. 577–598, Elsevier, 2024, https://doi.org/10.1016/B978-0-323-99474-3.00019-7.
110. Darie-Nita, R.N., Râpă, M., Visakh, P.M., Bio-Based Polyvinylchloride (PVC)-Related Blends, in: *Polyvinylchloride-based Blends: Preparation, Characterization and Applications*, V.P.M. & R.N. Darie-Nita (Ed.), pp. 211–234, Springer International Publishing, 2022, https://doi.org/10.1007/978-3-030-78455-3_10.
111. Ghosal, K. and Ghosh, S., Biodegradable polymers from lignocellulosic biomass and synthetic plastic waste: An emerging alternative for biomedical applications. *Mater. Sci. Eng.: R: Rep.*, *156*, 100761, 2023. https://doi.org/10.1016/j.mser.2023.100761.
112. Kang, J., Kim, J.Y., Sung, S., Lee, Y., Gu, S., Choi, J.-W., Yoo, C.-J., Suh, D.J., Choi, J., Ha, J.-M., Chemical upcycling of PVC-containing plastic wastes by thermal degradation and catalysis in a chlorine-rich environment. *Environ. Pollut.*, *342*, 123074, 2024. https://doi.org/10.1016/j.envpol.2023.123074.
113. Khalid, A.R., Shah, T., Asad, M., Ali, A., Samee, E., Adnan, F., Bhatti, M.F., Marhan, S., Kammann, C.I., Haider, G., Biochar alleviated the toxic effects of PVC microplastic in a soil-plant system by upregulating soil enzyme activities and microbial abundance. *Environ. Pollut.*, *332*, 121810, 2023. https://doi.org/10.1016/j.envpol.2023.121810.
114. Meng, J., Xu, B., Liu, F., Li, W., Sy, N., Zhou, X., Yan, B., Effects of chemical and natural ageing on the release of potentially toxic metal additives in commercial PVC microplastics. *Chemosphere*, *283*, 131274, 2021. https://doi.org/10.1016/j.chemosphere.2021.131274.
115. Nadali, E. and Naghdi, R., Effects of multiple extrusions on structure–property relationships of hybrid wood flour/poly (vinyl chloride) composites. *J. Thermoplast. Compos. Mater.*, *35*, 8, 1076–1093, 2022. https://doi.org/10.1177/0892705720930737.

116. Baptista, J., Simões, M., Borges, A., Effect of plant-based catecholic molecules on the prevention and eradication of *Escherichia coli* biofilms: A structure activity relationship study. *Int. Biodeterior. Biodegrad., 141*, 101–113, 2019. https://doi.org/10.1016/j.ibiod.2018.02.004.
117. Subhadra, B., Kim, D.H., Woo, K., Surendran, S., Choi, C.H., Control of Biofilm Formation in Healthcare: Recent Advances Exploiting Quorum-Sensing Interference Strategies and Multidrug Efflux Pump Inhibitors. *Materials, 11*, 9, 2018. Article 9. https://doi.org/10.3390/ma11091676.
118. Akinniyi, J.N., Atanda, S., Damilola, A., Ahmed, T., Ayanda, I., Badmos, F., Jimoh, M., Adegbola, R., *Perspective Chapter: Nanocomposites – Unlocking the Potentials for Diverse Applications*, 2024. https://doi.org/10.5772/intechopen.114914.
119. Abdelghany, A.M., Oraby, A.H., Asnag, G.M., Structural, thermal and electrical studies of polyethylene oxide/starch blend containing green synthesized gold nanoparticles. *J. Mol. Struct., 1180*, 15–25, 2019. https://doi.org/10.1016/j.molstruc.2018.11.095.
120. Quoc Pham, L., Uspenskaya, M.V., Olekhnovich, R.O., Olvera Bernal, R.A., A Review on Electrospun PVC Nanofibers: Fabrication, Properties, and Application. *Fibers*, 9, 2, 2021. Article 2. https://doi.org/10.3390/fib9020012.
121. Mandragutti, T., Jarso, T.S., Godi, S., Begum, S.S., K, B., Physicochemical characterization of polyhydroxybutyrate (PHB) produced by the rare halophile Brachybacterium paraconglomeratum MTCC 13074. *Microb. Cell Fact., 23*, 1, 59, 2024. https://doi.org/10.1186/s12934-024-02324-1.
122. Dutta, N., Newar, P., Narzary, M., Maji, T.K., Preparation and characterization of biobased hybrid nanocomposite of Polyvinyl chloride/thermoplastic starch with Mg–Al layered double hydroxide and waste rice husk ash. *J. Polym. Res., 29*, 9, 375, 2022. https://doi.org/10.1007/s10965-022-03226-9.
123. Dutta, N. and Maji, T.K., Synergic effect of montmorillonite and microcrystalline cellulose on the physicochemical properties of rice husk/PVC composite. *SN Appl. Sci., 2*, 3, 439, 2020. https://doi.org/10.1007/s42452-020-2255-9.
124. Liu, C., Luo, Y., Jia, Z., Particle configuration and properties of poly (vinyl chloride)/halloysite nanotubes nanocomposites *via in situ* suspension polymerization. *Polym. Composits.*, 35, 856–863, 2014.
125. Yang, L., Toughening and reinforcement of rigid PVC with silicone rubber/nano-$CaCO_3$ shell-core structured fillers. *Appl. Polymer Sci.*, 102, 2560–2567, 2006.
126. Yamak, H.B., Emulsion polymerization: effects of polymerization variables on the properties of vinyl acetate based emulsion polymers, polymer science (book), 2013.
127. Bouquey, M. and Loux, C., Morphological Study of Two-Phase Polymer Blends During Compounding in a Novel Compounder on the Basis of Elongational Flows. *Appl. Polym. Scienc.*, 119, 482–490, 2011.

128. Mousa, A., Heinrich, G., Wagenknecht, U., The potential of O-MMT as a reinforcing filler for uncured and dynamically cured PVC/XNBR composites. *J. Macromol. Science Part A Pure Appl. Chem.*, 45, 733–741, 2008.
129. Panda, S.K., Synthesis and Overall Migration Study of Chitosan-Encapsulated ZnO-Based ESO Bionanocomposite with Synergistic Antimicrobial Activity for Packaging Purpose. *Chem. Select*, 7, 2022. doi.org/10.1002/slct.202201433.
130. Mohanty, P., Behera, D., Panda, S.K., Bastia, T.K., Hybrid Composite Laminates from UPE/ESOA Blend Reinforced with Chitosan and Bamboo Fiber: A Study of Mechanical and Thermal Properties. *Asian J. Chem.*, 32, 6, 1321–1328, 2020.
131. Mohanty, P., Bastia, T.K., Behera, D., Panda, S.K., Chitosan Grafted Carbon Nanotubes Reinforced Vinyl Ester/UPE Blend Based Partially Bio Nanocomposite. *Asian J. Chem.*, 31, 9, 1943–1948, 2019.
132. Panda, S.K., Behera, D., Rath, P., Bastia, T.K., Enhanced properties of UPE/ESOA partially bio-nanocomposites reinforced with chitosan functionalized graphene nanoplatelets: An innovative approach. *Bull. Mater. Sci.*, 41, 4, 102, 2018.
133. Panda, S.K., Behera, D., Rath, P., Bastia, T.K., Advances in physical performances of UPE/ESOA partially bio-nanocomposites reinforced with graphene nanoplatelets. *Int. J. Res. Innovation Appl. Sci.*, 2, 8, 2017.
134. Panda, S.K. and Behera, D., Effect of red mud on mechanical and chemical properties of unsaturated polyester-epoxy bamboo fiber composites, in: *Materials Today: Proceeding*, Elsevier, 2017, ISSN:2214-7853.
135. Panda, S.K., Mohanty, P., Behera, D., Bastia, T.K., ESOA modified unsaturated polyester hybrid networks: A new perspective. *J. Appl. Polym. Sciences (Wiley)*, 133, 44345, 2016. DOI: 10.1002/APP.44345.

2

Polyvinyl Chloride (PVC)/Starch-Based Biocomposites and Bionanocomposites

Shiv Kumari Panda

Department of Chemistry, U.N. Autonomous College of Science & Technology, Adaspur, Cuttack, Odisha, India

Abstract

In current scenario, novel polymer materials like natural starch-reinforced polyvinyl chloride (PVC) biocomposites and bionanocomposites are the most interesting, demanding, and useful material in polymer industry. It possesses some significant influence in this field as concern has been raised up regarding the synthetic origin, sustainability, and some environmental issues for PVC. Introduction of natural starch as well as modified starch as reinforcing material for petroleum-based polymers through various physical and chemical processes carries great demand. It is because such innovative techniques can develop the physicochemical, functional, environmental, and mechanical characteristics in the synthetic material with expanding the range of application as a promising approach. Again the growing attention for nanostarch material boosts the revolutionary opportunity to improve the use of biomaterials in foodstuff, cosmetics, medication, and several other industrial fields. Additionally, the resulted bionanocomposites can demonstrate itself as a high-performance material for multiple sectors like packaging, agriculture, and medical devices.

***Keywords*:** Polysaccharide, starch, nanostarch, bionanocomposite, PVC/starch biocomposite

Email: shivkumaripanda@gmail.com

Shiv Kumari Panda and Visakh P. M (eds.) Polyvinyl Chloride-Based Biocomposites and Bionanocomposites, (29–52) © 2026 Scrivener Publishing LLC

2.1 Introduction

Polyvinyl chloride (PVC) is a broadly used inexpensive, synthetic, thermoplastic-based polymer made through the free radical polymerization of vinyl chloride monomeric units. These vinyl contents are the reason for the good tensile strength of this polymer. It has good durability with resistance to wear and tear. It shows chemical resistance towards acid, base, salts, and many solvents. Due to the chlorine monomeric unit, it is inherently flame retardant. Although it can be used for wiring and cable insulation, at high temperature, it is unstable. So, PVC has tremendous characteristics like good flexibility, high gloss, small lead content, flame retardancy, and finally the most important is multiplicity of manufacturing techniques. It has two varieties named as flexible and rigid polyvinyl chloride. Rigid PVC is non-plasticized, strong, and stiff with great utility in construction purpose. However, flexible PVCs contain plasticizer and can be used for cables, flooring, and medical purpose. PVC is generally safe for use, but its production and incineration can release harmful chemicals like dioxins. Modern recycling methods are helping reduce environmental concerns related to PVC. Generally, for flexible PVC, phthalates are the most widely used plasticizer that can enhance the flexibility, durability, and applicability of polyvinyl chloride. Plasticizers can lower the rigidity of PVC by lowering its glass transition temperature that can make the polymer more flexible and easier to handle. Most commonly used phthalate plasticizers are diethylhexyl phthalate (DEHP), dibutyl phthalate (DBP), benzyl butyl phthalate (BBP), and diisononyl phthalate (DINP). Again, use of plasticizer decides the utility of PVC-based products in various industries. Use of DBP plasticizer makes the PVC for its use as adhesives, paints, and coatings. Similarly, DINP-based PVC can be used in flooring, cables, and toys. DEHP plasticizer-based PVC can be used in automotive and wire industries. BBP plasticizer-based flexible PVC can be used in vinyl tiles, flooring, and sealants. But now, this phthalate plasticizer has become very contentious and now, it is the time to substitute them with various biodegradable plasticizers. Actually, phthalates create respiratory problems, cause developmental defects in newborns, interfere with the hormonal system, and generate infertility. It is also carcinogenic to health. Phthalates also create environmental pollution by leaking into water and soil bodies and persist for long periods. Such boundaries limit the potential applications of PVC. Three primary solutions have been proposed to address the growing environmental pollution created by the improper disposal of polymers. These include recycling, incineration, and the development and use of biodegradable polymers and their blends [1, 2]. However, ongoing research aims to address these challenges by developing innovative PVC products with improved properties, thereby expanding their applications and increasing their environmental as well as economic value across various industries. One effective method to mitigate these limitations involves incorporating various bio-fillers, such as clays, calcium carbonate, silica, talc, and also bio-based polysaccharides like chitin, chitosan, starch, glycogen, and cellulose, into PVC polymer nanocomposites [2]. This strategy provides several advantages, including improved mechanical strength, hardness, heat resistance, and barrier properties, along with better biodegradability [2]. Among all reinforcing biopolymers, starch is the most emerging one as that can satisfy all the necessary properties with increasing the biodegradability without sacrificing the inherent tendency of PVC.

Commonly abundant and naturally occurring starch is one of the polysaccharides that can be obtained from natural sources like corn, potato, and cassava. It is nontoxic, economical, renewable, and green polymer that is second to cellulose in occurrence. It is extensively available in the seeds, roots, and leaves of varieties of plants and some algae. Starch-based biocomposites and bionanocomposites have been widely considered because of their easy availability, inexpensiveness, simplicity of handling, and various developed physical and chemical properties [3, 4]. This globally available biopolymer can contribute to a healthy ecosystem when blended with any petroleum-based polymers like polyvinyl chloride [5]. Again, these starch-based bioproducts have drawn attention in various applications due to their properties like nanosized dimensions, high surface area, high absorptivity, biodegradability, nontoxicity, renewability, low density, and easy modification [6]. Natural starch has also some inadequacies like high hydrophilicity and low mechanical characteristics, for which it cannot be utilized alone in its regular natural or improved modified form. Modification of this natural polymer has also very important role for the present situation as it can turn into a suitable candidate for multiple applications. Starch can be modified through blending with other polymers, or it can be converted to nanoparticles by undergoing various physical and chemical modification [7, 8]. These physical modifications involve a combination of different types of polymers through milling, extrusion, introduction of plasticizers, and also chemical modifications that comprise esterification, cross-linking, substitution, oxidation, graft copolymerization, etherification, etc. to produce modified materials with innovative characteristics [9, 10].

This chapter explores a novel idea regarding the starch-based biocomposite and bionanocomposites reinforced with PVC and their applications in various fields with highlighting their potential as environmentally friendly alternatives to conventional PVC materials. Introduction of starch not only develops the ecological effect of PVC-based materials but also develops specific characteristics, such as biodegradability and lightweight properties. Again, it follows the twelve principles of green chemistry with special emphasis on the role and impact of starch as a biodegradable matrix component on composite performance, and also the future outlooks in the field of biocomposites from a non-biodegradable polymer. The twelve principles of green chemistry give priority to preventing waste or reintroduce this waste as a raw material for new synthesis is far more sustainable than cleaning it up afterward. In line with this philosophy, it is essential to transform synthetic polymeric materials to their corresponding biodegradable form by using biopolymers to improve their environmental impact.

2.2 Starch

Commonly, starch can be defined as a renewable, easily available plant-based polysaccharide with cheap and biodegradable properties. Generally, polysaccharides are of different types based on their structure and nature as given in Figure 2.1. Now-a-days, starch-based materials are receiving much more attention only due to its high availability from renewable resources. Starch is a homopolysaccharide that contains the same type

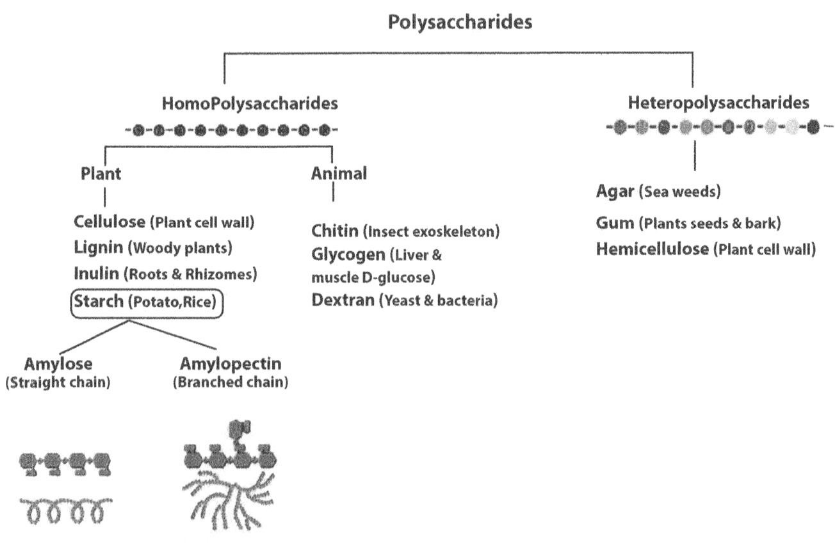

Figure 2.1 Types of polysaccharides.

of monosaccharide units. It can be formed by successive condensation of amylose and amylopectin. So, basically starch consists of two types of polymers, i.e., about 10 to 30% of mostly linear polymeric glucose units of amylose and another extremely branched polymer with several small chains of 70% to 90% of amylopectin. These linear polymeric D-glucose units of amylose are linked to each other by α-(1,4) glycosylic bonds; however, highly branched amylopectin is a structure with high molecular weight, and here, the D-glucose repeating units are simply connected by these same α-(1,4) glycosylic bonds but with about 5% of branching α-(1,6) glycosidic bonds [11].

Again, this amylopectin main backbone holds 10 to 60 glucose entities, and the splitting side chains contain 15 to 45 glucose units with approximately 5% of α-(1,6) branching points [11, 12]. Both these amylose and amylopectin are radially organized in an alternating way of concentric ring in the granules of starch as an amorphous and semicrystalline structure respectively [13]. So, simply, the single helices of amylose are circulated arbitrarily between the clusters of amylopectin. This amorphous region is the result of a combination of amylose with the branching point of amylopectin [14, 15]. Here, Figure 2.2 demonstrates various types of starches. Starch is a primary energy source in plants like corn, cassava, potato, rice, barley, wheat, peas, mung beans, corn, lentils, millet, sago, and banana [16–18]. It is deposited in the roots, seeds, stalks, and various storage organs of the plant [11].

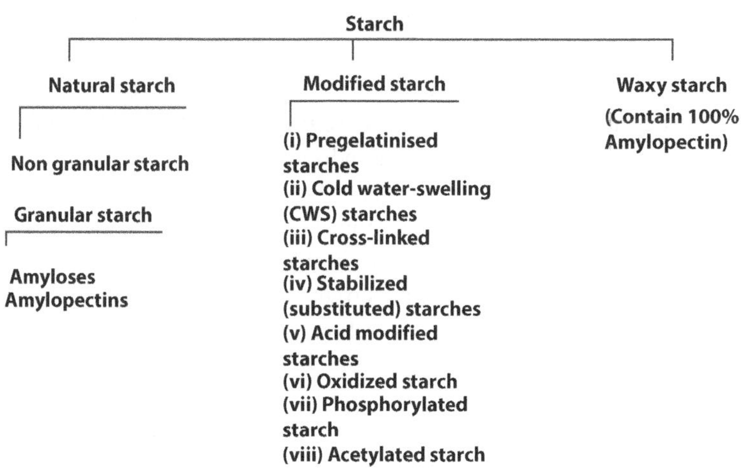

Figure 2.2 Classification of starch.

Starches from different sources show variation in their chemical composition like α-glucans, moisture, lipids, proteins, and phosphorylated residues; the structure of glucan components is amylose and amylopectin [19]. The size, nature, and shape of starch granule is the result of various ecological and hereditary factors which can be different with respect to the content, configuration, and arrangement of its two components, i.e., amylopectin and amylose [20]. So, granules of starch that are loaded with amylopectin and amylose are available in different magnitudes ranging from 2 μm to 150 μm.

In a broad way, different types of starches are available according to their source, structural arrangement, and functional properties. Again, in this current situation, new types of nanostarches are also available with more applicability. So, starches are of the following types:

2.2.1 Natural Starch

Such natural or native starches are naturally available in different types of plants like corn, wheat, and waxy maize. These starches may be granular or non-granular type. Granular starches are round or polygonal shape and present in various grains with different dimension range of 2 to 150 μm. These granular starches consist of two moieties, i.e., amylopectin and amylose that are involved by glycosidic linkages. Amylose is a stretched straight chain made up of thousands of units of glucose moieties. Here, the 1st carbon and 4th carbon of glucose units are attached, and results α-1, 4 glycosidic linkages. It can be converted to a 3D network when molecules associate with each other due to cooling and is responsible for the gelation of cooked, cooled starch pastes. The amylose content of cereal grains based starch is approximately 25–30% whereas roots possess 15–21%. Among these two molecules of granular starch, amylose molecules are present in 20% whereas amylopectin molecules are present in 80% of the polymers. Like amylose, the glucose chain of amylopectin contains linkage between the 1-carbon of the glucose and 6-carbon of the branch, i.e., α-1, 4 glycosidic linkages as well as α-1, 6 branching at every 15–30 glucose units of the chain. Generally, the starch chains are highly branched due to this amylopectin unit. These simple long chains make amylose more soluble in water in comparison to amylopectin. Therefore, starches having more concentration of amylopectin is responsible for the formation of a thick mixture, but here, gel cannot be formed. The higher the percentage of amylose, the stronger the gel, and the more the concentration of amylopectin, the more viscous the starch paste is. Figure 2.3 shows the detailed chemical structure of amylose and amylopectin. On the other hand, the non-granular type of

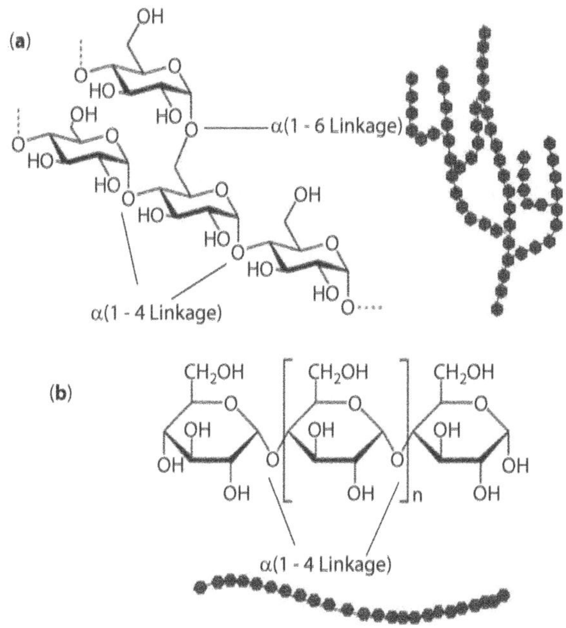

Figure 2.3 Chemical structure of granular starch: (a) amylopectin with branched chain and (b) amylose with long straight chain.

starch is generally utilized as a source for the formation of chemically processed starches for more innovative applications in various sectors.

2.2.2 Modified Starch

Functional modification of natural starches occurs through a physical, chemical, or enzymatic process that results in this modified starch with the introduction of various inter- and intramolecular bonds at various positions [21]. Again, starches contain a greater number of hydroxyl groups that make them hydrophilic in nature and low mechanical strength [22]. So, to avoid such bottle neck, the starch solution can be modified to various forms by introducing different functional groups. This modification can contribute to improve the stability, appearance, accessibility, and performance in various industrial applications. Such modifications are of various types:

(i) Pregelatinized Starches:
Such type of modified starches can be gelatinized instantly and then dried. It can easily swell in aqueous medium without any heat. So, the primary tendency

of pregelatinized starches is its easy dispersion in cold water. On reaction, it shows irreversible change as it cannot return back to the original form.

(ii) Cross-linked Starch:
These types of starches are chemically modified in the presence of chemical reagents like epichlorohydrin, sodium tripolyphosphate, phosphorus oxy chloride($POCl_3$), adipic acid, and acetic acid [21]. Such modification introduces covalent or noncovalent bonds among starch molecules with successful molecular interaction at selected hydroxyl (–OH) moieties of two adjacent starch molecules. These cross-linkings can enhance the stability of the starch by making it more pH resistant, acid resistant, and thermally stable than the native starch.

(iii) Stabilized Starches:
These starches are the result of chemical modification of native starch by succinic anhydride, propylene oxide, and acetic anhydride. Such starches show improved water-retaining capacity, binding and thickening ability, and thermal resistance. They are able to survive several times in freezing conditions.

(iv) Acid-Modified Starches:
Here, the starch is introduced to acid solutions like nitric acid, phosphoric acid, and citric acid for a specific time followed by heat treatment. Then, the starch solution undergoes neutralization followed by filtration. Such process reduces the inherent viscosity tendency of starch by breaking down the molecules into smaller form. It has wide industrial application in the food, fabric, and paper manufacturing.

(v) Oxidized Starch:
Starch on treatment with oxidizing reagents like hydrogen peroxide, nitrogen dioxide, and sodium hypochlorite can introduce carbonyl and carboxyl moieties in a starch molecule. Such modification can result to the decrease in the viscosity and increase in the film building capacity of the starches that have great applicability in paper coatings and adhesives.

(vi) Phosphorylated Starch:
Introduction of phosphate groups into starch in the presence of sodium tripolyphosphate or Sodium trimetaphosphate can accelerate the water-storing capacity and mechanical properties when it is used as a reinforcing material. In this phosphorylation process, starch metabolism occurs, where dikinases convert the phosphate groups to the glucosyl residues of amylopectin [23].

(vii) Acetylated Starch:
Chemical modification of starch by acetic anhydride or vinyl acetate for the introduction of acetyl functional moieties is called as acetylated starch. It has a great contribution for decreasing retrogradation and enhancing swelling capacity, granular size, and water absorption capacity as well as stability.

2.2.3 Waxy Starches

Waxy starches do not contain gel-forming amylose for which it cannot produce gels like ordinary starch. Also, it is less thick than native starch. These types of starches contain 100% amylopectin. So, waxy starches can be listed as proto pectin (methylated galacturonic acid polymer with high-molecular weight and insoluble in cold water and soluble in hot water), pectinic acid (soluble in water but cannot form gel), pectic acid (shorter chain derivative), etc. Due to some specific functional moieties, these starches possess high viscosity and better thermal resistance.

2.2.4 Thermoplastic Starches

Native natural starches are inherently water sensitive and do not possess thermoplastic properties. Ordinary starch can be converted to the highly useful synthetic thermoplastic starch through casting solutions or extrusion procedures by the introduction of plasticizers like water, glycerol, sorbitol, and other additives at high temperature [24]. Natural pristine starch gelatinizes and shows thermoplastic tendency with the addition of plasticizers followed by thermal processes. These thermoplastic starches show improved mechanical properties like tensile strength, elongation at break, elastic modulus, thermal stability, and flexibility and also reduce glass transition temperature (Tg). Actually, starch-based polymers have a variety of industrial uses, such as for food packaging, flexible films, injection-molded pots, candy trays, compost bags, plant pots, coatings, drug-delivery capsules, golf tees, cutlery, plates, and food containers [25]. Also, this thermoplastic starch can be prepared by introducing them as reinforcing filler with some biopolymers like polylactic acid and polyhydroxyalkanoate.

2.2.5 Nanostarch

The breaking down of starch granules from macro- to microscale and from micro- to nanoscale results in these nanostarches. These nano characteristics have drawn the attention of the current scenario not only because of their unique property like particle size, but also due to the enhanced functional characteristics [26]. Actually, the resulted nanostarches are of two different types: starch nanocrystals (SNCs) and starch nanoparticles (SNPs). SNCs are crystal-like portions that can be resulting from the disruption of amorphous areas in starch granules, whereas SNPs are produced from amorphous regions of gelatinized starch [27, 28]. Again, SNPs are considered as gifted biomaterials of starch as it is ornamented with various properties that can suggest numerous opportunities for improvement of quality for several innovative developments [29, 30]. SNPs are extensively used in active food packaging [31], bioactive compound encapsulation [32], and nanocomposite film [33]. SNPs are mainly manufactured by chemical approaches like enzymatic or acid hydrolysis and various physical processes like gamma radiation, high-pressure homogenization, milling, and ultra-sonication [34]. SNPs are widely used as reinforcing fillers for multiple polymer matrices to develop their mechanical, barrier, and environmental properties [28, 35]. These are generally atoxic and can be beneficial for preparing nanocomposite, absorbent, and emulsion stabilizers for food and non-food applications [34, 36, 37].

2.3 Starch-Based Biocomposites

Starch can successfully produce various types of biocomposites on introduction with numerous petroleum-based synthetic polymers as well as biopolymers through blending or as an effective reinforcing filler. So Figure 2.4 depicts that starch can act as a capable material to synthesize various types of biocomposite materials embedded with unique characteristics. During the period of primary evolution in composite industry in the year1970s, this starch was introduced as a green additive for synthetic artificial polymers to improve their biodegradability and also discover the hidden potential of this richly accessible natural resource [38]. Again, starch in its inherent form possesses high water sensitivity and poor material properties [39]. For which it cannot be used alone for any important application. Henceforth, various approaches have been introduced to overcome these issues over recent years including starch film, biocomposite, and starch as biofiller and bionanofiller also [40].

Starch is an appropriate reinforcing filler material due to its thermal stability and restricted interfering tendency with the normal melt flow characteristics of petroleum-based polymers [41]. So, blending of starch with petroleum-based nonbiodegradable polymers reduces their nonbiodegradability as starch can naturally be degraded through normal thermal oxidation or by various microorganisms by separating the frame of

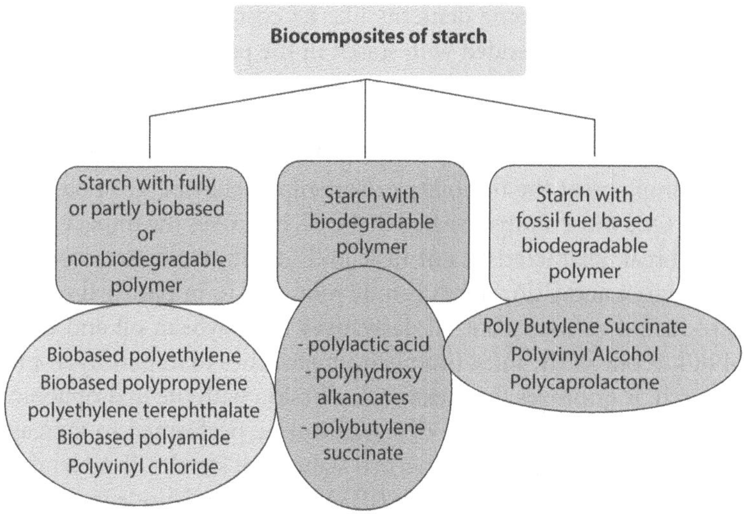

Figure 2.4 Various types of biocomposites from starch.

synthetic polymers [42]. The characteristics of a reinforcing polymer material are generally affected by the dimension and shape of the material and its compatibility with the polymer matrix [43]. Specifically, for the biopolymer like starch, material size plays a key role. But, the incompatibility arises due to the hydrophilic property of starch and the hydrophobic tendency of synthetic polymers that hinder the formation of a valid bond among the two polymers. So, various approaches have been taken to accelerate the overall performance of starch-reinforced polymers through the formation of blends and composites by using various compatibilizers and plasticizers or by introducing various functional groups that can enhance the miscibility between starch and synthetic polymer by improving the interaction through the formation of covalent bond among the two adjacent phases [44]. Starch is one of the most commonly used natural resin that has been used in different biocomposites as follows.

2.3.1 Starch/Biodegradable Polymer-Based Biocomposites

These types of biocomposites are fully eco-friendly as both the phases, i.e., resin as well as filler phases, are environmentally friendly, nontoxic, renewable, and naturally occurring.

Polylactic acid is a biodegradable, hydrophobic, and aliphatic polyester resin matrix resulting from lactic acid polymerization. Inherently, it carries excellent mechanical properties similar with synthetic petroleum-based polymers. But, it has some demerits like its cost and brittleness [45]. For this reason, it can be blended with starch in the presence of some compatibilizer or crosslinking agents that can enhance the mechanical properties of the resulted biocomposites. This compatibilizer binds the matrix (PLA) and filler (starch) molecules together. Again, here, another component can be added to implement the flexibility or softening behavior, i.e., polycaprolactone [46]. One more study revealed that PCL improves the impact and elongation at break characteristics of PLA [47]. Again, for better coordination between starch and PLA, starch can be modified to impart hydrophobicity by introducing various plant oil derivatives like soybean oil and castor oil [48, 49]. It is one of the most promising approaches to impart hydrophobicity in starch. It improves the interfacial adhesion with the hydrophobic PLA component. Again, acetylation of starch can be performed to increase the mechanical characteristics and thermal stability of starch–PLA blends [50].

Similarly, polyhydroxyalkanoate (PHA) is a natural biodegradable, nontoxic, hydrophobic, polyester of microbial origin and good film-forming capability [51]. Its limitation lies in its poor mechanical characteristics, more production cost, limited functionality, incompatibility with conventional thermal processing techniques, and weakness to thermal degradation. These bottlenecks can be avoided by introducing pure starch to make a blend and then a biocomposite, where both the matrix and filler gain some improvement than their neat counterpart. Now, the resulted biocomposite is technically a good candidate for various applications. The starch/PHA biocomposite was ornamented with better tensile strength, Young's modulus, and elongation to break as well as converted to cost effective by maintaining the normal biodegradability tendency.

Again, a new and interesting biocomposite of starch can be discussed where starch is blended with natural rubber in the presence of plasticizer cum compatibilizer, i.e., glycerol [52]. Here, glycerol is used to reduce the phase separation tendency of both the polymer and filler. Again, various chemical functionalizations have been performed in various literature to avoid this situation during the preparation of biocomposite.

2.3.2 Starch/Natural Fiber–Based Biocomposites

Presently, many works have been done on natural fiber–reinforced starch-based biocomposite by using nontoxic and chemically stable glycerine or glycerol as plasticizer [53, 54]. Here, for the preparation of this biocomposite, the starches may be from corn, potato, or cassava and natural fiber are in the form of pulp, powder, micro- or nanoparticle, or fibers of very small length. Results showed that introduction of about 15–20% fiber increased certain mechanical properties like tensile and flexural strength as well as elongation at break with decreased water absorbency. In the starch/fiber biocomposite, the addition of fiber did not affect the thermal properties of the composite positively though. There are so many articles based on the formation of okra fiber jute fiber-reinforced thermoplastic starch composite [55, 56].

2.3.3 Starch/Fossil Fuel–Based Biocomposite

Starch and fossil fuel–based ecofriendly polymers like polybutylene succinate [57], polyvinyl alcohol [58], and polycaprolactone [59] can produce biocomposites with inherent biodegradability as well as high mechanical strength and water resistance capacity. This innovative approach can effectively enhance the overall mechanical properties as well as thermal stability in comparison to pristine starch [58]. Here also, procedures were taken to overcome the poor compatibility among the starch and above matrix by including a compatibilizer [57] or plasticizers like calcium chloride [60] and different levels of plasticizers [59].

2.4 Starch-Based Bionanocomposites

Bionanocomposites can also be stated as green composites or biohybrids or bioplastics [5, 61]. Like a normal nanocomposite, also in starch-based bionanocomposites, starch is used in nanoform ranging from 1 nm to 100 nm. Therefore, nanostarch is blended with other ecofriendly and manmade synthetic polymers or combined with various nanomaterials to improve the mechanical, physical, and barrier properties [12, 62, 63]. Nanostarches are of two different types, i.e., starch nanocrystal and starch nanoparticles. Figure 2.5 shows different types of starches and their various preparation methods. Some starch-based bionanomaterials, are discussed below.

Currently, plant cell–based nanostarches and nanocellulose bionanocomposites have drawn much attention due to their excellent properties like improved biodegradability, enhanced crystallinity, renewability, biocompatibility, flexibility, as well as mechanical and thermal properties [64, 65]. About 5 wt.% of corn starch nanoparticles into 1.5% of nanocellulose fiber can make films with improved tensile properties by 85% and decreased water vapor permeability by 32% and the oxygen transmission rate by 80%. It also gives enhanced thermal stability and flexibility and diminish the contact angle from 38° to 69° [66].

Starch and chitin nanofiber-based bionanocomposites exhibited exceptional emulsion stability over a variety of ranges of temperatures and pHs

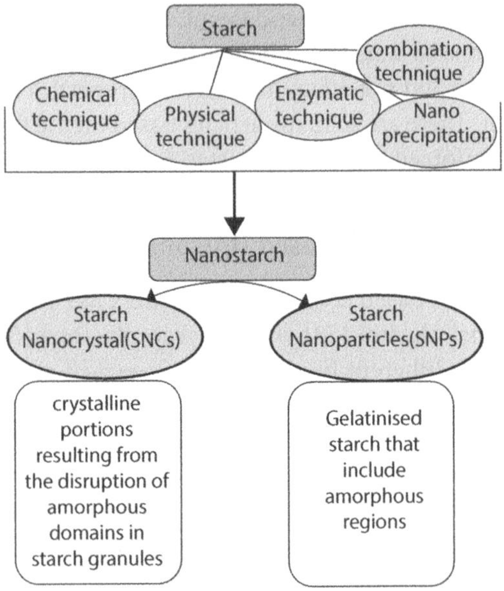

Figure 2.5 Scheme of nanostarch types and various preparation methods.

that can act as a perfect candidate as emulsion stabilizer. It can be used in a number of industries including medicines, food, cosmetics, paint, and coating [37]. Addition of 5 wt.% of chitin nanofibers with starch enhances the Young's modulus and the tensile strength and reduces the moisture absorption. Again, an effective combination of potato starch and chitin nanoparticles can result the bionanocomposite ornamented with improved tensile strength, storage modulus, glass transition temperature, and water vapor barrier properties [67]. These improved properties are the outcome of good interfacial interaction among starch polymeric resin and chitin nanoparticles. It can be used in pharmaceutical, packaging, green electronics, water treatment, and several other applications [68].

Introduction of montmorillonite nanoclay into corn starch can result into a bionanocomposite, with enhanced mechanical strength, swelling thickness, and reduced water or moisture absorption, oxygen permeation, as well as biodegradability in comparison to pristine corn starch. Moreover, the specific quantity of reinforcing montmorillonite nanoclay defines the structural features and applications of the nanocomposite [69]. Again, the addition of montmorillonite as a reinforcing filler in cross-linked polylactic acid and maleated functionalized starch bionanocomposite can successfully enhance the tensile strength and biodegradability [70].

2.5 PVC/Starch-Based Biocomposites

If in a synthesized material at least one component is derived from natural sources, then it is called as biocomposites. So here, biocomposites may be starch-reinforced biopolymer or natural fiber–reinforced starch polymer or starch-reinforced synthetic polymer-based biocomposites. Actually, polyvinyl chloride (PVC) is the mostly used synthetic thermoplastic polymer with various types of revolutionary features like simplicity of usage, exceptional insulating abilities, resistance to various chemicals, and fire resistance [71, 72]. Also, it has great utilization in construction [73], electronics [74], automotive [75, 76], packaging [77, 78], and healthcare [79]. But only the problem is that it remains chemically and physically inert for many years even after proper disposal due to the evil effect of chlorine which is the main culprit for environmental pollution. So, biodegradable filler like starch was introduced to produce biocomposites from petroleum-based polymer resin like PVC. Currently, a number of techniques were followed for introducing corn starch into PVC polymers to successfully produce a biocomposite [80]. According to earlier studies, there are various techniques used for the synthesis of biocomposites through dry mixing of starch and PVC, coprecipitation of cross-linked starch xanthate with a PVC latex as well as co-concentration of starch and PVC latex [81]. If we consider the characterization of the biocomposite here, then some mechanical studies presented the reduction in tensile properties and elongation at break only due to the introduction of starch in PVC [82]. It is because the greater the content of starch in a blend, the lesser the flexibility of the mixture and reduction in the mechanical properties of the biocomposite. Again, due to the combination of PVC and starch, some intermolecular force of interaction has been created and some type of intermolecular flow resistance between PVC chains has been generated. On the other hand, starch increases the Young's modulus due to the enhanced rigidity of the material. Actually, PVC is the plastic phase and starch is the rubbery phase and due to the combination, amorphous parts of starch undergo a transition from glassy to rubbery state [82]. Again, thermogravimetric analysis (TGA) indicated that the overall maximum degradation value of PVC gets positively influenced by the incorporation of starch. In addition to it, another important conclusion has been performed from a study that more concentration of starch can enhance the degree of dechlorination. Similarly, higher content of starch can successfully reduce the transition point of differential scanning calorimetry for the modified samples. This decrease in transition temperature indicates the valid interaction among

PVC polymer with starch filler through the formation of valid hydrogen bonding. Biodegradability study of the starch-reinforced PVC biocomposite showed a gradual weight loss during aging in simulated soil. Desired samples comprising higher quantity of starch resulted in little weight loss after 2, 4, and 6 months in simulated soil [82].

2.6 PVC/Starch-Based Bionanocomposites

PVC and starch-based bionanocomposites have gathered maximum attention only because of their prospective to combine the advantageous characteristics of both synthetic and natural polymers. The integration of starch into PVC matrices aims to enhance biodegradability and reduce environmental impact, while the incorporation of nanofillers seeks to improve mechanical, thermal, and barrier properties. Research has explored the incorporation of varying levels of starch into PVC resin matrix. A study showed that PVC compounds containing different starch levels like more starch content led to greater material stiffness but reduced tensile strength and elongation at break. The limited interfacial interaction between PVC and starch resulted in slight phase separation, affecting mechanical properties [82]. The bionanocomposite that is produced due to the addition of nanofillers like layered double hydroxides and rice husk ash into PVC and thermoplastic starch matrices has been investigated, and these bionanocomposites exhibited enhanced mechanical properties and thermal stability. It is the result of uniform dispersion of nanofillers and improved interfacial adhesion among the components [83]. Incorporating nanostarch with PVC can improve the overall biodegradability of the resulting bionanocomposite. In addition to that, samples with higher starch content showed negligible weight loss after prolonged exposure to simulated soil that is due to the encapsulation of starch by the PVC polymer matrix that can hinder the microbial degradation [82]. The development of thermoplastic starch-based bionanocomposites presents an excellent alternative to reduce dependence on petroleum-based polymers. The abundance, renewable origin, biodegradability, and low cost of starch contribute to the environmental benefits of these materials [84]. Again, now-a-days, some surface modification techniques and the use of a compatibilizer have been discovered to improve the compatibility and overall properties of the nanocomposites. It can become the answer to the challenge for the interfacial adhesion between a hydrophobic PVC polymer and hydrophilic starch biomaterial. Moreover, optimizing some advanced processing techniques is very effective for achieving uniform dispersion of nanofillers and consistent composite properties as well as scalability and commercial viability of these biomaterials.

2.7 Conclusion

The disposal of synthetic polymer-based products is becoming a matter of alarm day by day. So to get rid of such global issues, researchers have to give more emphasis on the development of biobased polymeric materials with the introduction of natural biopolymers with synthetic polymers. Also, according to the current trend, many research works have already been published related to various petroleum-based polymers that are finally resulted to biobased composites, nanocomposites, and bionanocomposites with multidimensional applications [85–91]. Here also, PVC is a synthetic nonbiodegradable polymer whereas starch is one of the most important natural polymers with inherently outstanding biocompatible properties. So, in this chapter, biomaterials of PVC and starch have been studied. Again here, this starch-based biopolymer was used as both a resin matrix as well as a reinforcing filler material for the development of new biocomposite and bionanocomposites. The exceptional biocompatibility, easy availability, inherently green tendency, affordable cost, and renewability of these bionanocomposites would open up many applications in agriculture, packaging, biomedicine, and many other fields [82]. Also, such synthesis of PVC and starch biomaterial shows very slight modification in the surface characteristics, reduction in molecular weight, and change in chemical structure proved by various morphological studies like SEM, FTIR, and NMR [92]. This synthesized PVC and starch-based biomaterial is expected to be a better construction material in the form of bioplastics than simple biocomposite or bionanocomposite. The starch-induced PVC biocomposite cum bioplastic is a cheap and eco-friendly, alternative of conventional polymers with potential applications in all major fields of polymeric applications.

AI disclosure statement: The author confirms that no artificial intelligence (AI)-generated content or AI-assisted writing tools were used in the preparation of this chapter (including summaries and the abstract). All text, analysis, and conclusions are the result of the author's original work.

References

1. Avella, M., Bonadies, E., Martuscelli, E., Rimedio, R., European current standardization for plastic packaging recoverable through composting and biodegradation. *Polym. Test.*, 20, 5, 517–521, 2001.
2. Bilck, A.P., Grossmann, M.V.E., Yamashita, F., Biodegradable mulch films for strawberry production. *Polym. Test.*, 29, 4, 471–476, 2010.
3. Ates, B., Koytepe, S., Ulu., A., Gurses, C., Thakur, V.K., Chemistry, Structures, and Advanced Applications of Nanocomposites from Biorenewable Resources. *Chem. Rev.*, 120, 9304–9362, 2020.
4. Saad, E.M., Elshaarawy, R.F., Mahmoud, S.A., El-Moselhy, K.M., New Ulva Lactuca Algae Based Chitosan Bio-Composites for Bioremediation of Cd(II) Ions. *J. Bioresour. Bioprod.*, 6, 3, 223–242, 2021.
5. Puiggali, J. and Katsarava, R., Chapter 7—Bionanocomposites, in: *Clay-Polymer Nanocomposites*, K. Jlassi, M.M. Chehimi, S. Thomas (Eds.), pp. 239–272, Elsevier, Amsterdam, The Netherlands, 2017.
6. Madhumitha, G., Fowsiya, J., Mohana Roopan, S., Thakur, V.K., Recent Advances in Starch–Clay Nanocomposites. *Int. J. Polym. Anal. Charact.*, 23, 331–345, 2018.
7. Be Miller, J.N., Chapter 5—Physical Modification of Starch, in: *Starch in Food*, 2nd, M. Sjoo and L. Nilsson (Eds.), pp. 223–253, Woodhead Publishing Series in Food Science, Technology and Nutrition, Woodhead Publishing, Cambridge, UK, 2018.
8. Gunawardene, O.H.P., Gunathilake, C.A., Amaraweera, A.P.S.M., Fernando, N.M.L., Manipura, A., Manamperi, W.A., Kulatunga, K.M.A.K., Rajapaksha, S.M., Gamage, A., Dassanayake, R.S., Removal of Pb(II) Ions from Aqueous Solution Using Modified Starch. *J. Compos. Sci.*, 5, 46, 2021.
9. Okoli, C.P. and Ofomaja, A.E., Degree of Time Dependency of Kinetic Coefficient as a Function of Adsorbate Concentration; New Insights from Adsorption of Tetracycline onto Monodispersed Starch-Stabilized Magnetic Nanocomposite. *J. Environ. Manag.*, 218, 139–147, 2018.
10. Zarski, A., Bajer, K., Kapu sniak, J., Review of the Most Important Methods of Improving the Processing Properties of Starch toward Non-Food Applications. *Polymers.*, 13, 832, 2021.
11. Arora, B., Bhatia, R., Attri, P., Bionanocomposites: Green Materials for a Sustainable Future, in: *New Polymer Nanocomposites for Environmental Remediation*, C.M. Hussain and A.K. Mishra (Eds.), pp. 699–712, Elsevier, Wilmington, NC, USA, 2018.
12. Garcia, N.L., Fama, L., D'Accorso, N.B., Goyanes, S., Biodegradable Starch Nanocomposites, in: *Eco-friendly Polymer Nanocomposites: Processing and Properties*, V.K. Thakur and M.K. Thakur (Eds.), pp. 17–77, Advanced Structured Materials; Springer, New Delhi, India, 2015.
13. Bertolini, A. (Ed.), *Starches: Characterization, Properties, and Applications*, 1st, CRC Press, Boca Raton, FL, USA, 2009.

14. Govindaraju, I., Zhuo, G.Y., Chakraborty, I., Melanthota, S.K., Mal, S.S., Sarmah, B., Baruah, V.J., Mahato, K.K., Mazumder, N., Investigation of Structural and Physico-Chemical Properties of Rice Starch with Varied Amylose Content: A Combined Microscopy, Spectroscopy, and Thermal Study. *Food Hydrocoll.*, 122, 107093, 2022.
15. Perez, S., Baldwin, P.M., Gallant, D.J., Chapter 5—Structural Features of Starch Granules I, in: *Starch*, 3rd, J. BeMiller and R. Whistler (Eds.), pp. 149–192, Food Science and Technology; Academic Press, San Diego, CA, USA, 2009.
16. Alves, Z., Abreu, B., Ferreira, N.M., Marques, E.F., Nunes, C., Ferreira, P., Enhancing the Dispersibility of Multiwalled Carbon Nanotubes within Starch-Based Films by the Use of Ionic Surfactants. *Carbohydr. Polym.*, 273, 118531, 2021.
17. Basiak, E., Lenart, A., Debeaufort, F., Effect of Starch Type on the Physico-Chemical Properties of Edible Films. *Int. J. Biol. Macromol.*, 98, 348–356, 2017.
18. Chaudhary, A.K. and Vijayakumar, R.P., Synthesis of Polystyrene/Starch/CNT Composite and Study on Its Biodegradability. *J. Polym. Res.*, 27, 187, 2020.
19. Copeland, L., Blazek, J., Salman, H., Tang, M.C., Form and Functionality of Starch. *Food Hydrocoll.*, 23, 1527–1534, 2009.
20. Cornejo-Ramirez, Y.I., Martínez-Cruz, O., Del Toro-Sanchez, C.L., Wong-Corral, F.J., Borboa-Flores, J., Cinco-Moroyoqui, F.J., The Structural Characteristics of Starches and Their Functional Properties. *CyTA—J. Food*, 16, 1003–1017, 2018.
21. Punia Bangar, S., Sunooj, K.V., Navaf, M., Phimolsiripol, Y., Whiteside, W.S., Recent advancements in cross-linked starches for food applications- a review. *Int. J. Food Prop.*, 27, 1, 411–430, 2024.
22. Bhat, P., Kumar, V., Goel, R., Sharma, S.K., Kaushik, S., Sharma, S., Tesema, M., Structural Modifications and Strategies for Native Starch for Applications in Advanced Drug Delivery. *Biomed Res. Int.*, 1–14, 2022.
23. Mahlow, S., Orzechowski, S., Fettke, J., Starch phosphorylation: insights and perspectives. *Cell. Mol. Life Sci.*, 73, 14, 2753–2764, 2016.
24. Ma, X.F., Yu, J.G., Wang, N., Fly ash reinforced thermoplastic starch composites. *Carbohydr. Polym.*, 67, 32–39, 2007.
25. Stepto, R.F.T., The processing of starch as a thermoplastic. *Macromol. Symp.*, 201, 203–212, 2003.
26. Alves, M.J.d.S., Chacon, W.D.C., Gagliardi, T.R., Agudelo Henao, A.C., Monteiro, A.R., Ayala Valencia, G., Food Applications of Starch Nanomaterials: A Review. *Starch-Starke*, 73, 2100046, 2021.
27. Qiu, C., Hu, Y., Jin, Z., McClements, D.J., Qin, Y., Xu, X., Wang, J., A Review of Green Techniques for the Synthesis of Size-Controlled Starch-Based Nanoparticles and Their Applications as Nanodelivery Systems. *Trends Food Sci. Technol.*, 92, 138–151, 2019.

28. Le Corre, D. and Angellier-Coussy, H., Preparation and Application of Starch Nanoparticles for Nanocomposites: A Review. *React. Funct. Polym.*, 85, 97–120, 2014.
29. Ahmad, M., Gani, A., Hassan, I., Huang, Q., Shabbir, H., Production and Characterization of Starch Nanoparticles by Mild Alkali Hydrolysis and Ultra-Sonication Process. *Sci. Rep.*, 10, 3533, 2020.
30. Kim, H.Y., Park, S.S., Lim, S.T., Preparation, Characterization and Utilization of Starch Nanoparticles. *Colloids. Surf. B Biointerfaces.*, 126, 607–620, 2015.
31. Basavegowda, N. and Baek, K.H., Advances in Functional Biopolymer-Based Nanocomposites for Active Food Packaging Applications. *Polymers*, 13, 4198, 2021.
32. Moran, D., Gutierrez, G., Blanco-Lopez, M.C., Marefati, A., Rayner, M., Matos, M., Synthesis of Starch Nanoparticles and Their Applications for Bioactive Compound Encapsulation. *Appl. Sci.*, 11, 4547, 2021.
33. Marta, H., Wijaya, C., Sukri, N., Cahyana, Y., Mohammad, M., A Comprehensive Study on Starch Nanoparticle Potential as a Reinforcing Material in Bioplastic. *Polymers*, 14, 4875, 2022.
34. Sandhu, K.S. and Nain, V., Starch Nanoparticles: Their Preparation and Applications, in: *Plant Biotechnology: Recent Advancements and Developments*, S.K. Gahlawat, R.K. Salar, P. Siwach, J.S. Duhan, S. Kumar, P. Kaur (Eds.), pp. 213–232, Springer, Singapore, 2017.
35. Santana, J.S., de Carvalho Costa, E.K., Rodrigues, P.R., Correia, P.R.C., Cruz, R.S., Druzian, J.I., Morphological, Barrier, and Mechanical Properties of Cassava Starch Films Reinforced with Cellulose and Starch Nanoparticles. *J. Appl. Polym. Sci.*, 136, 47001, 2019.
36. Campelo, P.H., Sant'Ana, A.S., Pedrosa Silva Clerici, M.T., Starch Nanoparticles: Production Methods, Structure, and Properties for Food Applications. *Curr. Opin. Food Sci.*, 33, 136–140, 2020.
37. Lee, Y.S., Tarte, R., Acevedo, N.C., Synergistic Effects of Starch Nanoparticles and Chitin Nanofibers on the Stability of Oil-in-Water Pickering Emulsions. *Food Chem.*, 363, 130301, 2021.
38. Otey, F.H., Current and Potential Uses of Starch Products in Plastics. *Polym. Plast. Technol. Eng.*, 7, 221–234, 1976.
39. Nawrath, C., Poirier, Y., Somerville, C., Plant Polymers for Biodegradable Plastics: Cellulose, Starch and Polyhydroxyalkanoates. *Mol. Breed.*, 1, 105–122, 1995.
40. Abe, M.M., Martins, J.R., Sanvezzo, P.B., Macedo, J.V., Brancifort, M.C., Halley, P., Botaro, V.R., Brienzo, M., Advantages and Disadvantages of Bioplastics Production from Starch and Lignocellulosic Components. *Polymers*, 13, 2484, 2021.
41. Fabunmi, O., Tabil, L., Panigrahi, S., Chang, P., *Developing Biodegradable Plastics from Starch*, ASABE, St. Joseph, MI, USA, 2007.

42. Rahmat, A.R., Rahman, W.A.W.A., Sin, L.T., Yussuf, A.A., Approaches to Improve Compatibility of Starch Filled Polymer System: A Review. *Mater. Sci. Eng., C*, 29, 2370–2377, 2009.
43. Willett, J.L., Chapter 19—Starch in Polymer Compositions 11, in: *Starch*, 3rd, J. BeMiller and R. Whistler (Eds.), pp. 715–743, Food Science and Technology; Academic Press, San Diego, CA, USA, 2009.
44. Liu, W., Liu, S., Wang, Z., Dai, B., Liu, J., Chen, Y., Zeng, G., He, Y., Liu, Y., Liu, R., Preparation and Characterization of Reinforced Starch-Based Composites with Compatibilizer by Simple Extrusion. *Carbohydr. Polym.*, 223, 115122, 2019.
45. Koh, J.J., Zhang, X., He, C., Fully Biodegradable Poly(Lactic Acid)/Starch Blends: A Review of Toughening Strategies. *Int. J. Biol. Macromol.*, 109, 99–113, 2018.
46. Collazo-Bigliardi, S., Ortega-Toro, R., Chiralt, A., Using Grafted Poly(ε-Caprolactone) for the Compatibilization of Thermoplastic Starch-Polylactic Acid Blends. *React. Funct. Polym.*, 142, 25–35, 2019.
47. Sarazin, P., L, G., Orts, W.J., Favis, B.D., Binary and Ternary Blends of Polylactide, Polycaprolactone and Thermoplastic Starch. *Polymer*, 49, 599–609, 2008.
48. Xiong, Z., Yang, Y., Feng, J., Zhang, X., Zhang, C., Tang, Z., Zhu, J., Preparation and Characterization of Poly (Lactic Acid)/Starch Composites Toughened with Epoxidized Soybean Oil. *Carbohydr. Polym.*, 92, 810–816, 2013.
49. Xiong, Z., Zhang, L., Ma, S., Yang, Y., Zhang, C., Tang, Z., Zhu, J., Effect of Castor Oil Enrichment Layer Produced by Reaction on the Properties of PLA/HDI-g-Starch Blends. *Carbohydr. Polym.*, 94, 235–243, 2013.
50. Nasseri, R., Ngunjiri, R., Moresoli, C., Yu, A., Yuan, Z., Xu, C., Charles., Poly (Lactic Acid)/Acetylated Starch Blends: Effect of Starch Acetylation on the Material Properties. *Carbohydr. Polym.*, 229, 115453, 2020.
51. Godbole, S., Gote, S., Latkar, M., Chakrabarti, T., Preparation and Characterization of Biodegradable Poly-3-Hydroxybutyrate–Starch Blend Films. *Bioresour. Technol.*, 86, 33–37, 2003.
52. Carvalho, A.J.F., Job, A.E., Alves, N., Curvelo, A.A.S., Gandini, A., Thermoplastic Starch/Natural Rubber Blends. *Carbohydr. Polym.*, 53, 95–99, 2003.
53. Chen, B. and Evans, J.R.G., Thermoplastic starch-clay nanocomposites and their characteristics. *Carbohydr. Polym.*, 61, 4, 455–46356, 2005.
54. Chang, P.R., Jian, R., Zheng, P., Yu, J., Ma, X., Preparation and properties of glycerol plasticized starch (GPS)/cellulose nanoparticle (CN) composites. *Carbohydr. Polym.*, 79, 2, 301–305, 2010.
55. Guleria, A., Singha, A.S., Rana, R.K., Mechanical, thermal, morphological, and biodegradable studies of okra cellulosic fiber reinforced starch-based biocomposites. *Adv. Polym. Technol.*, 37, 104–112, 2018.

56. Vilaseca, F., Mendez, J.A., Pelach, A., Llop, M., Canigueral, N., Girones, J., Turon, X., Mutje, P., Composite materials derived from biodegradable starch polymer and jute strands. *Process Biochem.*, 42, 329, 2007.
57. Yin, Q., Chen, F., Zhang, H., Liu, C., Fabrication and Characterisation of Thermoplastic Starch/Poly (Butylene Succinate) Blends with Maleated Poly (Butylene Succinate) as Compatibiliser. *Plast. Rubber Compos.*, 44, 362–367, 2015.
58. Guo, B., Zha, D., Li, B., Yin, P., Li, P., Polyvinyl Alcohol Microspheres Reinforced Thermoplastic Starch Composites. *Materials*, 11, 640, 2018.
59. Averous, L., Moro, L., Dole, P., Fringant, C., Properties of Thermoplastic Blends: Starch–Polycaprolactone. *Polymer*, 41, 4157–4167, 2000.
60. Jiang, X., Jiang, T., Gan, L., Zhang, X., Dai, H., Zhang, X., The Plasticizing Mechanism and Effect of Calcium Chloride on Starch/Poly (Vinyl Alcohol) Films. *Carbohydr. Polym.*, 90, 1677–1684, 2012.
61. Zafar, R., Zia, K.M., Tabasum, S., Jabeen, F., Noreen, A., Zuber, M., Polysaccharide Based Bionanocomposites, Properties and Applications: A Review. *Int. J. Biol. Macromol.*, 92, 1012–1024, 2016.
62. Singh, V., Yadav, P., Mishra, V., Recent Advances on Classification, Properties, Synthesis, and Characterization of Nanomaterials, in: *Green Synthesis of Nanomaterials for Bioenergy Applications*, pp. 83–97, John Wiley & Sons, Ltd, Hoboken, NJ, USA, 2020.
63. Othman, S.H., Bio-Nanocomposite Materials for Food Packaging Applications: Types of Biopolymer and Nano-Sized Filler. *Agric. Agric. Sci. Procedia*, 2, 296–303, 2014.
64. Reshmy, R., Philip, E., Paul, S.A., Madhavan, A., Sindhu, R., Binod, P., Pandey, A., Sirohi, R., Nanocellulose-Based Products for Sustainable Applications-Recent Trends and Possibilities. *Rev. Env. Sci. Biotechnol.*, 19, 779–806, 2020.
65. Trache, D., Tarchoun, A.F., Derradji, M., Hamidon, T.S., Masruchin, N., Brosse, N., Hussin, M.H., Nanocellulose: From Fundamentals to Advanced Applications. *Front. Chem.*, 8, 392, 2020.
66. Slavutsky, A.M. and Bertuzzi, M.A., Water Barrier Properties of Starch Films Reinforced with Cellulose Nanocrystals Obtained from Sugarcane Bagasse. *Carbohydr. Polym.*, 110, 53–61, 2014.
67. Heidari, M., Khomeiri, M., Yousefi, H., Rafieian, M., Kashiri, M., Chitin Nanofiber-Based Nanocomposites Containing Biodegradable Polymers for Food Packaging Applications. *J. Consum. Prot, Food Saf.*, 16, 237–246, 2021.
68. Chang, P.R., Jian, R., Yu, J., Ma, X., Starch-Based Composites Reinforced with Novel Chitin Nanoparticles. *Carbohydr. Polym.*, 80, 420–425, 2010.
69. Mohan, T., Devchand, K., Kanny, K., Barrier and Biodegradable Properties of Corn Starch-Derived Biopolymer Film Filled with Nanoclay Fillers. *J. Plast. Film Sheeting*, 33, 309–336, 2017.
70. Shayan, M., Azizi, H., Ghasemi, I., Karrabi, M., Effect of Modified Starch and Nanoclay Particles on Biodegradability and Mechanical Properties of Cross-Linked Poly Lactic Acid. *Carbohydr. Polym.*, 124, 237–244, 2015.

71. Ahmad, N., Kausar, A., Muhammad, B., Perspectives on polyvinyl chloride and carbon nanofiller composite: a review. *Polym. Plast. Technol. Eng.*, 55, 10, 1076–1098, 2016.
72. Mazhar, S., Promising PVC/MXene based flexible thin film nanocomposites with excellent dielectric, thermal and mechanical properties. *Ceram. Int.*, 46, 8, 12593–12605, 2020.
73. Quoc Pham, L., Uspenskaya, V.M., Olekhnovich, O.R., Olvera Bernal, A.R., A review on Electrospun PVC Nanofibers: fabrication, properties, and application. *Fibers*, 9, 2021.
74. Mohamed, A.T., Thermal experimental verification on effects of nanoparticles for enhancing electric and dielectric performance of polyvinyl chloride. *Measurement*, 89, 28–33, 2016.
75. Zattini, G., Ballardini, S., Benelli, T., Mazzocchetti, L., Giorgini, L., Safer plasticized polyvinyl chloride synthetic leathers for the automotive industry: evaluation of alternatives to antimony compounds as flame retardants. *Polym. Eng. Sci.*, 59, 12, 2488–2497, 2019.
76. Kim, J., Lee, J., Jo, C., Kang, C., Development of low cost carbon fibers based on chlorinated polyvinyl chloride(CPVC) for automotive applications. *Mater. Des.*, 204, 109682, 2021.
77. Moustafa, H., Morsy, M., Ateia, A.M., Abdel-Haleem, M.F., Ultrafast response humidity sensors based on polyvinyl chloride/graphene oxide nanocomposites for intelligent food packaging. *Sensor Actuator Phys.*, 331, 112918, 2021.
78. Hazer, B., Karahaliloglu, Z., Ashby, R.D., Poly (vinyl chloride) derived food packaging applications with Antioxidative and Anticancer properties. *ACS Food Sci. Technol.*, 3, 4, 761–771, 2023.
79. Alshammari, A.H. and Taha, T.A., Structure, thermal and dielectric insights of PVC/PVP/ZnFe2O4 polymer nanocomposites. *Eur. Phys. J. Plus*, 136, 12, 1201, 2021.
80. Otey, F.H., Westhoff, R.P., Mehltretter, C.L., Reactivity of Modified Starches with Isocyanates / Starch. *Staerke.*, 24, 107–110, 1972.
81. Richard, P.W., Felix, H.O., Charles, L.M., Charles, R.R., Starch-Filled Polyvinyl Chloride Plastics-Preparation and Evaluation. *Ind. Eng. Chem. Prod. Res. Develop.*, 13, 2, 1974.
82. Rosa, D.S., Rios, A.R., Viana, H.M., Tavares, M.I.B., Characterization of Poly (vinyl chloride) Compounds Containing Different Levels of Starch. *J. Vinyl Addit. Technol.*, 22, 4, 396–404, 2016.
83. Dutta, N., Newar, P., Narzary, M., Maji, T., Preparation and characterization of biobased hybrid nanocomposite of Polyvinyl chloride/thermoplastic starch with Mg–Al layered double hydroxide and waste rice husk ash. *J. Polym. Res.*, 29, 375, 2022.
84. de Freitas, A., de, S.M., da Silva, A.P.B., Montagna, L.S., Nogueira, I.A., Carvalho, N.K., de Faria, V.S., Lemes, A.P., Thermoplastic starch nanocomposites: sources, production and applications – a review. *J. Biomater. Sci., Polym. Ed.*, 33, 7, 900–945, 2022.

85. Panda, S.K., Synthesis and Overall Migration Study of Chitosan-Encapsulated ZnO-Based ESO Bionanocomposite with Synergistic Antimicrobial Activity for Packaging Purpose. *Chem. Select*, 7, 2022. doi.org/10.1002/slct.202201433.
86. Mohanty, P., Behera, D., Panda, S.K., Bastia, T.K., Hybrid Composite Laminates from UPE/ESOA Blend Reinforced with Chitosan and Bamboo Fiber: A Study of Mechanical and Thermal Properties. *Hybrid Compos. Laminates UPE/ESOA Blend Reinf. With Chitosan Bamboo Fiber A Study Mech. Therm. Prop.*, 32, 6, 1321–1328, 2020.
87. Mohanty, P., Bastia, T.K., Behera, D., Panda, S.K., Chitosan Grafted Carbon Nanotubes Reinforced Vinyl Ester/UPE Blend Based Partially Bio Nanocomposite. *Asian J. Chem.*, 31, 9, 1943–1948, 2019.
88. Panda, S.K., Behera, D., Rath, P., Bastia, T.K., Enhanced properties of UPE/ESOA partially bio-nanocomposites reinforced with chitosan functionalized graphene nanoplatelets: An innovative approach. *Bull. Mater. Sci.*, 41, 4, 102, 2018.
89. Panda, S.K., Behera, D., Rath, P., Bastia, T.K., Advances in physical performances of UPE/ESOA partially bio-nanocomposites reinforced with graphene nanoplatelets. *Int. J. Res. Innovation Appl. Sci.*, 2, 8, 2017.
90. Panda, S.K. and Behera, D., Effect of red mud on mechanical and chemical properties of unsaturated polyester-epoxy bamboo fiber composites, in: *Materials Today: Proceeding*, Elsevier, 2017, ISSN:2214-7853.
91. Panda, S.K., Mohanty, P., Behera, D., Bastia, T.K., ESOA modified unsaturated polyester hybrid networks: A new perspective. *J. Appl. Polym. Sciences (Wiley)*, 133, 44345, 2016. DOI: 10.1002/APP.44345.
92. Ali, M.I., Ahmed, S., Javed, I., Ali, N., Atiq, N., Hameed, A., Robson, G., Biodegradation of starch blended polyvinyl chloride films by isolated Phanerochaete chrysosporium PV1. *Int. J. Environ. Sci. Technol.*, 11, 339–348, 2014.

3

PVC/PLA Composites/Biocomposites

Rajni Garg[1], Rishav Garg[2]* and Nnabuk Okon Eddy[3]

[1]*Department of Applied Sciences and Humanities, Galgotias College of Engineering & Technology, Greater Noida, Uttar Pradesh, India*
[2]*Department of Civil Engineering, Galgotias College of Engineering and Technology, Greater Noida, Uttar Pradesh, India*
[3]*Department of Applied Science, Faculty of Natural Sciences, Walter Sisulu University, East London, South Africa*

Abstract
From the perspective of creating industrial ecology and circular economy, wastes are seen as valuable secondary raw materials. One way to reduce the impact of this pollution is by implementing recycling practices. Unfortunately, the recycled materials do not possess high-quality mechanical properties due to the incompatibility between different polymers. The sustainable way to counter pollution is development of biodegradable and eco-friendly materials. However, the physicochemical and structural properties of biodegradable materials are not as good as that of synthetic materials that are actually not biodegradable and add to the pollution issue. In this context, blending of biodegradable and non-biodegradable materials can provide an innovative composite material with desirable properties. This chapter explores the characteristics of such blends/composites obtained from suitable interaction of polyvinyl chloride (PVC), a synthetic polymer and polylactic acid (PLA), a biopolymer. The characteristics of both the polymers and their composites have been explored. The study reveals that blending of these polymers in presence of suitable plasticizer and compatibilizer can result in formation of biodegradable products with potential medical, packaging and other applications.

***Keywords*:** Composite, biodegradable, plastic, blend, sustainable, nanomaterials, biocompatible

**Corresponding author*: rajnigarg@science.org.in

Shiv Kumari Panda and Visakh P. M (eds.) Polyvinyl Chloride-Based Biocomposites and Bionanocomposites, (53–84) © 2026 Scrivener Publishing LLC

3.1 Introduction

Every day, the consumption of plastic materials continues to rise. Non-biodegradable plastics are disposed of in large amounts every year, both in industrial and residential settings. The issue of domestic, industrial, and agricultural solid waste is of significant concern and requires immediate attention [1]. Although plastics constitute a small fraction of these wastes, they face intense criticism because of their high visibility. First of all, disposal is a major challenge that comes with using synthetic polymers once their life cycle is complete [2]. Swapping out synthetic plastics with biodegradable polymers is another way to tackle this problem. The development of biodegradable plastics has the potential to make a significant contribution in addressing waste disposal challenges [3]. Due to the importance of waste reduction for the future of municipal solid waste, there are several significant challenges that must be addressed to ensure the growth of a new biodegradable plastics industry. Secondly, these items naturally decompose over time due to exposure to air, moisture, climate, and soil, eventually breaking down into the surrounding environment [4]. The products discarded by individuals will eventually turn into waste over time. The blending of polymers is a crucial method for creating innovative polymeric materials. These compositions often showcase properties that surpass those of any single polymer component [5]. Through the utilization of these blends, a diverse range of morphological states and chemical or physical qualities can be achieved by combining the advantageous characteristics of various molecular species [6].

Bioblends are a specific kind of polymer blend that includes at least one biodegradable polymer along with other types of polymers. Scientists are particularly interested in biodegradable polymer bioblends, as they have widespread applications, including packaging, medical devices, and drug delivery systems [7]. For the successful development of bioblends, it is crucial for the biodegradable polymers to be compatible with the other polymers that constitute the component. Extensive research has revealed that the degradation products of one polymer can significantly impact the degradation behavior of another polymer [8]. The compatibility of the polymer components in the blend serves as a determining factor in the structure and composition of the numerous products that are generated throughout the degradation process, as well as their interactions with one another in the ultimate mechanism of degradation. Within the category of biodegradable polymers, polylactic acid stands out as the thermoplastic polymer that possesses exceptional mechanical qualities [9]. The typical production steps for a large-scale production include fermentation of maize to produce lactic acid followed by chemical polymerization. In comparison to conventional petroleum plastics, which require between 500 and 1000 years to breakdown, pure PLA may be broken down into CO_2, CH_4, and H_2O in the environment in a period of time ranging from a few months to two years [10]. PLA offers the advantages of being biodegradable, having an annual renewable availability, and being processable. However, with the primary focus being on the biodegradability of the material, the thermomechanical behavior of a PLA-based material is equally important in its sustainable and effective utilization [11]. After undergoing typical melt procedures, PLA is melt blended with other elastomeric polymers such as poly(ecaprolactone), poly(butylene succinate), poly(hydroxyl butyrate), and poly butylene adipate-co-terephthalate. PLA is also melt blended with nonbiodegradable polymers including polypropylene (PP), polyacrylonitrile (ABS), polyethylene (PET), polystyrene (PS), polymethyl methacrylate (PMMA), polyvinyl alcohol (PVA), polyethylene (PET), and polyvinyl chloride (PVC) [12].

Another important consideration is of cost, as the price of biodegradable polymers continues to be higher than that of petroleum polymers, whereas the plastics industry is always looking for ways to cut costs. This is not only owing to the limited availability of the material, but also to the problems that are associated with their processing [13]. Under certain circumstances, such as when an expansive polymer may be over-engineered for a particular application, blending it with a less expensive polymer or filler and utilizing a compatibilizer that is equivalent to the polymer or filler can help lower costs [14]. In order to fulfill the expectations of the market and meet the technical criteria, it is necessary to question some fundamental elements. These fundamental essentials include competitive prices and mechanical properties that are either comparable or enhanced [15]. It is only possible for the prefix "bio" to be appealing if the costs of the materials are moderate and if there is an assurance that consumers will accept it. PVC goods are widely used because of their low costing and high performance, making it a desirable material for practical applications [16]. This has contributed to the widespread usage of PVC but also has resulted in concern for its non-biodegradable nature. However, PVC/PLA blends can be tailored to have desirable features of both the components in pursuit of sustainable practices [17].

The chapter provides an overview of the biodegradable polymers and their biocomposites. This chapter aims to explore the various characteristics PVC/PLA-based composites by adding additives or blending components in order to make it competitive with a high-end biomaterial.

The characteristics of both PVC and PLA have been explored in addition to the characteristics of their blends. Special emphasis has been given to the characterization techniques and the biodegradability of these innovative blends.

3.2 Overview of Biocomposites and Bionanocomposites

Composite materials with biopolymers, commonly known as "biocomposites," are currently generating a lot of interest. The concept of polymer reinforcement is quite old. Composites can be created by blending synthetic and natural polymers with suitable additives [18]. Biopolymers can be strengthened with various additives such as synthetic fibers or plastics that can be infused with natural fibers to enhance their compatibility with the environment and ability to break down naturally. Additional inorganic additives, such as clay and other inorganic minerals, are included for their reinforcing properties, affordability, accessibility, and environmental friendliness [19]. They are apprehended in a non-nanoparticulate phase, strengthening the plastics by adding significant percentages, and the products obtained are referred to as biocomposites and bioplastics [20]. Technically, these composites consist of a combination of bioplastics and synthetic plastics, which are then reinforced with either natural fibers, synthetic fibers, or a combination of both [21]. There are various types of composites made from a variety of biomaterials. It may involve the use of low-molecular-weight substances that play a crucial role in the lipid matrix of biological membranes, as well as microorganisms found in living organisms [22]. They can also be mineralized to create composites commonly referred to as "hybrids." Composites consist of synthetic plastics combined with microsized inorganic additives like carbon black, talc, calcium carbonate, layered silicates, and more [23]. For instance, carbon black has been utilized as fillers in the rubber industry since the early 20th century with the discovery of a reinforcing effect [24]. Various additives have been used to reinforce and modify natural rubber since ages. Bioplastics have been the primary composite materials prior to the rapid advancement of the petroleum-based chemical industry, including polymer production in 1950s. Wagner strengthened natural rubber by using silica nanoparticles with particle size ranging from 10 to 100 nm [25]. Ordered sheet-like composites were obtained by mixing cellulose with silica nanoparticles synthesized using the sol–gel technique in 1992. A hydrogel was developed by combining starch with montmorillonite in 1997 [26].

On the other hand, the composite materials of biopolymers with nanomaterials are referred to as "nanobiocomposites" or "bionanocomposites" as they contain components of biological origin and nanosized organic/inorganic materials [27]. The concept has gained popularity in early 1940s, and there has been a significant increase in the number of publications focused on bionanocomposites since then [32]. The biopolymer(s) present the organic component(s) and nanomaterials form the reinforcement part. It takes into consideration the existence of the reinforcing particles with at least one dimension in the nanoscale. These bionanocomposites are made up of materials that come from biological sources and include particles that are between 1 and 100 nm in size [30]. Thus, nanocomposites are distinct from regular composites due to the inclusion of fillers/additives in their nanosized form. There are a significant amount of reports available in the literature focusing on the synthesis, characterization, and specialized aspects of these novel materials, involving biopolymers and nanoparticles [21, 28, 29]. The concept is also applicable to the materials where the biocomponent may not be totally "constituent(s) of biological origin," but are biodegradable. With their biodegradability and eco-friendly nature, bionanocomposites are also termed as "green composites," "biohybrids," "biobased plastics," or "bioplastics" [31].

Bionanocomposites represent a cutting-edge category of composite materials. They vary in their thermal stability, solubility in water, biodegradability, and biocompatibility, which can be harnessed in various applications. Their functionalities, morphology, and properties such as mechanical strength, biocompatibility, and biodegradability can be altered significantly with change in components and synthetic techniques [33]. Bionanocomposites are distinct from biocomposites as the latter lack nanosized additives, despite being partially composed of biopolymers. This set of composite materials consists of various types of nano-scaled additives that can be tailored based on their morphological characteristics and synthetic techniques, which in turn determine their structure, properties, functionalities, and applications [34].

3.3 Lactic Acid

Scheele, the Swedish chemist, discovered lactic acid (2-hydroxy propionic acid), a naturally occurring organic acid. Its production is accomplished *via* chemical synthesis or fermenting carbohydrates like rice, corn, wheat, sugar beets, sugar cane, and sweet corn [35]. Figure 3.1 displays the chemical structure of monomer lactic acid. The compound possesses a chiral carbon atom within its molecular structure, which results in the phenomenon of enantiomerism. There are two distinct stereoisomers present: the dextrorotatory D (−) and the levorotatory L (+). The sign in the parenthesis denotes the orientation of the rotation of plane polarized light by the isomer. The L (+) isomer induces clockwise rotation of light, whereas the D (−) isomer induces anticlockwise rotation. Both isomers exhibit a symmetrical relationship, with one being the mirror image of the other [7]. These isomers are commonly known as L-LA and D-LA [35].

Lactic acid and its derivatives are sued to obtain lactic acid–based polymers that are biodegradable and eco-friendly. There are different techniques that can be used to create lactic acid–derived polymers in the form of homopolymers (PLA), copolymers (poly(lactide-co-glycolide)), and cross-linked polymers (polylactide). There are two primary pathways (Figure 3.2) for the synthesis of lactic acid–based polymers: polycondensation and ring-opening polymerization (ROP) [36].

Understanding the role of stereoisomerism in PLA is crucial for comprehending its thermal, mechanical, and biodegradation properties. The polycondensation of the two enantiomers of lactic acid leads to the formation of three different forms of PLA: poly(L-lactic acid) (PLLA), poly(D-lactic acid) (PDLA), and poly(D,L-lactic acid) (PDLLA). The crystalline structure of PLLA-α form has been confirmed by employing neutron diffraction and two-dimensional wide-angle X-ray techniques [37]. PLLA exhibits a comparatively elevated melting temperature (T_m). PDLA is classified as a semi-crystalline polymer, whereas PDLLA is categorized as an amorphous polymer. PDLLA is a racemic polymer with a unsystematic sequence of the enantiomers [37]. Several factors, such as the polymer's molecular weight, temperature, processing time, and duration of annealing treatment, can impact the degree of crystallinity [35].

On the other hand, the production of lactide involves the oligomerization and cyclization of the two enantiomers of lactic acid. The ROP of lactic acid to form lactide has been shown in Figure 3.2 [36]. The first step is for

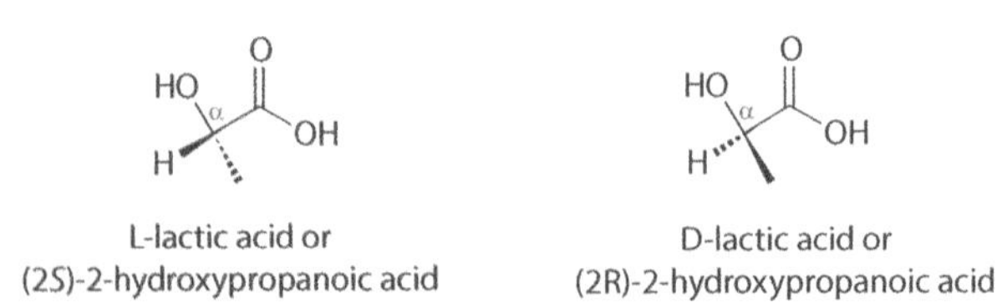

Figure 3.1 L- and D-isomers of lactic acid [35].

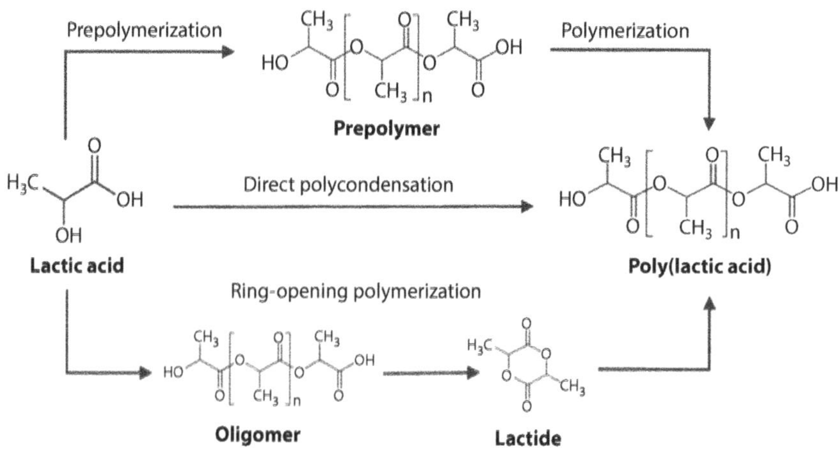

Figure 3.2 Polycondensation or ring-opening polymerization of lactic acid [36].

lactic acid to oligomerize into a low-molecular–weight prepolymer. Then, the prepolymer is cyclized into a cyclic dimer called a lactide [38]. After that, poly(lactide) is formed when the lactide goes through ring-opening. ROP is commonly utilized in commercial applications, resulting in the production of polylactide with a relatively high molecular weight. In addition, ROP enables precise control over the chemical reactions that occur during polymerization, allowing for the customization of desired properties [3].

3.4 Polylactic Acid

PLA is a widely used thermoplastic polymer that is derived from renewable resources such as maize, sugarcane, cassava, and corn. This biopolymer is highly compatible with the environment and can easily break down, making it an excellent choice for various applications. It is commonly utilized in food packaging applications [3]. It has a low-energy consumption and minimal greenhouse gas emissions during its production. PLA possesses a multitude of beneficial properties, such as its ability to biodegrade, wide availability, eco-friendly nature, and antibacterial capabilities, with impressive thermal and mechanical properties. These characteristics position it as a highly promising polymer for the development of sustainable and biodegradable products [12]. Certain drawbacks of PLA have been noted, including its limited toughness, thermal resistance, barrier properties,

Table 3.1 Comparative account of properties of PLA with other polymers [12].

Polymer	Elongation (%)	Flexural strength (MPa)	Tensile strength (MPa)	Tensile modulus (MPa)
PLA	2.5	70	49	3.2
PS	2.5	80	49	3.4
PVC	3	90	35.35	2.6
Nylon	5	95	71	2.9
PP	10	49	49	1.4

and glass transition temperature. A comparison of the strength properties of PLA with some synthetic polymers is shown in Table 3.1.

PLA had the highest tensile modulus compared to other polymers. It possesses greater flexural and tensile strength compared to PP and PVC. PLA is a highly compliant thermoplastic polymer that can be easily molded into desirable shapes and utilized for various uses. It demonstrates exceptional thermal stability, is highly resistant to oil and grease, and can withstand temperature variations. Its properties are well-suited for use in domestic appliances such as bottles, cups, trays, and more [6].

Efforts were made in the early 1930s to develop PLA by polymerizing and depolymerizing oligomeric lactides. Subsequently, Du Pont filed a patent in 1954 [39]. In the 1960s, researchers developed a method to create high molecular–weight PLA that had similar properties to plastics derived from petroleum. However, this method of production was not cost-effective [40]. In 1997, the process for commercial production of PLA was developed by Cargill Dow Corporation, marking a significant milestone in the industry. In this method, PLA was synthesized through the melting process instead of using a solution. In 2002, the production of PLA began with an annual capacity of 300 lbs in Blair (Nebraska). Through large-scale production, the cost of PLA production was significantly reduced [10]. Because of its biocompatibility, PLA was the foremost desirable polymer to be used for various applications in the field of biomedicine. As per Jem's law, the demand for PLA is growing to the point where it doubles every three to four years on the global market [41].

Nowadays, PLA is obtained by polycondensation of the two enantiomers of lactic acid. The properties of a polymer are determined by the chemical composition of the monomer constituting the polymer. The stereochemistry of PLA can be altered by altering the arrangement of L-LA and D-LA monomers during polymerization. This can be achieved through the use of appropriate catalysts such as 12-tungstophosphoric acid and optimal reaction conditions [42]. However, the process of polycondensation necessitates extended periods of polymerization, typically exceeding 30 hours. It is crucial to eliminate the water by-product to prevent any decrease in the molecular weight of the polymer. The typical semicrystalline structure of PLA can be modified by modifying the stereochemical composition of monomers to obtain polymers with amorphous or crystalline structures [3]. In addition, the physical and mechanical properties of PLA can be enhanced by adjusting D-LA and molecular weight. Table 3.2 presents the standard physical characteristics of the PLA [12].

In recent years, researchers and scientists have been increasingly interested in biocomposites. They are exploring ways to expand their applications across various sectors and make them more viable alternatives to petroleum-based products. This is driven by a growing global awareness of the need for sustainable solutions and the dwindling reserves of petroleum [13]. PLA is an excellent choice for replacing petroleum-derived products because of its abundance, ability to break down naturally, compatibility with living organisms, and lightweight nature [3]. These favorable characteristics make it suitable for applications in biomedicine, food packaging, and various other consumer goods. When compared to other biopolymers, the utilization of PLA offers several advantages as discussed below [10].

Table 3.2 Physical properties of PLA, PLLA/PDLA [3].

Property	PLA	PLLA or PDLA
Glass transition temperature (°C)	65–72	58
Melting temperature (°C)	220–230	170–190
Temperature for decomposition (°C)	-	310
Density (g/cm³)	-	1.25–1.29
Elasticity modulus (GPa)	8.6	7–10
Tensile strength (GPa)	0.88	0.12–2.3
Elongated at rupture (%)	30	4–7
Enthalpy of melting (J/g)	142-146	40–50

3.4.1 Eco-Friendly

PLA is a polymer derived from sustainable sources and is completely capable of undergoing biodegradation, making it environmentally friendly. Typically, PLA is sourced from plant-based materials like corn, wheat, or rice. PLA requires much less energy to produce compared to petroleum-based polymers [2]. Estimations indicate that in the future, this energy need can be further lowered to less than 10%. Reducing energy consumption also has the potential to make PLA production more cost-effective [39]. It is also biodegradable, recyclable, and compostable, making it eco-friendly. The production process also results in the consumption of carbon dioxide [35]. When PLA products undergo decomposition, they release CO_2 and H_2O, which are then used by plants to produce vegetable products through photosynthesis. These agricultural products are once again used to produce PLA. The life cycle of PLA (Figure 3.3) illustrates the process from its manufacture to its total decomposition [8].

These essential components contribute to the long-term sustainability of the PLA polymer. Each metric ton of PLA that is biodegrading releases about 1600 kg of CO_2 gas. When weighed against other polymers such as nylon at 7150 kg per metric ton, PET at 4150 kg, PP at 1850 kg, and PS at 2740 kg, this amount is comparatively lower [10]. Through the process of biodegradation, the emitted CO_2 from PLA contributes to the growth of agricultural products, leading to a decrease in prime greenhouse gas emission. Thus, as compared to synthetic polymers, the large-scale production of PLA from agricultural products can lead to a noteworthy decrease in greenhouse gas emissions [43]. A total of 75.4 MJ/kg of energy is required for the entire process, from maize cultivation to the manufacturing of PLA

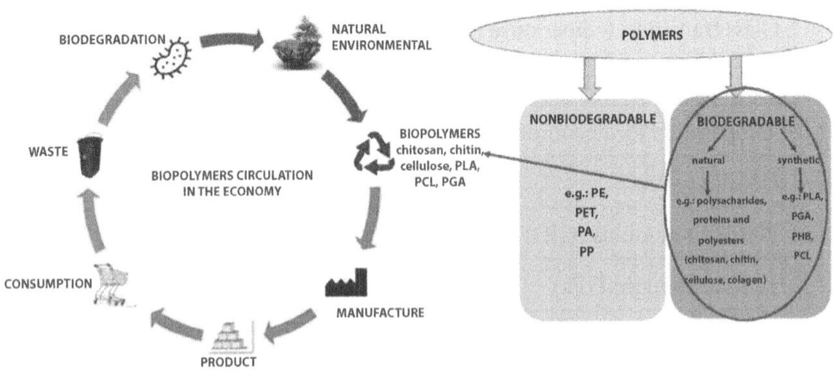

Figure 3.3 Life cycle of PLA [8].

pallets. This covers both 50.8 MJ/kg of fossil energy and 24.6 MJ/kg of renewable energy. Compared to the energy needed to make plastics made from petroleum, PLA requires far less fossil energy (25–55%) [11].

3.4.2 Biocompatible and Beneficial Properties

PLA is highly biocompatible, making it particularly appealing for use in biomedical applications. A biocompatible material should not cause any harmful effects in the surrounding tissues [12]. In addition, it is important to ensure that the byproducts of degradation do not hinder the process of tissue healing. PLA undergoes hydrolysis, in living organisms (including the human beings), converting to its constituent α-hydroxy acid. It is then processed through the tricarboxylic acid cycle and eliminated [38]. In addition, the degradation products of PLA are non-toxic at lesser compositions, which makes it a suitable option for biomedical applications. PLA has also been approved by the FDA for direct contact with biological fluids [7].

PLA exhibits superior thermal processibility when compared to other biopolymers like PHA, PEG, and PCL. It can be processed using various methods such as fiber spinning, injection molding, blow molding, film extrusion, thermoforming, and film forming [44]. Certain properties of PLA, such as its slow crystallization rate, low thermal resistance, poor ductility, brittleness, low impact strength, low elongation at break, and moisture sensitivity, impose limitations on its use in various engineering sectors [45]. However, despite the positive aspects mentioned earlier, PLA does have some limitations that restrict its use in certain applications with unappealing mechanical properties as compared to the synthetic polymers. These limitations include the following:

3.4.3 Low Toughness and Slow Degradation

PLA exhibits low toughness due to its brittle nature, characterized by less than 10% elongation at break. Although its tensile strength and elastic modulus closely resemble PET, its lack of toughness limits its usefulness in situations where plastic deformation is needed under higher stress levels, for instance in case of screws and fracture fixation plates [46]. Numerous variables affect the rate of PLA degradation, including molecular weight, crystallinity, morphology, water diffusion rate, and stereoisomeric content. In biomedical applications, the degradation rate is frequently regarded as a critical factor to consider [7]. The slow degradation rate results in a significantly extended lifespan within the body, lasting for several years in certain instances (typically 3–5 years) and poses a significant challenge when it comes to disposing of consumer commodities [34].

3.4.4 Slight Hydrophobicity and Difficulty in Modification

PLA exhibits a certain level of hydrophobicity, as indicated by its static water contact angle of around 80 °C. The PLA exhibits hydrophobic properties as a result of the CH_3 groups present in its chemical structure [38]. This can lead to a decrease in cell compatibility and, in certain situations, cause an inflammatory reaction when it comes into contact with bodily fluids [7].

PLA poses a challenge when it comes to modifying its surface and bulk due to its chemical inertness and lack of reactive side-chain groups. Recently, there have been several advancements in PLA-based technologies that focus on achieving chemical, mechanical, and biological properties that are on par with or even better than traditional polymers [47].

In the past decade, there has been a significant focus on addressing the need to modify PLA chemically or physically in order to ensure it possesses the desired properties for consumer and biomedical applications. In order for PLA to be successfully used in consumer and biomedical applications, it is important for its mechanical properties to be equal to or better than traditional plastics [45]. Additionally, the surface properties of PLA, such as roughness, hydrophilicity, and reactive functionalities, need to be carefully controlled. By incorporating PVC, the brittleness of PLA can be greatly improved, although this does come at the cost of reducing its modulus of elasticity [33]. Some researchers have proposed that incorporating nanoclay into PLA could potentially improve its performance. Due to hydrogen bond formation between the carbonyl group of PLA and the ammonium group of clay nanoparticles, there is complete blending of the composition [43]. Research has demonstrated that the characteristics of PLA can be enhanced by incorporating natural fibers, cellulose, and nanocellulose [48].

3.4.4.1 PVC

Polyvinyl chloride, also known as PVC, is a linear thermoplastic polymer. Typically, PVC has a low number of branches per molecule, usually less than 10. It is unlikely for there to be multiple long chains within a single molecule. When a chlorine atom is present, it enhances the attraction between molecules and contributes to the hardness and stiffness of the polymer [49]. It is worth noting that PVC has the ability to act as a weak proton donor. As a result, effective solvents for PVC are those that act as weak proton acceptors, such as cyclohexanone and tetrahydrofuran [50]. Based on X-ray studies, it has been observed that PVC is mostly amorphous, with a minor amount of crystallinity (< 5%) [51]. Research utilizing nuclear magnetic resonance (NMR) techniques indicate that traditional PVC consists of approximately 55% syndiotactic structure, with the remaining portion being predominantly atactic [24].

Polyvinyl chloride is highly commercially desirable due to its accessibility to basic raw materials and its possessed properties The high concentration of chlorine in the polymer provides it with flame retardant properties. However, it has certain limitations that need to be considered. Firstly, it is prone to brittleness at low temperatures. Additionally, its thermal stability is reduced, and it has a relatively narrow range for thermal processing [52]. By incorporating other monomers through copolymerization, numerous drawbacks can be mitigated. By combining the polymer with various additives, significant advancements have been made in enhancing the processing properties and quality of PVC [53]. These additives include thermal stabilizers, lubricants, impact modifiers, plasticizers, processing aids, and antioxidants. In most cases, these additive groups consist of micromolecular substances, with the exception of impact modifiers and processing aids [16]. PVC exhibits a higher heat distortion temperature compared to systems using liquid plasticizers. Nevertheless, the focus on the benefits provided by the polymer blends led to the substitution of micromolecular additives in PVC-based compositions with macromolecular compounds. Utilizing an additional polymer in conjunction with PVC enhances the processability and impact strength of PVC [54]. While most studies on PVC-based blends primarily examine their properties, there are also many investigations dedicated to determining the level of compatibility between components [45, 53, 55, 56]. The range of macromolecular compounds used for processability and utilization characteristics modification of PVC-based compositions is quite extensive. It includes both general use polymers such as polyethylene, polypropylene, polystyrene, and acrylic polymers, as well as engineering polymers like polyamides, thermoplastic polyesters, polycarbonates, and polyacetals [9].

3.4.4.2 Importance of PVC and PLA in the Development of Biocomposites and Bionanocomposites

Most of the plastics produced are derived from petroleum-based polymers and are non-biodegradable. Currently, a staggering billion tons of plastics have been produced worldwide. Only a small percentage of plastics, out of this, have been recycled, reused, and carbonized. Plastic waste is piling up in the environment, leading to significant environmental concerns [1]. Dealing with the environmental pollution caused by a significant amount of solid plastic waste is no easy task. Various methods have been employed to address this issue, with varying degrees of success. Plastics have a limited potential for reuse, but there are two types of recycling available commercially: mechanical and chemical recycling [57]. Recycling alone may not be a sufficient solution to effectively reduce and recycle solid plastic waste, as only a limited number of plastics can be recycled. So, another approach is to utilize a biodegradable polymer to partially or entirely substitute petroleum-based polymers [58]. However, many biopolymers are not widely utilized in various applications due to their higher cost and limited properties. As a result, there is a requirement to transform non-degradable polymers into biodegradable alternatives. Blending polymers has shown great potential in transforming non-biodegradable polymers into biodegradable ones. This tailoring of polymer properties makes a significant contribution to the field of polymer science and engineering [4].

Polyvinyl chloride (PVC) is considered one of the most efficient petroleum-based thermoplastic polymers. It holds a significant share among the six commonly used commercial polymers and is projected to experience a steady growth rate in the coming years. The global production capacity of PVC has reached millions of tons, which is used in a wide range of applications. Sadly, they often wind up in trash heaps after being used as lightweight, single-use products like cooking oil and packaging fruit juice, and they can create major environmental damage when not disposed of appropriately [5]. This pollution can have a wide-ranging impact on animal and plant life on Earth. Biodegradable polylactic acid is often blended with non-degradable polyvinyl chloride [55]. The non-degradable portion will undergo fragmentation into smaller pieces, potentially leading to the degradation of the non-degradable polymer [33].

3.5 Synthesis and Processing of PVC/PLA-Based Biocomposites and Bionanocomposites

The main objective of researchers across the globe is to develop innovative materials that could be used in waste management applications. The findings support the continuous endeavors in developing eco-friendly materials for waste management purposes [2]. PVC/PLA blend is a highly promising biodegradable polymer blend that has garnered significant attention from researchers in recent times. Literature reports the development and characterization of novel biomaterials (bioblends) made from poly(vinyl chloride) (PVC) and poly(lactic acid) (PLA) [59]. The thermal and mechanical properties, along with the thermal stability, of the binary blends were thoroughly examined through various scientific techniques such as tensile testing, microhardness measurements, thermal analysis, and scanning electron microscopy. However, the interaction between PVC and PLA is not favorable when they are combined directly [33]. This is because PVC and PLA polymers do not mix well, resulting in lower properties for the blends compared to the individual polymers. These types of issues are prompting researchers to conduct additional investigations into the modification of blends. They are exploring the use of a small amount of compatibilizer mixed in a polymeric solution and experimenting with various strategies, including the addition of copolymers [60].

Compatibilizers are used to improve the polymer blends since they interact with the components of the blends, altering the morphology and enhancing the mechanical properties. Compatibilizers can sort out the incompatibility between the two polymers [61]. These compatibilizers contain glycidyl methacrylate groups that can potentially interact with the functional end-groups of polylactide chains. Various reports emphasize the significant impact of maleic anhydride (MAH) as a compatibilizer in improving the compatibility and properties of these innovative biomaterials. MAH is a more effective agent for achieving compatibility between blends compared to phthalic anhydride (PAH) and succinic anhydride (SAH) [62]. Brito *et al.* explored the use of polypropylene grafted-MAH in various compositions of polylactic acid and polypropylene to obtain better wetting and impact strength with a significantly low melt flow index [44]. Hachemi, Belhaneche-Bensemra, and Massardier used melt mixing to obtain PVC/PLA bioblends with and without the addition of MAH as a compatibilizer. Dicumyl peroxide (DCP) was used as an initiator during *in situ* compatibilization with MAH. The incorporation of MAH improved the processability of the bioblends by reducing the mixing temperature [33].

Lyu *et al.* used ethylene-vinyl acetate–glycidyl methacrylate (EVM-GMA) random terpolymer using vinyl acetate (60 wt%) and glycidyl methacrylate (2.9 wt%) to enhance the compatibility of PVC/PLA blends to replace dioctyl phthalate (DOP), traditionally used as the polymeric plasticizer. Addition of PLA to the PVC matrix caused notable alterations in the rheological properties of the blends with improvement of the 3D printing performance as well as mechanical properties of the blends at a EVM–GMA/ PVC/PLA weight ratio of 5/45/50 [13]. Polymethyl methacrylate (PMMA) was utilized by Singh *et al.* as a compatibilizer to enhance the interfacial activity of PVC/PLA blends to further obtain a biodegradable film. A condensation method was used to synthesize PLA while solution casting technique was used to develop PVC/PLA blend films with and without the addition of PMMA as a compatibilizer [63]. Raksaksri *et al.* explored the synergistic effects of wood flour (WF) and poly(butylene adipate-co-terephthalate) (PBAT) in a binary/ternary blend system of PVC/PLA, analyzing the impact strength, Young's modulus, flexural properties, and thermal resistance, of the composite materials. The highest impact and tensile properties were exhibited by the PVC/PBAT samples, while the PVC/PLA samples displayed the greatest flexural and thermal properties. Average mechanical properties were exhibited by the PVC/PLA/PBAT ternary blended samples, suggesting a balance between the individual strengths of the components. The inclusion of WF to all the ternary blends further improved the overall performance of the composites. The PVC/PBAT/WF composite exhibited the highest impact strength, while the PVC/PLA/WF composite had the greatest flexural strength and thermal resistance. Likewise, PVC/PLA/PBAT/WF composites displayed the highest tensile strength and Young's modulus, demonstrating the synergistic effects of the ternary polymer blend and the wood filler. These findings suggest that the compatibility between the polymer matrix and the WF particles played a crucial role in determining the final properties of the composite materials. Thus, valuable insights into the optimization of PVC-based composites can be obtained by leveraging the unique characteristics of PLA, PBAT, and WF [60]. A twin-screw extrusion process (torque of 0.12 Nm, a screw temperature of 165°C, and a load of 5 kg) was used to prepare the fused deposition modeling (FDM) feedstock filament with an ideal combination for achieving optimal magnetization as PLA (52.5 wt%), PVC (15 wt%), wood dust (7.5 wt%), and Fe_3O_4 (25 wt%) [64].

3.6 Characterization of PVC/PLA Composites

The dynamics and harmonization of the two polymers in the blend are analyzed through the utilization of a comprehensive methodology. This involves conducting rheological, microscopic, and thermal analyses to comprehensively characterize the blends. Different analytical techniques such as Fourier-transform infrared spectroscopy (FTIR) [32], scanning electron microscopy (SEM) [65], thermogravimetric analysis (TGA) [66], vibrating sample magnetometer (VSM) [67], and differential scanning calorimetry (DSC) [68] have been used to characterize the polymer blends as well as study the underlying mechanisms that govern the properties of the blends.

A torque rheometer is generally utilized to the melt-mix of PVC/PLA in addition to employing various analytical techniques to study the phase morphology and compatibility of the blends. Generally, the compatibility of the two polymers is determined by studying the steady-state and dynamic rheological responses [47]. Systematic evaluation of the PVC/PLA blends' static and dynamic rheological properties offers valuable insights into their compatibility. The rheological analyses reveal important information regarding the viscoelastic behavior, melt viscosity, and other rheological properties of the polymer blends [69]. These findings play a critical role in understanding the processability and potential applications of these blends. Lyu *et al.* used static and dynamic rheological responses to examine the compatibility of PVC/PLA blends at a molar weight ratio of 50/50. By employing a range of rheological, microscopic, and thermal analysis techniques, the researchers were able to gain a deeper insight into the compatibility of the PLA/PVC blends. They offered valuable insights for the development and optimization of materials based on PLA/PVC, as well as their various applications [13].

The thermal properties and potential phase separation within the blends are studied through DSC measurements. For instance, TGA results revealed the enhanced thermal stability of the PLA/PVC bioblends in a molar composition of 50/50 [13]. On the other hand, during the DSC analysis, no clear glass transition temperatures (Tg) were found for the individual components indicating that a miscible polymer blend was formed at a EVM–GMA/PVC/PLA weight ratio of 5/45/50 [5]. SEM analysis is used to examine the morphological characteristics of the blends as a visual evidence for the compatibility of the PVC/PLA blends. In a study by Hachemi, Belhaneche-Bensemra, and Massardier, PLA chains were found evenly distributed within the PVC matrix of PVC/PLA, during the SEM analysis, suggesting the elimination of the phase separation in the presence of MAH leading to enhanced compatibility between the PVC and PLA components with the inclusion of MAH [33]. From the SEM micrographs (Figure 3.4), Lyu *et al.* found the uniform distribution of PLA within the PVC matrix in the presence of DOP and Ca–Zn stabilizers, suggesting a strong interfacial adhesion and a high degree of compatibility between these two polymers [13]. Lyu *et al.* obtained a non-linear relationship for the storage modulus and complex viscosity of the blends with the PLA content indicating the existence of interactions between the two polymer components. The successful incorporation of PLA into the PVC matrix was accredited to the observed non-linear rheological behavior and the homogeneous phase

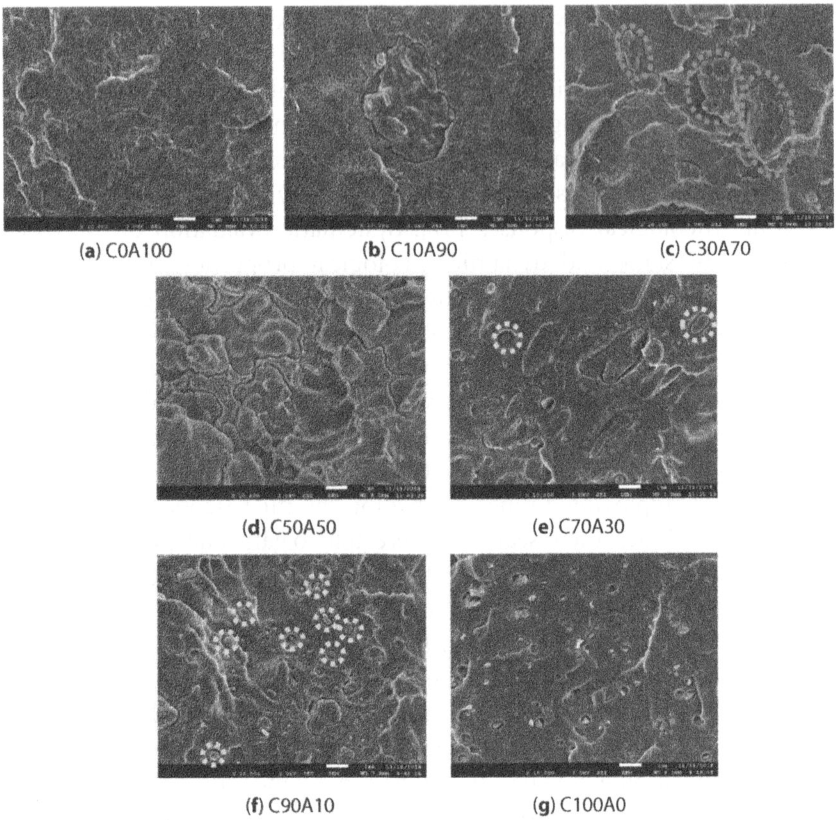

Figure 3.4 SEM images of PVC/PLA blends; composition represented as CxAy with x and y as mass fraction of PVC and PLA, respectively; (a) matrix structure of C0A100; (b) a typical "sea-island" structure of C10A90; (c) an "island" phase structure of C30A70 with size ranging from 3 to 8 μm outlined in green circles; (d) a distinct co-continuous phase structure of C50A50; (e) "island structure of C70A30 with microparticles of Ca–Zn stabilizers having a diameter less than 1 μm outlined in yellow circles; (f) uniformly distributed structure of C90A10; (g) matrix structure of C100A0, reproduced with permission from [13].

morphology [45]. Singh *et al.* examined an immiscible morphology of PVC/PLA blend without a compatibilizer evidenced by the presence of large domain sizes. This indicated a lack of compatibility between the two polymer components, further confirmed with two distinct glass transition temperatures. However, the introduction of PMMA as a compatibilizer reduced the domain size thereby improving the miscibility of the blend augmented with a single and intermediate glass transition temperature [63].

VSM is used to ascertain the superparamagnetic behavior of samples containing Fe_3O_4. The incorporation of Fe_3O_4 particles into the PLA matrix induces the magnetic characteristics while having a minimal impact on the mechanical properties. Studying the magnetic and mechanical properties of the polymer blends containing Fe_3O_4 particles offers valuable insights for creating self-assembly materials. Understanding the relationship between material composition, processing conditions, and resulting properties is crucial for the practical application of these materials in fields like robotics, electronics, and biomedical engineering. Kumar *et al.* explored the magnetic properties of a polymer blend consisting of PVC/PLA/wood dust in the presence of Fe_3O_4 particles for self-assembly purposes. Thoroughly characterizing and optimizing the TSE parameters contributes to this understanding [64]. Thus, various analysis techniques are valuable to gain a deeper understanding of the compatibility between PVC and PLA in blends.

3.7 Properties of PVC/PLA Composites

Thermoplastics have a wide range of applications due to their flexibility in incorporating various reinforcing materials into the core framework. Various filler materials have been discovered to be effective in strengthening polymeric bases and achieving specific properties [70]. For example, ceramic particles can enhance hardness and tensile strength, while magnetite particles (Fe_3O_4) can introduce magnetic properties [67]. Additionally, alumina particles can be used to reduce the melt flow index (MFI) of the composite [71]. Polymeric composites have a wide range of applications in the electronics and electrical industry. They are known for their insulating properties and low electrical conductance, making them ideal for use in printed circuit boards and holders for various devices. Researchers have utilized a range of fillers, including aluminum oxide, silica powder, iron powder, and graphene powder, to enhance the thermal conductivity of the polymeric matrix [23]. These blends have great potential for a wide range of applications. However, optimization of the process parameters is required to achieve desired mechanical properties.

3D-printed parts with 4D properties, such as the ability to change shape or function over time, have a broad range of applications in both structural and non-structural engineering fields. Different processing parameters and composition of the blends impact the properties of these functional 3D-printed prototypes. This information is valuable for potential structural and non-structural engineering applications [14]. Kumar *et al.* examined the mechanical properties of PVC/PLA/wood dust/Fe_3O_4 composite material created through 3D printing. Statistical analysis was used to optimize the 3D printing conditions, and a 100% infill density, 45° infill angle, and 70 mm/s infill speed yielded the optimal results for peak strength and break strength. The sample with maximum density and low infill angle had the least amount of porosity, as confirmed in the morphological analysis, resulting in improved mechanical properties. These findings were backed by photomicrographs that demonstrated the impact of porosity on the fractured samples during the mechanical characterization of this unique composite material created through 3D printing [72].

3.8 Biodegradability of PVC/PLA Composites

PVC is a synthetic polymer that is commonly used in various applications such as packaging, roofing, flooring, curtains, hoses, and toys. It has a high Tg (80–85°C) and is non-biodegradable [73]. It has been blended with a number of different polymers in order to improve their mechanical properties and broaden their applications in commercial and industrial settings. Despite the fact that PVC may have several potential advantages, its use is not recommended. When exposed to light, PVC waste continues to have a long-lasting impact on the environment, as it results in the formation of harmful products during its degradation (Figure 3.5). Scientists across the globe are perusing to mitigate the environmental impact of plastic waste and promote the adoption of more environmentally friendly packaging solutions with increased level of biodegradability [5].

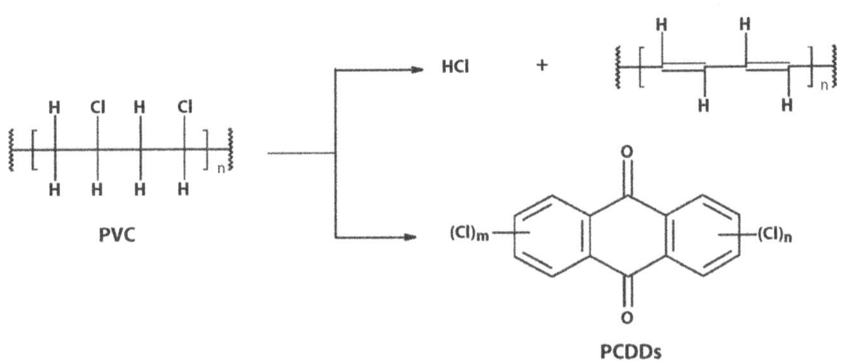

Figure 3.5 Formation of harmful products by degradation of PVC [5].

There are three main categories of biodegradable polymers. The first category includes biopolymers that are naturally produced by plants, animals, and microorganisms. Examples of these biopolymers include chitin, cellulose, starch, and polyhydroxyalkanoates (PHAs) [74]. The second category consists of synthetic polymers like polylactic acid (PLA), poly(e-caprolactone) (PCL), poly(ethylene/butylene adipate), and poly(ethylene/butylene succinate). There are several commercially available products such as Biopol, Eco-PLA, and Bionolle [75]. Lastly, there are polymer blends that are obtained by blending/mixing natural and synthetic polymers. Various aliphatic polyesters, including PHAs, PCL, and PLLA, are environmentally friendly and biodegradable polymers that have been used to obtain these blends with synthetic polymers [15]. The biobased and biodegradable polymers have garnered significant attention as a potential replacement for synthetic, non-biodegradable plastics. However, the brittleness and high cost of these polymers have led to the exploration of blending it with other polymers, such as PVC, to improve its mechanical properties and economic viability [76]. PLA is a biodegradable polymer that has a suitable T_g (60 °C) and T_m (130–180 °C). It finds applications in various medical implants and coating of controlled-release capsules [44]. It is a significant polymer because of its ease of generation, excellent biocompatibility and biodegradability, and impressive mechanical properties. It can undergo hydrolysis in the human body due to the presence of ester groups in the chain [77]. The challenges faced in processing PLA have continued to be the primary obstacle in its application resulting in the need of blend with suitable modifier. These blends have been used in the development and characterization of biodegradable polymeric films, demonstrating the feasibility of utilization as a sustainable alternative to conventional plastic materials [78].

Santana *et al.* examined the biodegradation of PVC/PLA blend under soil burial (up to 180 days) through use of a range of analytical techniques, including FTIR, TGA, and SEM, to closely monitor the physical and chemical changes occurring in the PLA/PVC blend during the biodegradation process. The degree of biodegradation was evaluated through measurements of mass loss, changes in mechanical properties, and the identification of specific functional group alterations. There were distinct changes in the chemical structure of the blend, indicating the gradual decomposition of the PLA portion. The mechanical properties of the blend, such as tensile strength and elongation at break, decreased over the burial period, further corroborating the ongoing biodegradation process [55].

Raksaksri *et al.* aimed to enhance the mechanical properties, thermal resistance, and biodegradability of PVC/PLA blends through the incorporation of PBAT and WF. PVC components were found to be non-biodegradable in the SEM analysis even after 6 months. However, addition of the PBAT improved the biodegradability of the samples in the soil burial tests. The biodegradability of the samples was further enhanced with the addition of WF. The enhancement was attributed to the increase in water uptake by the specimens in presence of WF containing hydrophilic hydroxyl groups forming hydrogen bonds with the water molecules [60]. Singh *et al.* conducted a comprehensive investigation into the development and characterization of a biodegradable polymeric film composed of PVC/PLA blends in the presence of PMMA to address the pressing global issue of environmental contamination caused by the accumulation of non-degradable plastic waste. PVC/PLA blend with 30% PLA was found to exhibit the highest level of biodegradability as around 50% of the film degraded within 8 months. This finding highlights the potential utilization of PMMA/PLA blends as better environmentally-friendly alternative to conventional, non-degradable plastic films. The shelf-life of tomatoes stored under the film developed by using these blends was extended with its quality attributes maintained up to 21 days when stored at ambient temperature (25°C) suggesting the practical applications of the developed film in the food packaging industry, where the preservation of perishable goods is of paramount importance [63].

3.9 Challenges and Future Perspectives

The demands of the present and those of the future are dependent upon the production of sustainable resources such as using natural polymers that are eco-friendly being biodegradable but have some drawbacks that can be altered by forming composites with additives [44]. Biocomposites produced from PLA have promising applications in a wide range of high-performance industries, including transportation, clothing, medicine, home appliances, and more [33]. Sustainable biocomposites have the potential to become a market leader in the production of long-lasting goods. However, the production methods, natural fiber supply, reinforcing content, chemical treatments, and interfacial bonding all have a significant impact on the characteristics of PLA-based biocomposites [60]. The blends/composites of a natural and synthetic polymer are not totally biodegradable as the synthetic plastic is almost non-biodegradable. For instance, PLA/PVC blends have been found to exhibit a relatively slow rate of biodegradation in the soil environment as the PVC component of the blend acts as a barrier, slowing down the biodegradation of the PLA [8].

Thus, comprehensive analysis of the biodegradation of PLA/PVC blends is necessary to provide valuable insights for the development of more sustainable and environmentally friendly polymer materials.

Globally, 3D printers rely on PLA and its biocomposites. Reaching performance levels comparable to more conventional methods, such as compression molding and injection molding, is a significant challenge [79]. The benefits of 4D printing include the ability to create items with autonomous behavior and the ability to manufacture complex designs [17]. Because of their form memory effect and other beneficial features, PLA biocomposites could be extremely useful for 4D printing [6]. Further research is required to optimize the composition and processing of such blends to enhance their biodegradability, ultimately contributing to the advancement of the circular economy and the reduction of plastic waste accumulation in the environment [64]. The results have important implications for the development of sustainable and high-performance materials in various applications, such as packaging, construction, and automotive industries.

3.10 Conclusion

Biocomposites and green composites based on PLA have been more popular as a result of the growing interest in using eco-friendly materials in engineering. As the world's population becomes more environmentally conscious, PLA-based biocomposites are becoming acknowledged as a better alternative to non-biodegradable petroleum-derived products. Composites like this offer an attractive substitute for petroleum-based products that don't break down in the environment because of their good mechanical and thermal qualities. The properties, advantages, disadvantages, characterization, difficulties, applications, and future prospects of PLA/PVC biocomposites are covered extensively in this chapter. The study reveals that optimal blending can result in highly tailored and desirable thermomechanical properties along with good biodegradability. Numerous uses have been discovered for PLA-based biocomposites; however, more research and development are needed to lower processing costs, improve performance, and increase industrial adoption.

AI disclosure statement: The author acknowledges the use of AI-based tools for minor language editing or grammatical correction. However, all intellectual contributions, analyses, and conclusions presented in this chapter are entirely the author's own.

References

1. Amelia, T.S.M., Kee, S.H., Senathirajah, K., Shao, Y.T., Ridzuan, N.A.M., Azmeerizal, A.S., Vigneswari, S., Lakshiminarayanan, R., Ramakrishna, S., Bhubalan, K., Solution or Pollution? A paradigm shifts in understanding the fate and threats of biodegradable plastics in the marine environment. *J. Hazard. Mater. Adv.*, 20, 100902, 2025, ISSN 2772-4166. https://doi.org/10.1016/j.hazadv.2025.100902.
2. Gopalan, G.P. and Anas, S., Structural, Morphological, and Textural Properties of Biopolymers, in: *Handbook of Biopolymers*, S. Thomas, A. AR, C. Jose Chirayil, B. Thomas (Eds.), pp. 323–363, Springer Nature Singapore, Singapore, 2023, https://doi.org/10.1007/978-981-19-0710-4_56.
3. Khouri, N.G., Bahú, J.O., Blanco-Llamero, C., Severino, P., Concha, V.O.C., Souto, E.B., Polylactic Acid (PLA): Propearties, Synthesis, and Biomedical Applications – A Review of the Literature. *J. Mol. Struct.*, 1309, 138243, 2024. https://doi.org/https://doi.org/10.1016/j.molstruc.2024.138243.
4. Singh, P., Pandey, V.K., Singh, R., Singh, K., Dash, K.K., Malik, S., Unveiling the Potential of Starch-Blended Biodegradable Polymers for Substantializing the Eco-Friendly Innovations. *J. Agric. Food Res.*, 15, 101065, 2024. https://doi.org/https://doi.org/10.1016/j.jafr.2024.101065.
5. Kudzin, M.H., Piwowarska, D., Festinger, N., Chruściel, J.J., Risks Associated with the Presence of Polyvinyl Chloride in the Environment and Methods for Its Disposal and Utilization. *Materials*, 17, 1, 2024. https://doi.org/10.3390/ma17010173.
6. Haq, R.H.A., Taib, I., Rahman, M.N.A., Haw, H.F., Abdullah, H., Ahmad, S., Ariffin, A.M.T., Hassan, M.F., Mechanical Properties of PCL/PLA Composite Sample Produced from 3D Printer and Injection Molding. *Int. J. Integr. Eng.*, 11, 5, 102–108, 2019. https://doi.org/10.30880/ijie.2019.11.05.014.
7. Singhvi, M.S., Zinjarde, S.S., Gokhale, D.V., Polylactic Acid: Synthesis and Biomedical Applications. *J. Appl. Microbiol.*, 127, 6, 1612–1626, 2019. https://doi.org/10.1111/jam.14290.
8. Dziuba, R., Kucharska, M., Madej-Kiełbik, L., Sulak, K., Wiśniewska-Wrona, M., Biopolymers and Biomaterials for Special Applications within the Context of the Circular Economy. *Materials*, 14, 24, 2021. https://doi.org/10.3390/ma14247704.
9. Belukhichev, E.V., Sitnikova, V.E., Samuylova, E.O., Uspenskaya, M.V., Martynova, D.M., Films Based on a Blend of PVC with Copolymer of 3-Hydroxybutyrate with 3-Hydroxyhexanoate. *Polymers*, 12, 2, 2020. https://doi.org/10.3390/polym12020270.
10. Vink, E.T.H., Rábago, K.R., Glassner, D.A., Gruber, P.R., Applications of Life Cycle Assessment to NatureWorks™ Polylactide (PLA) Production. *Polym. Degrad. Stab.*, 80, 3, 403–419, 2003. https://doi.org/https://doi.org/10.1016/S0141-3910(02)00372-5.

11. Vink, E.T.H., Glassner, D.A., Kolstad, J.J., Wooley, R.J., O'Connor, R.P., Original Research: The Eco-Profiles for Current and near-Future NatureWorks® Polylactide (PLA) Production. *Ind. Biotechnol.*, 3, 1, 58–81, 2007. https://doi.org/10.1089/ind.2007.3.058.
12. Kühnert, I., Spörer, Y., Brünig, H., Tran, N.H.A., Rudolph, N., *Processing of Poly(Lactic Acid) BT - Industrial Applications of Poly(Lactic Acid)*, M.L. Di Lorenzo and R. Androsch (Eds.), pp. 1–33, Springer International Publishing, Cham, 2018, https://doi.org/10.1007/12_2017_30.
13. Lyu, Y., Pang, J., Gao, Z., Zhang, Q., Shi, X., Characterization of the Compatibility of PVC/PLA Blends by Aid of Rheological Responses. *Polymer*, 176, 20–29, 2019. https://doi.org/10.1016/j.polymer.2019.05.016.
14. Joseph, T.M., Kallingal, A., Suresh, A.M., Mahapatra, D.K., Hasanin, M.S., Haponiuk, J., Thomas, S., 3D Printing of Polylactic Acid: Recent Advances and Opportunities. *Int. J. Adv. Manuf. Technol.*, 125, 3, 1015–1035, 2023. https://doi.org/10.1007/s00170-022-10795-y.
15. Maiti, T.K., Singh, J., Dixit, P., Majhi, J., Chattopadhyay, S., Enhancing Mechanical and Interfacial Properties of PEEK/Epoxy/SWCNT Composites Employing Aromatic Hydroxyl and Amine-Functionalized SWCNTs. *Chem. Eng. J. Adv.*, 10, February, 100274, 2022. https://doi.org/10.1016/j.ceja.2022.100274.
16. Ait-Touchente, Z., Khellaf, M., Raffin, G., Lebaz, N., Elaissari, A., Recent Advances in Polyvinyl Chloride (PVC) Recycling. *Polym. Adv. Technol.*, 35, 1, e6228, 2024. https://doi.org/https://doi.org/10.1002/pat.6228.
17. Kumar, S., Singh, R., Singh, T.P., Batish, A., On Investigations of Thermal Conductivity, Circumferential Compressive Strength, and Surface Characterization of 3D-Printed Hybrid Blended Magnetostrictive PLA Composite. *J. Thermoplast. Compos. Mater.*, 35, 5, 631–650, 2022. https://doi.org/10.1177/0892705720907651.
18. Garg, R. and Garg, R., Performance Evaluation of Polypropylene Fiber Waste Reinforced Concrete in Presence of Silica Fume. *Mater. Today Proc.*, 43, xxxx, 809–816, 2020. https://doi.org/10.1016/j.matpr.2020.06.482.
19. Garg, R., Garg, R., Eddy, N.O., Applications of Nanocoatings, in: *Sustainable Approach to Protective nanocoatings*, pp. 28–56, 2024, https://doi.org/10.4018/979-8-3693-3136-1.ch002.
20. Havstad, M.R., *Chapter 5 - Biodegradable Plastics*, Editor(s): Trevor M. Letcher, Plastic Waste and Recycling, pp. 97–129, Academic Press, 2020. ISBN 9780128178805, https://doi.org/10.1016/B978-0-12-817880-5.00005-0.
21. Alves, Z., Ferreira, N.M., Ferreira, P., Nunes, C., Design of Heat Sealable Starch-Chitosan Bioplastics Reinforced with Reduced Graphene Oxide for Active Food Packaging. *Carbohydr. Polym.*, 291, 119517, 2022. https://doi.org/10.1016/j.carbpol.2022.119517.
22. Šuput, D., Popović, S., Hromiš, N., Ugarković, J., Degradable Packaging Materials: Sources, Application and Decomposition Routes. *J. Process. Energy Agric.*, 25, 2, 37–42, 2021. https://doi.org/10.5937/jpea25-30971.

23. Lule, Z. and Kim, J., Thermally Conductive and Highly Rigid Polylactic Acid (PLA) Hybrid Composite Filled with Surface Treated Alumina/Nano-Sized Aluminum Nitride. *Composites, Part A*, 124, 105506, 2019. https://doi.org/https://doi.org/10.1016/j.compositesa.2019.105506.
24. Gilbert, M., Plasticized PVC Elastomers, Editor(s): K.H. Jürgen Buschow, Robert W. Cahn, Merton C. Flemings, Bernhard Ilschner, Edward J. Kramer, Subhash Mahajan, Patrick Veyssière, in: *Encyclopedia of Materials: Science and Technology*, pp. 1–3, Elsevier, 2001, ISBN 9780080431529, https://doi.org/10.1016/B0-08-043152-6/01259-6.
25. Wagner, M.P., Non Black Fillers in Elastomers. *Rubber World*, 164, 46, 1941.
26. Perry, C.C. and Lu, Y., Preparation of Silicas from Silicon Complexes: Role of Cellulose in Polymerisation and Aggregation Control. *J. Chem. Soc.{,} Faraday Trans.*, 88, 19, 2915–2921, 1992. https://doi.org/10.1039/FT9928802915.
27. Eddy, N., Eddy, N., Eddy, N., Nkereuwem, U.E., Oladele, J.O., Kelle, H.I., Ogoko, E.C., Odiongenyi, A., Ameh, P., Ukpei, R.A., Ogbodo, R., Garg, R., Garg, R., Freundlich, M., Ebenso, E.E., Emori, W., Synthesis , Characterization and Application of Novel Microporous Ca-Al-Si Oxide Nanocomposite for Photocatalytic Degradation of Methyl Orange Dye in Aqueous Media.
28. Merino, D., Ollier, R., Lanfranconi, M., Alvarez, V., Preparation and Characterization of Soy Lecithin-Modified Bentonites. *Appl. Clay Sci.*, 127–128, 17–22, 2016. https://doi.org/10.1016/j.clay.2016.04.006.
29. Anjum, A., Anjum, A., Kashif, M., Garg, R., Eddy, N.O., Jahan, I., Kumari, A., Ersan, G., Wabaidur, S.M., *Food Chemistry Advances Nano-Scale Innovations: Properties, Types, and Applications of Nanomaterials in Packaging.*
30. Yu, J., Yang, J., Liu, B., Ma, X., Preparation and Characterization of Glycerol Plasticized-Pea Starch/ZnO-Carboxymethylcellulose Sodium Nanocomposites. *Bioresour. Technol.*, 100, 11, 2832–2841, 2009. https://doi.org/10.1016/j.biortech.2008.12.045.
31. Filiciotto, L. and Rothenberg, G., Biodegradable Plastics: Standards, Policies, and Impacts. *ChemSusChem*, 14, 1, 56–72, 2021. https://doi.org/10.1002/cssc.202002044.
32. Ahmed, J., Arfat, Y.A., Bher, A., Mulla, M., Jacob, H., Auras, R., Active Chicken Meat Packaging Based on Polylactide Films and Bimetallic Ag–Cu Nanoparticles and Essential Oil. *J. Food Sci.*, 83, 5, 1299–1310, 2018. https://doi.org/10.1111/1750-3841.14121.
33. Hachemi, R., Belhaneche-Bensemra, N., Massardier, V., Elaboration and Characterization of Bioblends Based on PVC/PLA. *J. Appl. Polym. Sci.*, 131, 7, 1–7, 2014. https://doi.org/10.1002/app.40045.
34. Mathew, S., S, S., Mathew, J., Radhakrishnan, E.K., Biodegradable and Active Nanocomposite Pouches Reinforced with Silver Nanoparticles for Improved Packaging of Chicken Sausages. *Food Packag. Shelf Life*, 19, August 2018, 155–166, 2019. https://doi.org/10.1016/j.fpsl.2018.12.009.

35. de França, J.O.C., da Silva Valadares, D., Paiva, M.F., Dias, S.C.L., Dias, J.A., Polymers Based on PLA from Synthesis Using D,L-Lactic Acid (or Racemic Lactide) and Some Biomedical Applications: A Short Review. *Polymers*, 14, 12, 2022. https://doi.org/10.3390/polym14122317.
36. Hu, Y., Daoud, W.A., Cheuk, K.K.L., Lin, C.S.K., Newly Developed Techniques on Polycondensation, Ring-Opening Polymerization and Polymer Modification: Focus on Poly(Lactic Acid). *Materials*, 9, 3, 2016. https://doi.org/10.3390/ma9030133.
37. Othman, N., Xu, C., Mehrkhodavandi, P., Hatzikiriakos, S.G., Thermorheological and Mechanical Behavior of Polylactide and Its Enantiomeric Diblock Copolymers and Blends. *Polymer*, 53, 12, 2443–2452, 2012. https://doi.org/https://doi.org/10.1016/j.polymer.2012.03.068.
38. Castro-Aguirre, E., Iñiguez-Franco, F., Samsudin, H., Fang, X., Auras, R., Poly(Lactic Acid)—Mass Production, Processing, Industrial Applications, and End of Life. *Adv. Drug Delivery Rev.*, 107, 333–366, 2016. https://doi.org/https://doi.org/10.1016/j.addr.2016.03.010.
39. Ramakrishna, S., Rangaswamy, V., Jain, D., Jagdambalal, R., Patel, P., Kar, D., Ramachandran, S., Ganeshpure, P., Satpathy, U., Process for the Production of Polylactic Acid (PLA) from Renewable Feedstocks, 2005. https://patents.google.com/patent/US20060036062A1/en.
40. Dorgan, J.R., Braun, B., Wegner, J.R., Knauss, D.M., Poly(Lactic Acids): A Brief Review, in: *Degradable Polymers and Materials*, vol. 939, pp. 102–125 American Chemical Society, 2006, https://doi.org/doi:10.1021/bk-2006-0939.ch007.
41. Jem, K.J. and Tan, B., The Development and Challenges of Poly (Lactic Acid) and Poly (Glycolic Acid). *Adv. Ind. Eng. Polym. Res.*, 3, 2, 60–70, 2020. https://doi.org/https://doi.org/10.1016/j.aiepr.2020.01.002.
42. Chafran, L.S., Paiva, M.F., França, J.O.C., Sales, M.J.A., Dias, S.C.L., Dias, J.A., Preparation of PLA Blends by Polycondensation of D,L-Lactic Acid Using Supported 12-Tungstophosphoric Acid as a Heterogeneous Catalyst. *Heliyon*, 5, 5, e01810, 2019. https://doi.org/https://doi.org/10.1016/j.heliyon.2019.e01810.
43. Mistretta, M.C., Botta, L., La Mantia, F.P., Di Fiore, A., Cascone, M., Film Blowing of Biodegradable Polymer Nanocomposites for Agricultural Applications. *Macromol. Mater. Eng.*, 306, 9, 1–8, 2021. https://doi.org/10.1002/mame.202100177.
44. Brito, G.F., Agrawal, P., Araújo, E.M., De Mélo, T.J.A., Polylactide/Biopolyethylene Bioblends. *Polimeros*, 22, 5, 427–429, 2012. https://doi.org/10.1590/S0104-14282012005000072.
45. Lyu, Y., Yin, H., Chen, Y., Zhang, Q., Shi, X., Structure and Evolution of Multiphase Composites for 3D Printing. *J. Mater. Sci.*, 55, 16, 6861–6874, 2020. https://doi.org/10.1007/s10853-020-04505-w.

46. Pardo-Figuerez, M., López-Córdoba, A., Torres-Giner, S., Lagaron, J.M., Superhydrophobic Bio-Coating Made by Co-Continuous Electrospinning and Electrospraying Polyethylene Terephthalate Films Proposed as Easy Emptying Transparent Food Packaging. *Coatings*, 8, 10, 2018. https://doi.org/10.3390/coatings8100364.
47. Xie, D., Zhao, Y., Li, Y., LaChance, A.M., Lai, J., Sun, L., Chen, J., Rheological, Thermal, and Degradation Properties of PLA/PPG Blends. *Mater. (Basel Switzerland),* 12, 21, 2019. https://doi.org/10.3390/ma12213519.
48. Benini, K.C.C. de C., Bomfim, A.S.C. de, Voorwald, H.J.C., Cellulose-Reinforced Polylactic Acid Composites for Three-Dimensional Printing Using Polyethylene Glycol as an Additive: A Comprehensive Review. *Polymers*, 15, 19, 2023. https://doi.org/10.3390/polym15193960.
49. DE Vries, A.J., Bonnebat, C., Carrega, M., Dilute Solution Properties and Molecular Characterization of Polyvinyl Chloride. *Pure Appl. Chem.*, 26, 2, 209–240, 1971. https://doi.org/10.1351/pac197126020209.
50. Wu, Q., Xie, W., Wu, H., Wang, L., Liang, S., Chang, H., Liu, B., Effect of Volatile Solvent and Evaporation Time on Formation and Performance of PVC/PVC-g-PEGMA Blended Membranes. *RSC Adv.*, 9, 59, 34486–34495, 2019. https://doi.org/10.1039/c9ra05454e.
51. Patil, S.S. and Jena, H.M., Performance Assessment of Polyvinyl Chloride Films Plasticized with Citrullus Lanatus Seed Oil Based Novel Plasticizer. *Polym. Test.*, 101, 107271, 2021. https://doi.org/https://doi.org/10.1016/j.polymertesting.2021.107271.
52. Sim, M.-J., Cha, S.-H., Lee, J.-C., Enhancement of Flame Retardancy and Physical Property for Poly(Vinyl Chloride) Having Renewable Cardanol-Based Self-Polymerizable Phosphonate under Heat Treatment Process. *Polym. Test.*, 100, 107266, 2021. https://doi.org/https://doi.org/10.1016/j.polymertesting.2021.107266.
53. Abdollahi Boraei, S.B., Bakhshandeh, B., Mohammadzadeh, F., Haghighi, D.M., Mohammadpour, Z., Clay-Reinforced PVC Composites and Nanocomposites. *Heliyon*, 10, 7, e29196, 2024. https://doi.org/10.1016/j.heliyon.2024.e29196.
54. Zhang, Y., Ding, C., Zhang, N., Wang, Y., Zhang, Y., Li, H., Zhang, Y., Formulation Optimization and Performance Enhancement of Polyvinyl Chloride-Silica Micropowder Composites with a Promising Flooring Application. *J. Mater. Res. Technol.*, 26, 8944–8960, 2023. https://doi.org/https://doi.org/10.1016/j.jmrt.2023.09.168.
55. Santana, V.T., Gonçalves, S.P.C., Agnelli, J.A.M., Martins-Franchetti, S.M., Biodegradation of a Polylactic Acid/Polyvinyl Chloride Blend in Soil. *J. Appl. Polym. Sci.*, 125, 1, 536–540, 2012. https://doi.org/10.1002/app.35685.
56. Li, X., Li, W., Jiang, Y., Ding, Y., Yun, J., Tang, Y., Zhang, P., Effect of Nano-ZnO-Coated Active Packaging on Quality of Fresh-Cut "Fuji" Apple. *Int. J. Food Sci. Technol.*, 46, 9, 1947–1955, 2011. https://doi.org/10.1111/j.1365-2621.2011.02706.x.

57. Ragaert, K., Delva, L., Van Geem, K., Mechanical and Chemical Recycling of Solid Plastic Waste. *Waste Manage. (Oxford)*, 69, 24–58, 2017. https://doi.org/https://doi.org/10.1016/j.wasman.2017.07.044.
58. Samir, A., Ashour, F.H., Hakim, A.A.A., Bassyouni, M., Recent Advances in Biodegradable Polymers for Sustainable Applications. *npj Mater. Degrad.*, 6, 1, 68, 2022. https://doi.org/10.1038/s41529-022-00277-7.
59. Saikia, A. and Karak, N., Chapter 1 - Introduction to Biodegradable and Biocompatible Polymer Nanocomposites: Synthesis, Structure, Fundamental Properties, Biocompatibility, and Biodegradability, Editor(s): Kalim Deshmukh, Mayank Pandey, in: *Biodegradable and Biocompatible Polymer Nanocomposites*, pp. 1–34, Elsevier, 2023, ISBN 9780323916967, https://doi.org/10.1016/B978-0-323-91696-7.00007-6.
60. Raksaksri, L., Chaiwutthinan, P., Larpkasemsuk, A., Chuayjuljit, S., Boonmahitthisud, A., Properties of Wood-Plastic Composites Based on PVC/PLA/PBAT Ternary Blend. *J. Met. Mater. Miner.*, 30, 2, 124–135, 2020. https://doi.org/10.14456/jmmm.2020.29.
61. Koh, H.C., Park, J.S., Jeong, M.A., Hwang, H.Y., Hong, Y.T., Ha, S.Y., Nam, S.Y., Preparation and Gas Permeation Properties of Biodegradable Polymer/Layered Silicate Nanocomposite Membranes. *Desalination*, 233, 1–3, 201–209, 2008. https://doi.org/10.1016/j.desal.2007.09.043.
62. Lopera-Valle, A. and Elias, A., Amine Responsive Poly(Lactic Acid) (PLA) and Succinic Anhydride (SAh) Graft-Polymer: Synthesis and Characterization. *Polymers*, 11, 9, 2019. https://doi.org/10.3390/polym11091466.
63. Singh, S.K., Chowdhury, A., Thakur, R.R., Mangaraj, S., Sami, R., Aljahani, A.H., Aljuraide, N.I., Helal, M., Development and Characterization of Polyvinyl Chloride/Poly Lactic Acid Blend Based Biodegradable Polymeric Films. *Mater. Express*, 13, 4, 632–643, 2023. https://doi.org/10.1166/mex.2023.2388.
64. Kumar, S., Singh, R., Singh, T.P., Batish, A., Investigations for Magnetic Properties of PLA-PVC-Fe3O4-Wood Dust Blend for Self-Assembly Applications. *J. Thermoplast. Compos. Mater.*, 34, 7, 929–951, 2021. https://doi.org/10.1177/0892705719857778.
65. Eyholzer, C., Tingaut, P., Zimmermann, T., Oksman, K., Dispersion and Reinforcing Potential of Carboxymethylated Nanofibrillated Cellulose Powders Modified with 1-Hexanol in Extruded Poly(Lactic Acid) (PLA) Composites. *J. Polym. Environ.*, 20, 4, 1052–1062, 2012. https://doi.org/10.1007/s10924-012-0508-4.
66. Khammassi, S., Tarfaoui, M., Škrlová, K., Měřínská, D., Plachá, D., Erchiqui, F., Poly(Lactic Acid) (PLA)-Based Nanocomposites: Impact of Vermiculite, Silver, and Graphene Oxide on Thermal Stability, Isothermal Crystallization, and Local Mechanical Behavior. *J. Compos. Sci.*, 6, 4, 2022. https://doi.org/10.3390/jcs6040112.

67. Vega-García, A., Mariño-Castellanos, P., Fernández-Santiesteban, E., Velázquez-Infante, J., Guerrero, F., Peña-Garcia, R., Structural, Magnetic, and Electrical Properties of a SrFe12O19/PLA Composite. *Ceram. Int.*, 49, 23, Part B, 38877–38884, 2023. https://doi.org/https://doi.org/10.1016/j.ceramint.2023.09.224.
68. Chieng, B.W., Ibrahim, N.A., Yunus, W.M.Z.W., Hussein, M.Z., Poly(Lactic Acid)/Poly(Ethylene Glycol) Polymer Nanocomposites: Effects of Graphene Nanoplatelets. *Polymers*, 6, 1, 93–104, 2014. https://doi.org/10.3390/polym6010093.
69. Ramli, H., Zainal, N.F.A., Hess, M., Chan, C.H., Basic Principle and Good Practices of Rheology for Polymers for Teachers and Beginners. *Chem. Teach. Int.*, 4, 4, 307–326, 2022. https://doi.org/10.1515/cti-2022-0010.
70. Mumtaz, N., Li, Y., Artiaga, R., Farooq, Z., Mumtaz, A., Guo, Q., Nisa, F.-U., Fillers and Methods to Improve the Effective (Out-Plane) Thermal Conductivity of Polymeric Thermal Interface Materials – A Review. *Heliyon*, 10, 3, e25381, 2024. https://doi.org/https://doi.org/10.1016/j.heliyon.2024.e25381.
71. Doronin, F., Rudakova, A., Rytikov, G., Nazarov, V., A Novel Determination of the Melt Flow Index of Composite Filaments Used in Extrusion Additive Manufacturing. *Polym. Test.*, 133, 108376, 2024. https://doi.org/https://doi.org/10.1016/j.polymertesting.2024.108376.
72. Kumar, S., Singh, R., Singh, T.P., Batish, A., On Mechanical Characterization of 3-D Printed PLA-PVC-Wood Dust-Fe3O4 Composite. *J. Thermoplast. Compos. Mater.*, 35, 1, 36–53, 2022. https://doi.org/10.1177/0892705719879195.
73. Tomaszewska, J., Sterzyński, T., Woźniak-Braszak, A., Banaszak, M., Review of Recent Developments of Glass Transition in PVC Nanocomposites. *Polymers*, 13, 24, 2021. https://doi.org/10.3390/polym13244336.
74. Maréchal, F., *Biodegradable Plastics BT - Biodegradable Polymers and Plastics*, E. Chiellini and R. Solaro (Eds.), pp. 67–71, Springer US, Boston, MA, 2003.
75. Nilsen-Nygaard, J., Fernández, E.N., Radusin, T., Rotabakk, B.T., Sarfraz, J., Sharmin, N., Sivertsvik, M., Sone, I., Pettersen, M.K., Current Status of Biobased and Biodegradable Food Packaging Materials: Impact on Food Quality and Effect of Innovative Processing Technologies. *Compr. Rev. Food Sci. Food Saf.*, 20, 2, 1333–1380, 2021. https://doi.org/https://doi.org/10.1111/1541-4337.12715.
76. Abdullah, Z.W. and Dong, Y., Biodegradable and Water Resistant Poly(Vinyl) Alcohol (PVA)/Starch (ST)/Glycerol (GL)/Halloysite Nanotube (HNT) Nanocomposite Films for Sustainable Food Packaging. *Front. Mater.*, 6, April, 1–17, 2019. https://doi.org/10.3389/fmats.2019.00058.
77. Li, K., Bian, S., Zhen, W., Li, H., Zhao, L., Performance, Crystallization and Rheological Behavior of Poly(Lactic Acid)/N-(2-Hydroxyl) Propyl-3-Trimethyl Ammonium Chitosan Chloride Intercalated Vermiculite Grafted Poly(Acrylamide) Nanocomposites. *React. Funct. Polym.*, 158, 104791, 2021. https://doi.org/10.1016/j.reactfunctpolym.2020.104791.

78. Vasile, C., Râpă, M., Ştefan, M., Stan, M., Macavei, S., Darie-Niţă, R.N., Barbu-Tudoran, L., Vodnar, D.C., Popa, E.E., Ştefan, R., Borodi, G., Brebu, M., New PLA/ZnO:Cu/Ag Bionanocomposites for Food Packaging. *eXPRESS Polym. Lett.*, 11, 7, 531–544, 2017. https://doi.org/10.3144/expresspolymlett.2017.51.
79. Kumar, S., Singh, R., Singh, T.P., Batish, A., Mechanical and Morphological Properties Correlation of PLA-PVC-Wood Powder-Fe3O4 Composite Matrix for 3D Printing, in: *Additive, Subtractive, and Hybrid Technologies: Recent Innovations in Manufacturing*, C. Prakash, S. Singh, S. Ramakrishna (Eds.), pp. 51–58, Springer International Publishing, Cham, 2022, https://doi.org/10.1007/978-3-030-99569-0_4.

4

Cellulose Nanocrystals and Nanofibers Bionanocomposite in Medical Applications

Wan Norfazilah Wan Ismail[1,2*] and Nurul Hidayah Abu Bakar[1]

[1]*Faculty of Industrial Sciences and Technology, Universiti Malaysia Pahang Al-Sultan Abdullah, Lebuh Persiaran Tun Khalil Yaakob, Kuantan, Pahang, Malaysia*
[2]*Bioaromatic Research Centre, Universiti Malaysia Pahang Al-Sultan Abdullah, Lebuh Persiaran Tun Khalil Yaakob, Kuantan, Pahang, Malaysia*

Abstract

Cellulose, the most abundant natural polymer on Earth, has emerged as a promising resource for the development of sustainable bionanocomposites. Its nanoscale derivatives, cellulose nanocrystals (CNCs) and cellulose nanofibers (CNFs), offer remarkable properties such as high surface area, mechanical strength, biocompatibility, and renewability. This chapter explores the extraction methods, fabrication techniques, and biomedical applications of CNC- and CNF-based bionanocomposites. Various extraction strategies, including acid hydrolysis, TEMPO-mediated oxidation, mechanical processing, and enzymatic treatments, are discussed with a focus on optimizing yield, morphology, and environmental impact. CNCs, with their rigid crystalline structure, and CNFs, with their flexible fibrillar network, serve as reinforcing agents in biopolymer matrices to enhance mechanical, barrier, and functional properties. Advanced fabrication methods, including electrospinning, solvent casting, 3D printing, and freeze drying, enable the design of customized bionanocomposites for biomedical applications. These materials show significant potential in drug delivery, tissue engineering, wound healing, biosensing, and antimicrobial systems. CNCs are particularly valuable in controlled drug delivery, implant coatings, and biosensor platforms, while CNFs are ideal for wound dressings, hydrogels, and tissue scaffolds. Despite their promise, challenges remain in achieving uniform dispersion, scaling up production, and meeting regulatory standards. Future research directions focus on green processing, functionalization, and integration with digital manufacturing to advance clinical

**Corresponding author*: norfazilah@umpsa.edu.my

Shiv Kumari Panda and Visakh P. M (eds.) Polyvinyl Chloride-Based Biocomposites and Bionanocomposites, (85–124) © 2026 Scrivener Publishing LLC

translation. Overall, nanocellulose-based bionanocomposites represent a versatile, eco-friendly solution for next-generation medical materials, supporting the global shift toward sustainable and high-performance biomedical technologies.

Keywords: Cellulose nanocrystals, cellulose nanofibers, bionanocomposites, biomedical applications, extraction, fabrication

4.1 Introduction to Cellulose Nanocrystals and Nanofibers

Imagine a resource so abundant on Earth that it can be found almost everywhere: in the towering trees of forests, the softness of cotton, the tiniest bacteria, the green expanse of algae, and even in marine creatures like tunicates, the only animals known to produce it. This remarkable material is cellulose, nature's building block, which forms microscopic fibers called fibrils. These fibrils contain a unique structure: tightly packed crystalline regions interspersed with flexible, amorphous segments. This duality gives cellulose an extraordinary range of applications, from structural reinforcement to advanced medical innovations. More than just a structural component, cellulose holds immense potential. Scientists have unlocked its secrets to creating cellulose nanomaterials with remarkable properties: lightweight but strong [5, 11, 113], flexible yet robust [11, 81], and environmentally friendly [17, 103, 125]. Derived from renewable sources like plants and bacteria, cellulose exhibits an elastic strength [3, 100, 104] comparable to some metals while being biodegradable and sustainable.

When processed further, cellulose can be transformed into cellulose nanocrystals (CNCs) or cellulose nanofibers (CNFs). CNCs and CNFs have garnered significant attention in recent years due to their remarkable properties and potential applications in various fields, particularly in medical applications [38]. CNCs are rod-like nanoparticles derived from the crystalline regions of cellulose, while CNFs are flexible, long fibers comprising both crystalline and amorphous regions (Figure 4.1). These nanomaterials have sizes ranging from just a few nanometers to over a micron, depending on how they are prepared [106]. Factors like the type of cellulose source (be it a tree, algae, or bacteria) and the extraction process play a big role in shaping their final form and properties.

Something even more fascinating is unique characteristics of these nanocelluloses, including high surface area, excellent mechanical

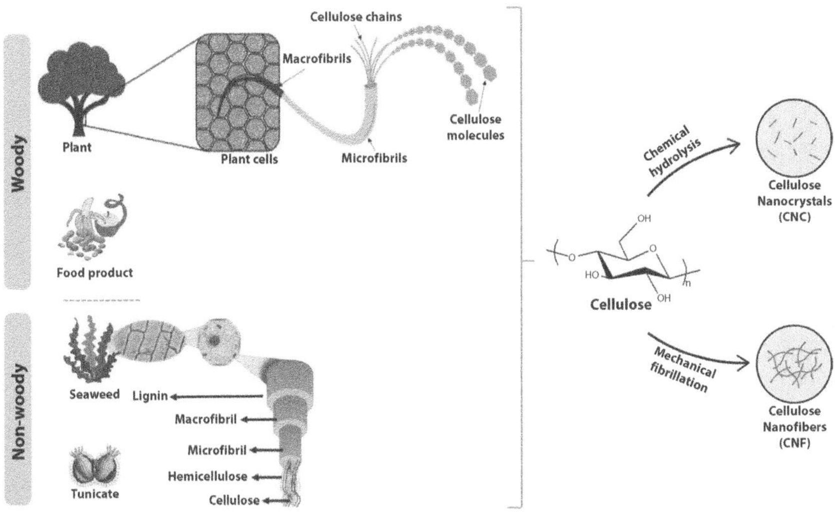

Figure 4.1 Sources and extraction pathways of CNCs and CNFs. Adapted and modified from [53, 62, 141].

properties, renewability, and biocompatibility [8, 49, 68, 99], making them ideal candidates for bionanocomposite materials. Their nanoscale dimensions and ability to form strong hydrogen bonds allow them to enhance the physical and chemical properties of composite materials. They are used to make stronger, lighter materials; improve barriers in packaging; create advanced hydrogels; enhance electronics; insulate against heat; and even deliver medicines precisely where they are needed. Additionally, their biodegradability aligns with the increasing demand for sustainable and eco-friendly materials in medical applications. It is like nature's way of giving us a sustainable toolkit for solving modern problems. Moreover, the need for a new socioeconomic transformation narrative to achieve the United Nations' Sustainable Development Goals (SDGs) by 2030 has emphasized the link between sustainable biomass utilization, environmental conservation, and ecosystem resilience, driving researchers to focus on biodegradable and sustainable renewable resources like natural fiber, which is highly valued for enhancing the mechanical and barrier properties of bioplastics [76].

The rapid increase in the number of publications related to CNCs, CNFs, and biocomposites for medical applications over time highlights the growing scientific interest in this field (Figure 4.2). Early research on

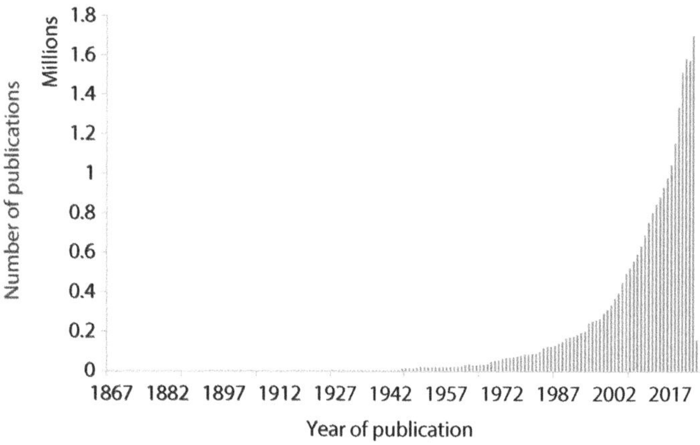

Figure 4.2 Trend of research publications over time. Exported from Scopus (1867–2026), accessed on 30th January 2025.

these materials was relatively scarce, as seen by the minimal publications before the mid-20th century. However, from the early 2000s onwards, there has been an exponential rise in publications, indicating a surge in both research and technological advancements. This trend is likely driven by the increasing awareness of sustainable materials, advancements in nanotechnology, and the broadening scope of medical applications utilizing nanocellulose. The peak in recent years suggests an accelerating momentum in research, with CNCs and CNFs being recognized as key materials in biomedical innovations, including drug delivery, tissue engineering, and biocomposites.

Figure 4.3 showcases the top 10 countries contributing to research in CNCs and CNFs for related medical applications. The United States (7,212,281), China (3,862,545), and the United Kingdom (2,022,163) lead in the number of publications, suggesting that these countries have invested significantly in nanocellulose research. The prominence of the United States may be attributed to strong funding in materials science and biotechnology [120], as well as an established network of research institutions. China follows closely, reflecting the country's increasing emphasis on nanotechnology and biomaterials in recent years [89]. The United Kingdom, with its leading universities and collaborative research initiatives, also plays a significant role in advancing CNC and CNF research [32]. Other countries, including Japan, Germany, India, Canada, Italy,

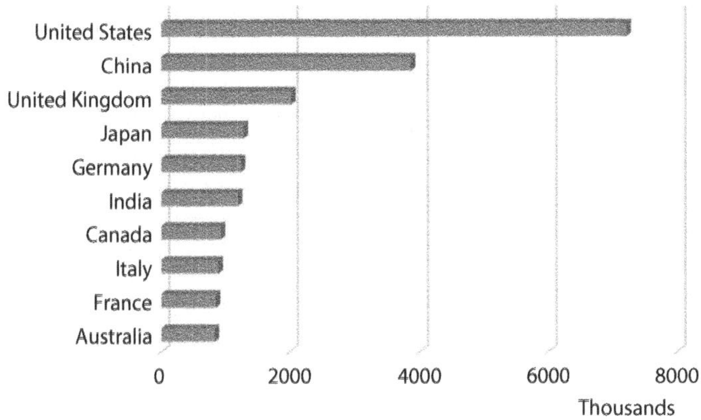

Figure 4.3 Leading countries in publications. Exported from Scopus (1867–2026), accessed on 30th January 2025.

France, and Australia, have also made substantial contributions, demonstrating global interest in nanocellulose applications. The distribution of research efforts suggests that nanocellulose is a worldwide area of interest, with diverse applications spanning biomedicine, packaging, and industrial uses.

4.2 Extraction Methods for CNCs and CNFs

The extraction methods of CNCs and CNFs from various biomass sources are a crucial step in harnessing their remarkable properties for biomedical and industrial applications. The process typically begins with the removal of non-cellulosic components (pretreatment and purification). This step will remove lignin, hemicellulose, and other impurities. This is often achieved using alkaline pretreatment and bleaching [43]. During alkaline pretreatment, sodium hydroxide or potassium hydroxide dissolves hemicellulose and some lignin, reducing fiber stiffness and improving fibrillation efficiency. Next, sodium chlorite combined with acetic acid is applied at controlled temperatures for several hours to eliminate residual lignin, further whitening the fibers and enhancing their crystallinity [67].

After pretreatment, two distinct routes for CNC extraction are evident: direct extraction from cellulose *via* acid hydrolysis or an indirect route where CNF is first produced through chemical, mechanical, or enzymatic treatments before undergoing acid hydrolysis to yield CNCs (Figure 4.4). The acid hydrolysis of cellulose sources, which selectively removes the amorphous regions, leaves behind the crystalline domains. Commonly used acids include sulfuric acid (H_2SO_4), hydrochloric acid (HCl), and phosphoric acid (H_3PO_4) [21, 43, 66]. H_2SO_4 is particularly popular due to its ability to introduce sulfate ester groups onto the CNC surface, enhancing colloidal stability [37]. In an acid hydrolysis approach, purified cellulose is hydrolyzed with concentrated acid under control conditions. This selective process breaks down amorphous regions while preserving crystalline domains, resulting in rod-like nanocrystals. The hydrolysis process is quenched by adding cold water, followed by centrifugation and dialysis to remove excess acid (Figure 4.5). The resulting colloidal suspension is then ultrasonicated to ensure uniform dispersion of CNCs. Depending on the cellulose source, CNCs vary in dimensions, typically ranging from 3 to 50 nm in width and 50 to 1160 nm in length [10]. Alternative acids, such as HCl or H_3PO_4, have been explored to modify CNC surface properties.

Meanwhile, the isolation of CNFs can be achieved by various extraction methods. The chemical approach includes 2,2,6,6-tetramethylpiperidine-1-oxyl (TEMPO)-mediated oxidation [67, 85, 131], ionic liquids [40, 134], and electrospinning [36, 91, 94, 102]. TEMPO-mediated oxidation (Figure 4.6) introduces carboxyl groups ($-COO^-$) onto cellulose, enhancing fibrillation and dispersibility, while ionic liquids employ non-volatile solvents (e.g., 1-butyl-3-methylimidazolium chloride) to selectively dissolve amorphous cellulose, breaking hydrogen bonds, aiding nanofibril dispersion without generating harmful waste. Both chemical approaches create a dispersed cellulose solution, which is then fed into a high-voltage electrospinning system, where it undergoes electrical stretching; the solvent evaporates and lastly leaves behind continuous CNFs. Cellulose dispersion can also be blended with polymers (e.g., polyvinyl alcohol, polyethylene oxide, polylactic acid) or additives to optimize fiber formation and mechanical properties before electrospinning.

Emerging techniques, such as mechanical process and enzymatic hydrolysis, offer more sustainable alternatives. Mechanical techniques such as cryo-crushing, high-pressure homogenization, microfluidization, ultrasonication, grinding, and compression/roller crushing use shear, impact, and cavitation forces to break down cellulose into CNFs. High-pressure homogenization involves breaking down cellulose fibers into nanofibers through a narrow nozzle at extremely high pressures (up to 2000 bar) that

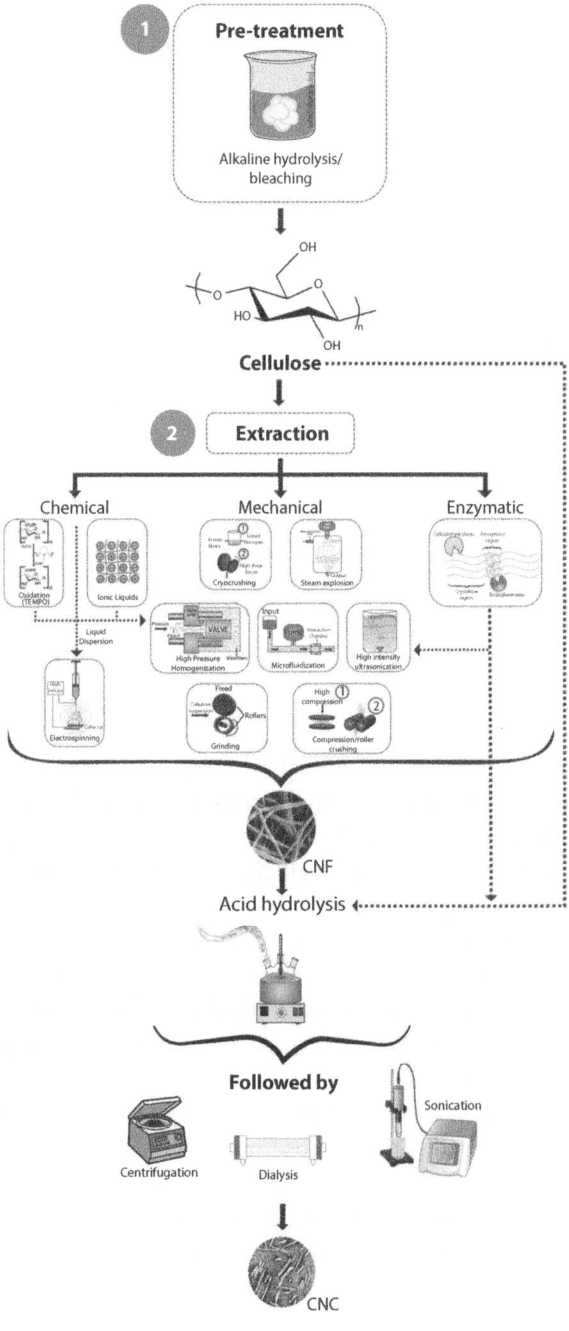

Figure 4.4 Extraction methods to obtain CNCs and CNFs. Adapted and modified from [73, 97, 123].

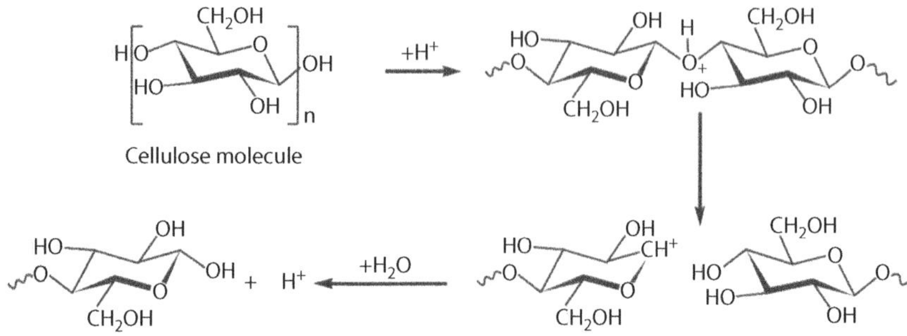

Figure 4.5 Mechanism of acid hydrolysis of cellulose [88].

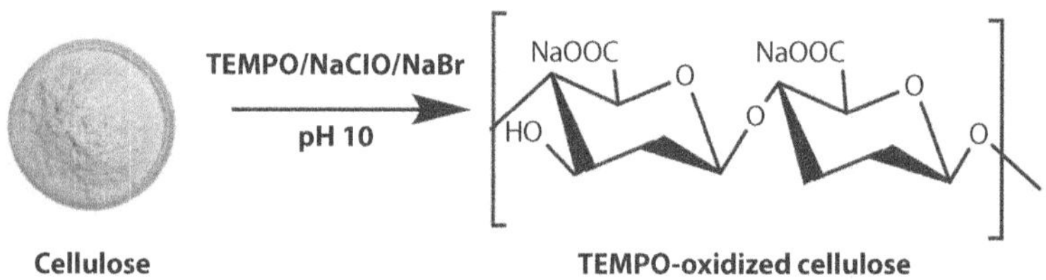

Figure 4.6 TEMPO-mediated oxidation of CNF reaction scheme. Adapted and modified from Madivoli *et al.* [63].

generate high shear forces [96, 124, 133]. This method results in well-dispersed CNFs with high aspect ratios but requires multiple passes to achieve uniform fibrillation. Microfluidization is able to produce highly uniform CNFs with improved fibrillation efficiency by a high-pressure fluidizer where cellulose slurry is forced through a microchannel system, causing high shear, impact, and cavitation forces [16]. Additionally, grinding and refining methods involve mechanical shearing using a rotating disk grinder, progressively breaking down cellulose into nanofibers. Although effective, this method is energy-intensive and requires multiple passes to achieve fine CNFs [55].

Cryo-crushing and ultrasonication are another mechanical technique to separate nanofibers from cellulose molecules. Using high mechanical impact, cryo-crushing causes the formation of ice crystal that disrupts the fiber network which promotes fibrillation [105, 107]. This technique is useful for achieving high yields of ultrafine CNFs. In the meantime, ultrasonication techniques are often used in combination with chemical pretreatments to enhance CNF dispersion. It involves the application of high-intensity ultrasonic waves (20–40 kHz) to a cellulose suspension, generating acoustic cavitation that helps break down fiber bundles. Advances in hybrid methods combining mechanical and chemical approaches, such as TEMPO-mediated oxidation followed by microfluidization, have demonstrated enhanced yield and improved control over fiber morphology.

Those mechanical and chemical processes typically require high-energy input and environmental concerns, respectively, making them less economically viable for large-scale applications [27]. Enzymatic hydrolysis, involving cellobiohydrolases, endoglucanases, xylanase, and laccase enzymes are employed during enzymatic pretreatment to selectively degrade hemicellulose and lignin without damaging the cellulose structure, leading to increased fiber purity while minimizing environmental impact [58]. Figure 4.4 also illustrates the role of enzymatic pretreatment in facilitating both mechanical fibrillation (ultrasonication) and acid hydrolysis. When followed by high-intensity ultrasonication, the enzymatic approach reduces the energy required for mechanical fibrillation, as the partially hydrolyzed cellulose fibers are more susceptible to cavitation-induced fragmentation, leading to finer, more uniform CNFs.

Simultaneously, enzymatic treatment also serves as a preparatory step for acid hydrolysis, where the selective breakdown of amorphous cellulose exposes more crystalline regions, improving the efficiency and selectivity of CNC extraction. This enzymatic-assisted hydrolysis reduces the required acid concentration and hydrolysis time, minimizing over-degradation and environmental impact while producing CNCs with enhanced crystallinity, aspect ratio, and thermal stability. By integrating enzymatic hydrolysis before either ultrasonication or acid hydrolysis processing, this approach enhances nanocellulose yield, quality, and sustainability, offering an energy-efficient and environmentally friendly alternative for CNF and CNC production. This pretreatment, also known as green extraction method, has gained popularity due to its ability to reduce chemical usage and lower environmental impact [43]. After CNF formation, acid hydrolysis, followed by centrifugation, sonication, and dialysis, removes amorphous cellulose to yield high-crystallinity CNCs with enhanced mechanical properties. Table 4.1 summarizes the recent extraction methods to isolate CNCs and CNFs.

Table 4.1 Summary of recent extraction methods to isolate CNCs and CNFs and their fabrication techniques.

Type of nanocellulose	Sources of cellulose	Extraction method	Extraction conditions	Fabrication technique	References
CNC	*Dictyota bartayresiana* (brown seaweed from *Dictyotaceae*)	Acid hydrolysis using H_2SO_4	Acid concentration: 51 wt% Reaction time: 30 min Temperature: 45°C	Freeze drying	Murugesan *et al.* [71]
	Miscanthus X Giganteus	Ionic liquid using N,N-dimethyl-N-butylamine and H_2SO_4 followed by oxidation using alkaline H_2O_2	**Ionic liquid** Ratio: 80% IL:20% water Temperature: 120 and 170°C Reaction time: 45 min and 6 h **Oxidation** NaOH concentration: 10 wt% H_2O_2 concentration: 30 wt% Reaction time: 60 min Temperature: 80°C	Freeze drying	Rafat *et al.* [90]
	Grape pomace	Acid hydrolysis using H_2SO_4	Acid concentration: 64 wt% Reaction time: 30 min Temperature: 45°C	Freeze drying	Taymaz & Uslu [114]

(*Continued*)

Table 4.1 Summary of recent extraction methods to isolate CNCs and CNFs and their fabrication techniques. (*Continued*)

Type of nanocellulose	Sources of cellulose	Extraction method	Extraction conditions	Fabrication technique	References
CNC	Hemp stalks, wheat straws, and wheat husks	Acid hydrolysis using H_2SO_4	Acid concentration: 60 wt% Solid-liquid ratio: 1:10 Reaction time: 3~4 h Temperature: 45°C Stirring rate: 170 rpm	Electrospinning and freeze drying	Wu *et al.* [129]
	Hemp stalks	Grinding and microfluidization	**Grinding** Sieve size: 4.00 mm, 0.25 mm and 0.12 mm **Microfluidization** Chamber nozzle size: 200 µm Pressure: 30,000 psi Repetition: 5 times	3D printing	Bruvere *et al.* [15]

(*Continued*)

Table 4.1 Summary of recent extraction methods to isolate CNCs and CNFs and their fabrication techniques. (*Continued*)

Type of nanocellulose	Sources of cellulose	Extraction method	Extraction conditions	Fabrication technique	References
CNC	Poplar wood	Steam explosion followed by acid hydrolysis using H_2SO_4	**Steam explosion** Temperature: 200°C Pressure: 15.5 bar Reaction time: 5 min **Acid hydrolysis** Acid concentration: 62 wt% Reaction time: 2 h Temperature: 45°C Stirring rate: 200 rpm	Freeze drying	Haddis *et al.* [42]
	Northern Bleached Softwood Kraft pulp	Acid catalyzed-subcritical water hydrolysis	Subcritical water temperature: 120-150°C Reaction time: 60 and 120 min Reaction pressure: 2,600 and 2,800 psi Acid concentration: 0.1 M orthophosphoric acid Water to pulp ratio: 100:1 mL/g	Freeze drying	Osei-Bonsu *et al.* [78]

(*Continued*)

Table 4.1 Summary of recent extraction methods to isolate CNCs and CNFs and their fabrication techniques. (*Continued*)

Type of nanocellulose	Sources of cellulose	Extraction method	Extraction conditions	Fabrication technique	References
CNC	*Lemna minor L.* (duckweed)	Ionic liquid using triethylammonium and H_2SO_4 followed by acid hydrolysis using H_2SO_4	**Ionic liquid (IL)** Ratio: 90% IL:10% water Temperature: 120°C Reaction time: 4 h **Acid hydrolysis** Acid concentration: 64 wt% Reaction time: 30 min Temperature: 45°C	Freeze drying	Puglia *et al.* [86]
	Enset ventricosum	Acid hydrolysis using H_2SO_4	Acid concentration: 55 wt% Reaction time: 45 min Temperature: 50°C	Freeze drying	Tenaye *et al.* [115]
	Aloe vera waste leaves	Acid hydrolysis using H_2SO_4	Acid concentration: 64 wt% Temperature: 50°C	Freeze drying	Triantafyllou *et al.* [118]
	Alpinia nigra rhizome	Acid hydrolysis using H_2SO_4	Acid concentration: 30 wt% Reaction time: 60 min Temperature: 80°C	Freeze drying	Baruah *et al.* [12]

(*Continued*)

Table 4.1 Summary of recent extraction methods to isolate CNCs and CNFs and their fabrication techniques. (*Continued*)

Type of nanocellulose	Sources of cellulose	Extraction method	Extraction conditions	Fabrication technique	References
CNF	Willow bark	Hot water extraction (HWE) and microfluidization	**HWE** Temperature: 80°C Time: 30 min **Microfluidization** Chamber nozzle: 100-400 μm Repetition: 6-12 times	Freeze drying	Dou *et al.* [29]
	Sawdust	Mechanical grinding using a Supermass collider	Speed: 1500 rpm Time: 20 min	3D printing	John *et al.* [47]
	Softwood-bleached kraft pulp	Horizontal bead mill	NA	Melting extrusion technique	Ren *et al.* [95]
	Rice husk	High-pressure homogenization	Pressure: 120 MPa Cycle: 6, 9 and 12	Freeze drying	Samsalee *et al.* [101]

(*Continued*)

Table 4.1 Summary of recent extraction methods to isolate CNCs and CNFs and their fabrication techniques. (*Continued*)

Type of nanocellulose	Sources of cellulose	Extraction method	Extraction conditions	Fabrication technique	References
CNF	Ginger fiber, wheat straw, bamboo pulp, aspen wood chips, and flax fiber	Acid hydrolysis using H_2SO_4	Acid concentration: 64 wt% Reaction time: 30 min Temperature: 40°C	Freeze drying	Wang *et al.* [122]
	Ramie fibers	DES pretreatment followed by ball milling	**DES pretreatment** Reaction time: 4 h Temperature: 100°C **Ball milling** Agate ball size: 10-20 mm Agate pot size: 500 mL Ball-to-pulp slurry ratio: 2.5:1 Rotation speed: 1100 rpm Reaction time: 6 h	Solvent casting	Yu *et al.* [135]

*NA: not available

4.3 Incorporation of Nanoparticles in Bionanocomposites

Bionanocomposites are materials where natural, biodegradable polymers (like cellulose, chitosan, or alginate) are incorporated with nanomaterials (such as nanoparticles), also known as fillers, to improve their properties [35, 75]. CNCs and CNFs are not considered "bionanocomposites" themselves, but rather they are the building blocks for creating bionanocomposites when combined with other biopolymers like polylactic acid (PLA) or starch, forming a composite material with enhanced properties due to the nanosized cellulose particles dispersed within the matrix [6, 51, 64, 119]. Essentially, the "nano" part of the bionanocomposite comes from the CNCs or CNFs [69, 117].

One of the most interesting aspects of incorporating CNC and CNF into biopolymer matrices lies in their distinct morphology and the specific reinforcement mechanisms they provide. CNCs, due to their rigid, rod-like crystalline structure, impart significant stiffness and tensile strength to the host polymer [56]. They are particularly effective in enhancing barrier properties and mechanical performance in films, coatings, and implantable devices. On the other hand, CNFs, with their entangled fibrillar network and flexible nature, offer improvements in toughness, ductility, and impact resistance [127]. Their ability to form continuous, web-like networks within the polymer matrix also aids in stress transfer and energy dissipation [13], making them excellent candidates for soft biomedical materials such as wound dressings and injectable hydrogels.

What further elevates the value of CNCs and CNFs in bionanocomposites is their ability to be processed using advanced alignment and patterning techniques. Methods such as shear-induced orientation, magnetic field alignment, and 3D printing allow researchers to control the orientation of nanocellulose within the biopolymer matrix, tailoring anisotropic properties that better match the mechanical behavior of biological tissues [9, 34]. For instance, aligned CNF structures have been used to guide neural growth, while CNC-reinforced PLA matrices have demonstrated enhanced mechanical performance when aligned along the axis of stress.

Furthermore, the incorporation of CNCs and CNFs into biopolymers contributes not only to mechanical enhancement but also to the multifunctionality of the final material. These nanocellulose components can act as carriers for drugs, antimicrobials, or growth factors when modified appropriately, transforming simple structural materials into dynamic, therapeutic platforms. For example, CNF-alginate hydrogels loaded with antibiotics or silver nanoparticles have shown significant efficacy in wound care, combining moisture retention, mechanical protection, and antibacterial activity [93]. Similarly, CNCs functionalized with targeting ligands or responsive moieties have been used in controlled drug delivery systems, especially in tumor-targeted formulations [41].

Achieving uniform dispersion of the nanocellulose and ensuring adequate compatibility with the matrix are key factors for attaining optimal material properties. However, nanocellulose has a natural tendency to agglomerate within the polymer system, which can hinder effective dispersion and interfacial bonding [84]. This challenge is compounded by the hydrophilic nature of nanocellulose and the often-hydrophobic character of many biopolymer matrices such as PLA. To address this issue, several fabrication techniques have been devised to enhance the distribution and interaction of nanocellulose within biopolymer matrices.

4.4 Fabrication Techniques for the Bionanocomposites

The fabrication processes of cellulose-based bionanocomposites, particularly those involving CNC and CNF, significantly influence the properties and application potential of the resulting materials. Notable techniques for incorporating CNC and CNF into polymer matrices include electrospinning, solvent casting, freeze drying, and 3D printing (Figure 4.7) [50]. Electrospinning has proven highly effective for producing cellulose-based nanofibers. CNC and CNF, after surface modification or blending with compatible polymers, are processed *via* electrospinning, leading to nanofibers with enhanced mechanical properties. The orientation and uniform distribution of CNC and CNF within these fibers contribute to improved structural integrity and biocompatibility, making them particularly suitable for biomedical applications such as tissue engineering scaffolds.

Solvent casting is another prevalent fabrication method, where aqueous suspensions of CNC and CNF are combined with polymer solutions to form homogeneous mixtures that are subsequently cast into films [126]. This technique allows for precise control over the composite's thickness and uniformity, which is crucial for enhancing mechanical properties [77].

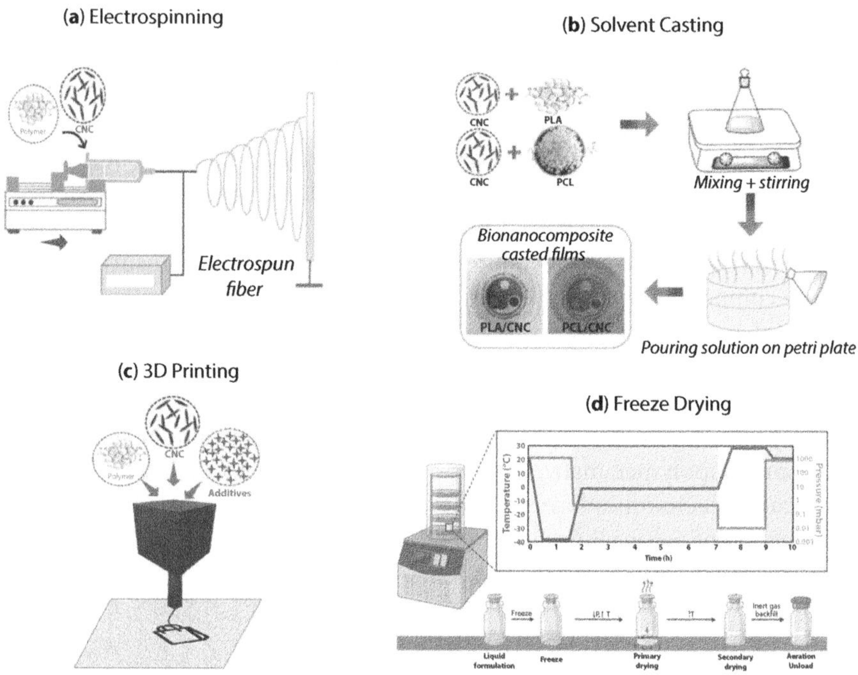

Figure 4.7 Different fabrication techniques of nanocellulose-based bionanocomposites: (a) electrospinning, (b) solvent casting, (c) 3D printing, and (d) freeze drying technique. Adapted and modified from Ghosh *et al.* [39] and Xiang *et al.* [130].

Surface modification of CNC and CNF, typically through chemical functionalization, is frequently applied to improve their dispersion within hydrophobic matrices, thereby achieving composites with superior mechanical properties and stability. Meanwhile, 3D printing has emerged as an advanced technique for fabricating CNC and CNF nanocomposites with complex and tailored geometries. The adaptability of 3D printing allows precise layer-by-layer construction of composite materials. CNC and CNF can be effectively integrated as reinforcing additives or as primary components of bio-inks, significantly enhancing the mechanical performance and functionality of printed constructs [52]. This method holds substantial promise for personalized medicine and complex tissue engineering applications, including cartilage and bone regeneration.

Finally, freeze drying, or lyophilization, enables the production of porous aerogel structures from CNC and CNF composites [19]. This method involves freezing the aqueous suspension and subsequently subliming the solvent, yielding aerogels characterized by low density and high porosity [65]. Such structures are advantageous in biomedical fields due to their lightweight nature and controllable porosity, which are beneficial for drug delivery and regenerative medicine. Table 4.1 also includes an additional column indicating the corresponding fabrication techniques used for the cellulose-based bionanocomposites.

Apart from the conventional techniques, recent research has introduced melt-compounding as a promising green fabrication technique [95]. This method combines acetylation surface modification with melt extrusion to efficiently disperse hydrophobically modified CNFs into PLA. The melt-compounding technique offers scalability and environmental advantages by eliminating solvent usage, making it particularly attractive for industrial applications. The melt-compounding method specifically leverages chemical modification through acetylation to significantly enhance chemical interactions between hydrophobic-modified CNFs and the PLA matrix, reducing nanoparticle aggregation and improving dispersion. This approach not only enhances the dispersion and compatibility of CNFs within hydrophobic polymer matrices but also significantly improves the rheological, crystallization, and foaming properties of the resulting bionanocomposites.

4.5 Various Applications in Medical Field

The integration of nanocellulose into bionanocomposite matrices has garnered increasing attention in recent years due to its significant potential in various medical applications. The biocompatibility and structural similarity of nanocellulose to the extracellular matrix (ECM) make it a suitable material for medical use. In this context, nanocellulose-based bionanocomposites have been employed in a wide range of biomedical applications, including wound healing, tissue engineering, drug delivery, biosensing, and antimicrobial platforms (Figure 4.8).

One of the principal applications of CNC-based composites is in drug delivery systems. The ability to functionalize CNCs with hydrophilic or hydrophobic moieties allows for encapsulation and sustained release of therapeutic agents [80]. Their excellent colloidal stability in aqueous systems ensures effective distribution and retention at the target site. Studies have reported the development of pH-sensitive CNCs capable of delivering anticancer drugs specifically to tumor microenvironments, thereby

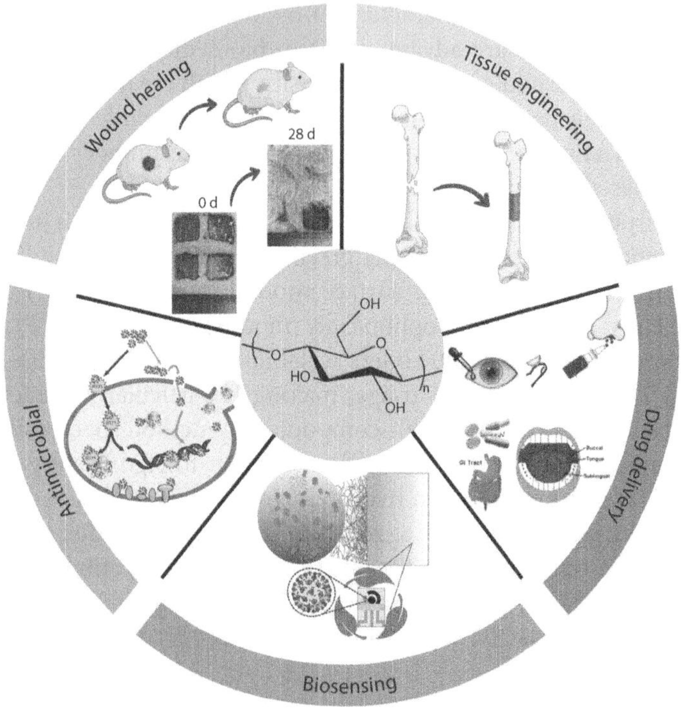

Figure 4.8 Applications of cellulose-based bionanocomposites in selected medical fields. Adapted and modified from Abazari *et al.* [1], Benoit *et al.* [14], De Lima *et al.* [23], Liu *et al.* [60], Sun *et al.* [108] and Wright *et al.* [128].

minimizing systemic toxicity [31, 57, 138]. Additionally, CNCs have been incorporated into implant coatings and injectable scaffolds where they enhance mechanical integrity while providing a biocompatible interface for cellular adhesion [20, 139]. Their rigidity and high aspect ratio contribute to improved structural stability in soft or load-bearing tissues [109–111]. Furthermore, CNCs have been successfully blended with conductive polymers and metal nanoparticles to develop biosensors for diagnostic purposes [25]. These composites offer enhanced sensitivity and stability for electrochemical detection of glucose [28, 33, 116], cholesterol [4, 7], or hormone biomarkers [30, 70].

CNC-based bionanocomposites also find utility in antimicrobial materials. When coupled with metal oxides such as zinc oxide or silver nanoparticles, CNCs act as dispersing and stabilizing agents that maximize antimicrobial activity. These formulations are being evaluated for application in surgical masks, wound dressings, and hospital textiles where the prevention of microbial growth is essential [2, 61, 72]. The crystallinity and nanoscale dimensions of CNCs not only improve the dispersion of active agents but also increase the surface interaction with microbial membranes, amplifying their antibacterial efficacy. Moreover, the integration of CNCs into 3D printing bioinks such as PLA [26, 74], cellulose [44], and poly(ethylene glycol) diacrylate-acrylated epoxidized soybean oil [98] has also been explored for constructing customized biomedical implants. Their rheological properties, when optimized, support extrusion-based bioprinting, and the resultant structures demonstrate favorable cell viability and mechanical performance. Overall, CNC-based bionanocomposites exhibit multi-functional capabilities, making them highly adaptable for modern biomedical applications.

CNF-based hydrogels have shown substantial potential in wound healing due to their excellent water retention capacity, flexibility, and ability to maintain a moist environment [22, 45, 132, 140]. These hydrogels form conformable barriers over wounds, promoting tissue regeneration while preventing microbial invasion. CNFs can be crosslinked with natural polymers such as alginate or gelatin to enhance mechanical stability and biological performance. Moreover, CNF-based dressings loaded with bioactive agents or antimicrobial nanoparticles exhibit accelerated healing and reduced inflammation in chronic wound models. In the context of tissue engineering, CNF composites are used to fabricate scaffolds for skin, cartilage, and even neural tissues [54]. The inherent shear-thinning property of CNF suspensions facilitates their use in injectable formats, allowing minimally invasive administration. Once implanted, CNF scaffolds support cell infiltration and proliferation due to their fibrous morphology and large surface area [18]. Studies have demonstrated that blending CNFs with collagen or hyaluronic acid yields hybrid scaffolds that better support tissue regeneration, especially in dermal and subcutaneous layers [59, 82, 136, 137].

Drug delivery is another field where CNFs have proven their relevance. Their high surface area and the ability to form dense networks allow the entrapment of therapeutic agents within gel matrices [24, 48, 87]. CNF-based carriers can be engineered for topical, oral, or transdermal delivery systems. Their biodegradability and low toxicity are particularly beneficial for long-term or sustained-release formulations. Furthermore, oxidized CNFs exhibit improved mucoadhesive properties, which can be exploited in gastrointestinal or nasal drug delivery routes [79]. Similar to CNCs, CNFs also serve as functional components in bioinks for 3D bioprinting [46, 83, 112]. Their structural integrity and tunable viscoelastic properties support cell-laden printing, enabling the development of patient-specific grafts and constructs [92]. CNFs combined with cell-compatible hydrogels have been used to create tissue-like architectures that maintain mechanical resilience and biological performance over extended culture periods.

Moreover, CNF-based films and membranes have been evaluated for barrier applications in implantable devices [25]. Their low permeability towards gases and fluids, coupled with mechanical durability, provides a means of isolating sensitive tissues or protecting bioelectronic interfaces [121]. These applications reflect the versatility of CNF-based composites as they transition from fundamental research into translational biomedical solutions. The applications of cellulose-based bionanocomposites in medical fields are summarized in Table 4.2.

4.6 Future Prospects and Challenges

The integration of CNCs and CNFs into biopolymer matrices continues to show great promise in developing sustainable, high-performance bionanocomposites for biomedical applications. However, several important challenges remain before their full clinical and commercial potential can be realized.

One of the most persistent obstacles is achieving uniform dispersion of CNCs and CNFs within hydrophobic biopolymer matrices, such as polylactic acid, polycaprolactone, or polyhydroxyalkanoates. The strong tendency of nanocellulose to self-agglomerate, driven by extensive hydrogen bonding and high surface energy, compromises interfacial adhesion and leads to heterogeneous structures. This in turn limits the mechanical performance and functional consistency of the final composite. Although surface functionalization methods, such as TEMPO oxidation, acetylation, or graft copolymerization, have been developed to improve compatibility, the scalability, reproducibility, and regulatory acceptability of these chemical modifications require further investigation.

Another key challenge is the lack of standardized processing techniques for producing CNC- and CNF-based bionanocomposites at industrial scales. Techniques such as solvent casting, extrusion, or electrospinning each influence nanocellulose alignment, matrix distribution, and end-use performance differently. Optimization of these methods for

Table 4.2 Biomedical applications of cellulose-based bionanocomposites.

Type of nanocellulose	Polymer matrix	Applications	References
CNC	Graphene/cellulose	Electrochemical biosensors	Dong *et al.* [28]
	Catechol-modified chitosan	Wound healing	Lu *et al.* [61]
	Polylactic acid	Bone tissue engineering	N'Gatta *et al.* [74]
	Poly(ethylene glycol) diacrylate and acrylated epoxidized soybean oil	Tissue engineering	Rosa *et al.* [98]
	Cellulose	Anticancer activity against colorectal cancer cells	Yusefi *et al.* [138]
	Polyacrylamide	Drug delivery of doxorubicin	Ehsanimehr *et al.* [31]
	Polylactic acid	Custom-made implants, tissue and cell printing	Dinesh Kumar *et al.* [26]
	Poly(glycidylmethacrylate-co ethylene glycol dimethacrylate)	Detection of cortisol in human sweat	Mugo & Alberkant [70]
	Cellulose	Dermal and oral blood glucose monitoring	Tracey *et al.* [116]
	Polyaniline	Electrochemical enzymatic biosensor of cholesterol	Abdi *et al.* [4]

(*Continued*)

Table 4.2 Biomedical applications of cellulose-based bionanocomposites. (*Continued*)

Type of nanocellulose	Polymer matrix	Applications	References
CNC	Poly(3-hydroxybutyrate-co-3-hydroxy valerate	UV absorber and antibacterial agents	Abdalkarim *et al.* [2]
	Cellulose	Electrochemical sensing of cholesterol	Anirudhan *et al.* [7]
	Polypyrrole	Biosensing detection of glucose	Esmaeili *et al.* [33]
CNF	Polyvinyl alcohol	Wound healing	Zhang *et al.* [140]
	Poly(lactic-co-glycolic acid)	Bone regeneration and guided tissue regeneration	Chen *et al.* [18]
	Polyvinyl alcohol-polyethylene glycol/glycerin	Wound dressing	Jafri *et al.* [45]
	Alginate	Antibiotic delivery	Putra *et al.* [87]
	Collagen	Drug delivery	Yue *et al.* [137]
	Chitosan	Tissue engineering	Tamo *et al.* [112]
	Polyvinyl alcohol-borax	Wound dressing	Yang *et al.* [132]
	Collagen	Tissue engineering and regenerative medicine	Yue *et al.* [136]

(*Continued*)

Table 4.2 Biomedical applications of cellulose-based bionanocomposites. (*Continued*)

Type of nanocellulose	Polymer matrix	Applications	References
CNF	Sodium alginate/chitosan	Drug delivery	Deng *et al.* [24]
	Alginate	Bioink for human meniscus tissue engineering	Lan *et al.* [54]
	Pectin	Biofabrication and tissue engineering	Pitton *et al.* [83]
	Chitosan	Tissue engineering	Tamo *et al.* [111]
	κ-carrageenan oligosaccharide	Periodontitis treatment	Johnson *et al.* [48]
	Alginate	Tissue engineering	Tamo *et al.* [110]
	Cellulose	Dopamine detection	Durairaj *et al.* [30]
	Cellulose	Antibacterial activity against pathogenic microorganisms	Mwafy *et al.* [72]
	Gelatin	Tissue repair	Jiang *et al.* [46]
	Collagen	Scaffold materials	Liu *et al.* [59]

biomedical-grade materials remains largely empirical and labor-intensive. Furthermore, sterilization processes, such as autoclaving or gamma irradiation, may alter the nanostructure or compromise the composite's properties and must therefore be carefully tailored for specific formulations.

From a biomedical perspective, more comprehensive *in vivo* studies are needed to validate long-term safety, degradation behavior, and immunological response. While CNCs and CNFs are generally regarded as biocompatible and non-toxic, the effects of prolonged exposure, accumulation in tissues, and interaction with biological fluids under clinical conditions are still underexplored. Regulatory guidelines for cellulose-based nanomaterials remain vague, which poses a hurdle for product approval and market entry.

Despite these challenges, significant opportunities lie ahead. Advancements in green synthesis and low-energy processing are making CNC and CNF production more sustainable and cost-effective. The development of hybrid composites, combining CNCs or CNFs with functional fillers like metal oxides, conductive polymers, or antimicrobial agents, has introduced new functionalities, such as responsive drug release, antibacterial behavior, or electrical conductivity. These multifunctional materials hold strong potential in targeted therapy, smart wound care, and implantable sensors.

Furthermore, the rising use of digital manufacturing tools, such as 3D bioprinting and microfluidics, offers exciting avenues for designing customizable bionanocomposite structures with hierarchical organization and patient-specific geometries. The alignment of CNFs or CNCs through shear, magnetic, or flow-based methods can replicate tissue anisotropy, enhancing biomimicry and mechanical integration with host tissues. Looking ahead, interdisciplinary research combining material science, bioengineering, and regulatory science will be essential to unlock the next generation of nanocellulose-based bionanocomposites. As understanding deepens regarding structure–function relationships and biological interactions, CNC and CNF materials are expected to play a central role in the future of sustainable, functional, and intelligent biomaterials.

4.7 Conclusion

In conclusion, CNCs and CNFs offer great potential as sustainable, high-performance materials for biomedical applications due to their renewability, biocompatibility, and excellent mechanical properties. Their incorporation into biopolymer matrices has enabled the development of multifunctional bionanocomposites for drug delivery, wound healing, tissue engineering, and antimicrobial systems. While challenges remain in achieving uniform dispersion, scaling production, and meeting regulatory standards, advances in green extraction, surface modification, and digital manufacturing offer promising solutions. Continued interdisciplinary research and process optimization are essential to fully realize the clinical and commercial potential of nanocellulose-based bionanocomposites for future medical innovations.

AI disclosure statement: The author confirms that no artificial intelligence (AI)-generated content or AI-assisted writing tools were used in the preparation of this chapter (including summaries and the abstract). All text, analysis, and conclusions are the result of the author's original work.

References

1. Abazari, M.F., Gholizadeh, S., Karizi, S.Z., Birgani, N.H., Abazari, D., Paknia, S., Derakhshankhah, H., Allahyari, Z., Amini, S.M., Hamidi, M., Delattre, C., Recent Advances in Cellulose-Based Structures as the Wound-Healing Biomaterials: A Clinically Oriented Review. *Appl. Sci.*, *11*, 17, 7769, 2021. https://doi.org/10.3390/app11177769.
2. Abdalkarim, S.Y.H., Yu, H.-Y., Wang, C., Yang, L., Guan, Y., Huang, L., Yao, J., Sheet-like Cellulose Nanocrystal-ZnO Nanohybrids as Multifunctional Reinforcing Agents in Biopolyester Composite Nanofibers with Ultrahigh UV-Shielding and Antibacterial Performances. *ACS Appl. Bio Mater.*, *1*, 3, 714–727, 2018. https://doi.org/10.1021/acsabm.8b00188.
3. Abdelhamid, H.N. and Mathew, A.P., Cellulose-Based Nanomaterials Advance Biomedicine: A Review. *Int. J. Mol. Sci.*, *23*, 10, 5405, 2022. https://doi.org/10.3390/ijms23105405.
4. Abdi, M.M., Razalli, R.L., Tahir, P.M., Chaibakhsh, N., Hassani, M., Mir, M., Optimized fabrication of newly cholesterol biosensor based on nanocellulose. *Int. J. Biol. Macromol.*, *126*, 1213–1222, 2019. https://doi.org/10.1016/j.ijbiomac.2019.01.001.
5. Abouzeid, R., Cellulose Nanomaterials and its Applications: Mini Review. *Nanomed. Nanotechnol. Open Access*, *9*, 2, 1–9, 2024. https://doi.org/10.23880/nnoa-16000301.
6. Agbakoba, V.C., Hlangothi, P., Andrew, J., John, M.J., Preparation of cellulose nanocrystal (CNCs) reinforced polylactic acid (PLA) bionanocomposites filaments using biobased additives for 3D printing applications. *Nanoscale Adv.*, *5*, 17, 4447–4463, 2023. https://doi.org/10.1039/D3NA00281K.
7. Anirudhan, T.S., Deepa, J.R., Binussreejayan, Electrochemical sensing of cholesterol by molecularly imprinted polymer of silylated graphene oxide

and chemically modified nanocellulose polymer. *Mater. Sci. Engineering C*, *92*, 942–956, 2018. https://doi.org/10.1016/j.msec.2018.07.041.

8. Arockiasamy, F.S., Manoharan, B., Santhi, V.M., Prakalathan, K., Periasamy, D., Dhandapani, A., Natarajan, V., Krishnasamy, S., Thiagamani, S.M.K., Ilyas, R.A., Navigating the nano-world future: Harnessing cellulose nanocrystals from green sources for sustainable innovation. *Heliyon*, *11*, 1, e41188, 2025. https://doi.org/10.1016/j.heliyon.2024.e41188.
9. Babaei-Ghazvini, A., Cudmore, B., Dunlop, M.J., Acharya, B., Bissessur, R., Ahmed, M., Whelan, W.M., Effect of magnetic field alignment of cellulose nanocrystals in starch nanocomposites: Physicochemical and mechanical properties. *Carbohydr. Polym.*, *247*, 116688, 2020. https://doi.org/10.1016/j.carbpol.2020.116688.
10. Babaei-Ghazvini, A., Vafakish, B., Patel, R., Falua, K.J., Dunlop, M.J., Acharya, B., Cellulose nanocrystals in the development of biodegradable materials: A review on CNC resources, modification, and their hybridization. *Int. J. Biol. Macromol.*, *258*, 128834, 2024. https://doi.org/10.1016/j.ijbiomac.2023.128834.
11. Bader, Q.A., Al-Sharify, Z.T., Dhabab, J.M., Zaidan, H.K., Rheima, A.M., Athair, D.M., Joseph, T.M., Kianfar, E., Cellulose nanomaterials in oil and gas industry and bio-manufacture: Current situation and future outlook. *Case Stud. Chem. Environ. Eng.*, *10*, 100993, 2024. https://doi.org/10.1016/j.cscee.2024.100993.
12. Baruah, R., Goswami, M., Das, A.M., Nath, D., Talukdar, K., Multifunctional ZnO Bionanocomposites in the Treatment of Polluted Water and Controlling of Multi-drug Resistant Bacteria. *J. Mol. Struct.*, *1283*, 135251, 2023. https://doi.org/10.1016/j.molstruc.2023.135251.
13. Benítez, A.J., Lossada, F., Zhu, B., Rudolph, T., Walther, A., Understanding Toughness in Bioinspired Cellulose Nanofibril/Polymer Nanocomposites. *Biomacromolecules*, *17*, 7, 2417–2426, 2016. https://doi.org/10.1021/acs.biomac.6b00533.
14. Benoit, D.S.W., Overby, C.T., Sims Jr., K.R., Ackun-Farmmer, M.A., Drug Delivery Systems, in: *Biomaterials Science*, pp. 1237–1266, Elsevier, 2020, https://doi.org/10.1016/B978-0-12-816137-1.00078-7.
15. Bruvere, B.B., Gromova, A., Jurinovs, M., Platnieks, O., Rižikovs, J., Pa&xxxx0305;že, A., Godiņa, D., Mieriņa, I., Heinmaa, I., Smits, K., Rjabovs, V., Gaidukovs, S., Valorizing Biopolyester Suberin: Modification of Cellulose Nanocrystals and Performance Assessment in 3D-Printed Biobased Acrylates. *ACS Omega*, *9*, 42, 42786–42798, 2024. https://doi.org/10.1021/acsomega.4c04419.
16. Bucci, A.J., Van Hekken, D.L., Tunick, M.H., Renye, J.A., Tomasula, P.M., The effects of microfluidization on the physical, microbial, chemical, and coagulation properties of milk. *J. Dairy Sci.*, *101*, 8, 6990–7001, 2018. https://doi.org/10.3168/jds.2017-13907.

17. Carolin, C.F., Kamalesh, T., Kumar, P.S., Hemavathy, R.V., Rangasamy, G., A critical review on sustainable cellulose materials and its multifaceted applications. *Ind. Crops Prod.*, *203*, 117221, 2023. https://doi.org/10.1016/j.indcrop.2023.117221.
18. Chen, J., Ma, A., Zhang, Y., Sun, L., Yang, K., Vanegas Sáenz, J.R., Hong, G., Mechanical and biological properties of cellulose nanofibers as a dental biomaterial. *J. Dent. Sci.*, S1991790225000388, 2025. https://doi.org/10.1016/j.jds.2025.02.006.
19. Chen, S., Ren, J., Chen, R., Cryopreservation and Desiccation Preservation of Cells, in: *Comprehensive Biotechnology*, pp. 157–166, Elsevier, 2019, https://doi.org/10.1016/B978-0-444-64046-8.00451-1.
20. Chimpibul, W., Nakaji-Hirabayashi, T., Yuan, X., Matsumura, K., Controlling the degradation of cellulose scaffolds with Malaprade oxidation for tissue engineering. *J. Mater. Chem. B*, *8*, 35, 7904–7913, 2020. https://doi.org/10.1039/D0TB01015D.
21. Da Costa, F.A.T., Dufresne, A., Song, T., Parra, D.F., Exploring acid hydrolysis conditions and extended mechanical processing for producing cellulose nanocrystal and nanofibrils from pineapple leaf fibers. *Int. J. Biol. Macromol.*, *306*, 141755, 2025. https://doi.org/10.1016/j.ijbiomac.2025.141755.
22. Datta, D., Bandi, S.P., Colaco, V., Dhas, N., Saha, S.S., Hussain, S.Z., Singh, S., Cellulose-Based Nanofibers Infused with Biotherapeutics for Enhanced Wound-Healing Applications. *ACS Polym. Au*, *5*, 2, 80–104, 2025. https://doi.org/10.1021/acspolymersau.4c00092.
23. De Lima, L.F., Ferreira, A.L., Ranjan, I., Collman, R.G., De Araujo, W.R., De La Fuente-Nunez, C., A bacterial cellulose-based and low-cost electrochemical biosensor for ultrasensitive detection of SARS-CoV-2. *Cell Rep. Phys. Sci.*, *4*, 8, 101476, 2023. https://doi.org/10.1016/j.xcrp.2023.101476.
24. Deng, W., Tang, Y., Mao, J., Zhou, Y., Chen, T., Zhu, X., Cellulose nanofibril as a crosslinker to reinforce the sodium alginate/chitosan hydrogels. *Int. J. Biol. Macromol.*, *189*, 890–899, 2021. https://doi.org/10.1016/j.ijbiomac.2021.08.172.
25. Dias, O.A.T., Konar, S., Leão, A.L., Yang, W., Tjong, J., Sain, M., Current State of Applications of Nanocellulose in Flexible Energy and Electronic Devices. *Front. Chem.*, *8*, 420, 2020. https://doi.org/10.3389/fchem.2020.00420.
26. Dinesh Kumar, S., Venkadeshwaran, K., Aravindan, M.K., Fused deposition modelling of PLA reinforced with cellulose nano-crystals. *Mater. Today Proc.*, *33*, 868–875, 2020. https://doi.org/10.1016/j.matpr.2020.06.404.
27. Djafari Petroudy, S.R., Chabot, B., Loranger, E., Naebe, M., Shojaeiarani, J., Gharehkhani, S., Ahvazi, B., Hu, J., Thomas, S., Recent Advances in Cellulose Nanofibers Preparation through Energy-Efficient Approaches: A Review. *Energies*, *14*, 20, 6792, 2021. https://doi.org/10.3390/en14206792.
28. Dong, L., Ren, S., Zhang, X., Yang, Y., Wu, Q., Lei, T., *In-situ* synthesis of Pt nanoparticles/reduced graphene oxide/cellulose nanohybrid for

nonenzymatic glucose sensing. *Carbohydr. Polym.*, *303*, 120463, 2023. https://doi.org/10.1016/j.carbpol.2022.120463.

29. Dou, J., Vuorinen, T., Koivula, H., Forsman, N., Sipponen, M., Hietala, S., Self-Standing Lignin-Containing Willow Bark Nanocellulose Films for Oxygen Blocking and UV Shielding. *ACS Appl. Nano Mater.*, *4*, 3, 2921–2929, 2021. https://doi.org/10.1021/acsanm.1c00071.
30. Durairaj, V., Wester, N., Etula, J., Laurila, T., Lehtonen, J., Rojas, O.J., Pahimanolis, N., Koskinen, J., Multiwalled Carbon Nanotubes/Nanofibrillar Cellulose/Nafion Composite-Modified Tetrahedral Amorphous Carbon Electrodes for Selective Dopamine Detection. *J. Phys. Chem. C*, *123*, 40, 24826–24836, 2019. https://doi.org/10.1021/acs.jpcc.9b05537.
31. Ehsanimehr, S., Najafi Moghadam, P., Dehaen, W., Shafiei-Irannejad, V., Synthesis of pH-sensitive nanocarriers based on polyacrylamide grafted nanocrystalline cellulose for targeted drug delivery to folate receptor in breast cancer cells. *Eur. Polym. J.*, *150*, 110398, 2021. https://doi.org/10.1016/j.eurpolymj.2021.110398.
32. Eichhorn, S.J., Etale, A., Wang, J., Berglund, L.A., Li, Y., Cai, Y., Chen, C., Cranston, E.D., Johns, M.A., Fang, Z., Li, G., Hu, L., Khandelwal, M., Lee, K.-Y., Oksman, K., Pinitsoontorn, S., Quero, F., Sebastian, A., Titirici, M.M., Frka-Petesic, B., Current international research into cellulose as a functional nanomaterial for advanced applications. *J. Mater. Sci.*, *57*, 10, 5697–5767, 2022. https://doi.org/10.1007/s10853-022-06903-8.
33. Esmaeili, C., Abdi, M., Mathew, A., Jonoobi, M., Oksman, K., Rezayi, M., Synergy Effect of Nanocrystalline Cellulose for the Biosensing Detection of Glucose. *Sensors*, *15*, 10, 24681–24697, 2015. https://doi.org/10.3390/s151024681.
34. Ferreira, F.V., Souza, A.G., Ajdary, R., De Souza, L.P., Lopes, J.H., Correa, D.S., Siqueira, G., Barud, H.S., Rosa, D.D.S., Mattoso, L.H.C., Rojas, O.J., Nanocellulose-based porous materials: Regulation and pathway to commercialization in regenerative medicine. *Bioact. Mater.*, *29*, 151–176, 2023. https://doi.org/10.1016/j.bioactmat.2023.06.020.
35. Fura, S., Hakim (Assala), Q., Toledo, D., Agatemor, C., Biomedical applications of bionanocomposites, in: *Advances in Bionanocomposites*, pp. 351–364, Elsevier, 2024, https://doi.org/10.1016/B978-0-323-91764-3.00015-2.
36. Gan, Y.X., Processing of nanofibers for oil-spill cleaning application, in: *Polymeric Nanofibers and their Composites*, pp. 69–89, Elsevier, 2025, https://doi.org/10.1016/B978-0-443-14128-7.00004-3.
37. Garg, M., Linares, M., Zozoulenko, I., Theoretical Rationalization of Self-Assembly of Cellulose Nanocrystals: Effect of Surface Modifications and Counterions. *Biomacromolecules*, *21*, 8, 3069–3080, 2020. https://doi.org/10.1021/acs.biomac.0c00469.
38. Ghilan, A., Nicu, R., Ciolacu, D.E., Ciolacu, F., Insight into the Latest Medical Applications of Nanocellulose. *Materials*, *16*, 12, 4447, 2023. https://doi.org/10.3390/ma16124447.

39. Ghosh, T., Roy, S., Khan, A., Mondal, K., Ezati, P., Rhim, J.-W., Agricultural waste-derived cellulose nanocrystals for sustainable active food packaging applications. *Food Hydrocolloids, 154,* 110141, 2024. https://doi.org/10.1016/j.foodhyd.2024.110141.
40. Glińska, K., Gitalt, J., Torrens, E., Plechkova, N., Bengoa, C., Extraction of cellulose from corn stover using designed ionic liquids with improved reusing capabilities. *Process Saf. Environ. Prot., 147,* 181–191, 2021. https://doi.org/10.1016/j.psep.2020.09.035.
41. Guo, Y., He, X., Williams, G.R., Zhou, Y., Liao, X., Xiao, Z., Yu, C., Liu, Y., Tumor microenvironment-responsive hyperbranched polymers for controlled drug delivery. *J. Pharm. Anal., 14,* 12, 101003, 2024. https://doi.org/10.1016/j.jpha.2024.101003.
42. Haddis, D.Z., Chae, M., Asomaning, J., Bressler, D.C., Evaluation of steam explosion pretreatment on the cellulose nanocrystals (CNCs) yield from poplar wood. *Carbohydr. Polym., 323,* 121460, 2024. https://doi.org/10.1016/j.carbpol.2023.121460.
43. Hu, M., Lv, X., Wang, Y., Zhang, Y., Dai, H., Guideline for the extraction of nanocellulose from lignocellulosic feedstocks. *Food Biomacromolecules, 1,* 1, 9–17, 2024. https://doi.org/10.1002/fob2.12011.
44. Iglesias-Mejuto, A., Malandain, N., Ferreira-Gonçalves, T., Ardao, I., Reis, C.P., Laromaine, A., Roig, A., García-González, C.A., Cellulose-in-cellulose 3D-printed bioaerogels for bone tissue engineering. *Cellulose, 31,* 1, 515–534, 2024. https://doi.org/10.1007/s10570-023-05601-1.
45. Jafri, N.H.S., Jimat, D.N., Wan Nawawi, W.M.F., Ahmad Nor, Y., Amid, A., Effect of Incorporating Cellulose Nanofibers and Lemongrass Essential Oil in Polyvinyl Alcohol-Polyethylene Glycol/Glycerin Hydrogel for Wound Dressing. *IIUM Eng. J., 25,* 2, 99–115, 2024. https://doi.org/10.31436/iiumej.v25i2.3240.
46. Jiang, Y., Xv, X., Liu, D., Yang, Z., Zhang, Q., Shi, H., Zhao, G., Zhou, J., Preparation of cellulose nanofiber-reinforced gelatin hydrogel and optimization for 3D printing applications. *BioResources, 13,* 3, 5909–5924, 2018. https://doi.org/10.15376/biores.13.3.5909-5924.
47. John, M.J., Dyanti, N., Mokhena, T., Agbakoba, V., Sithole, B., Design and Development of Cellulosic Bionanocomposites from Forestry Waste Residues for 3D Printing Applications. *Materials, 14,* 13, 3462, 2021. https://doi.org/10.3390/ma14133462.
48. Johnson, A., Kong, F., Miao, S., Lin, H.-T.V., Thomas, S., Huang, Y.-C., Kong, Z.-L., Therapeutic effects of antibiotics loaded cellulose nanofiber and κ-carrageenan oligosaccharide composite hydrogels for periodontitis treatment. *Sci. Rep., 10,* 1, 18037, 2020. https://doi.org/10.1038/s41598-020-74845-9.
49. Jose, S.A., Cowan, N., Davidson, M., Godina, G., Smith, I., Xin, J., Menezes, P.L., A Comprehensive Review on Cellulose Nanofibers, Nanomaterials, and Composites: Manufacturing, Properties, and Applications. *Nanomaterials, 15,* 5, 356, 2025. https://doi.org/10.3390/nano15050356.

50. Joseph, B., K, S.V., Sabu, C., Kalarikkal, N., Thomas, S., Cellulose nanocomposites: Fabrication and biomedical applications. *J. Bioresour. Bioprod.*, 5, 4, 223–237, 2020. https://doi.org/10.1016/j.jobab.2020.10.001.
51. Khairnar, Y., Hansora, D., Hazra, C., Kundu, D., Tayde, S., Tonde, S., Naik, J., Chatterjee, A., Cellulose bionanocomposites for sustainable planet and people: A global snapshot of preparation, properties, and applications. *Carbohydr. Polym. Technol. Appl.*, *2*, 100065, 2021. https://doi.org/10.1016/j.carpta.2021.100065.
52. Khalid, M.Y., Arif, Z.U., Noroozi, R., Hossain, M., Ramakrishna, S., Umer, R., 3D/4D printing of cellulose nanocrystals-based biomaterials: Additives for sustainable applications. *Int. J. Biol. Macromol.*, *251*, 126287, 2023. https://doi.org/10.1016/j.ijbiomac.2023.126287.
53. Kim, M. and Doh, H., Upcycling Food By-products: Characteristics and Applications of Nanocellulose. *Chem. – An Asian J.*, *19*, 6, e202301068, 2024. https://doi.org/10.1002/asia.202301068.
54. Lan, X., Ma, Z., Szojka, A.R.A., Kunze, M., Mulet-Sierra, A., Vyhlidal, M.J., Boluk, Y., Adesida, A.B., TEMPO-Oxidized Cellulose Nanofiber-Alginate Hydrogel as a Bioink for Human Meniscus Tissue Engineering. *Front. Bioeng. Biotechnol.*, *9*, 766399, 2021. https://doi.org/10.3389/fbioe.2021.766399.
55. Lee, H. and Mani, S., Mechanical pretreatment of cellulose pulp to produce cellulose nanofibrils using a dry grinding method. *Ind. Crops Prod.*, *104*, 179–187, 2017. https://doi.org/10.1016/j.indcrop.2017.04.044.
56. Lee, W.J., Clancy, A.J., Kontturi, E., Bismarck, A., Shaffer, M.S.P., Strong and Stiff: High-Performance Cellulose Nanocrystal/Poly(vinyl alcohol) Composite Fibers. *ACS Appl. Mater. Interfaces*, *8*, 46, 31500–31504, 2016. https://doi.org/10.1021/acsami.6b11578.
57. Li, Y., Liu, W., Wang, Y., Lu, S., Cellulose Based Nano-Scaffolds for Targeted Cancer Therapies: Current Status and Future Perspective. *Int. J. Nanomed.*, *20*, 199–213, 2025. https://doi.org/10.2147/IJN.S500261.
58. Lima, A.R., Cristofoli, N.L., Rosa Da Costa, A.M., Saraiva, J.A., Vieira, M.C., Comparative study of the production of cellulose nanofibers from agro-industrial waste streams of Salicornia ramosissima by acid and enzymatic treatment. *Food Bioprod. Process.*, *137*, 214–225, 2023. https://doi.org/10.1016/j.fbp.2022.11.012.
59. Liu, C.-Y., Goto, D., Hongo, C., Matsumoto, T., Nishino, T., Collagen/Cellulose Nanofiber Blend Scaffolds Prepared at Various pH Conditions. *ACS Appl. Bio Mater.*, *1*, 5, 1362–1368, 2018a. https://doi.org/10.1021/acsabm.8b00302.
60. Liu, Q., Li, J., Zhong, X., Dai, Z., Lu, Z., Yang, H., Chen, R., Enhanced antibacterial activity and mechanism studies of Ag/Bi2O3 nanocomposites. *Adv. Powder Technol.*, *29*, 9, 2082–2090, 2018b. https://doi.org/10.1016/j.apt.2018.05.015.
61. Lu, H., Li, X., Zhang, M., Xu, C., Li, W., Wan, L., Antibacterial Cellulose Nanocrystal-Incorporated Hydrogels With Satisfactory Vascularization for

Enhancing Skin Regeneration. *Front. Bioeng. Biotechnol.*, *10*, 876936, 2022. https://doi.org/10.3389/fbioe.2022.876936.

62. Lu, Y., Li, J., Ge, L., Xie, W., Wu, D., Pickering emulsion stabilized with fibrous nanocelluloses: Insight into fiber flexibility-emulsifying capacity relations. *Carbohydr. Polym.*, *255*, 117483, 2021. https://doi.org/10.1016/j.carbpol.2020.117483.
63. Madivoli, E.S., Kareru, P.G., Gachanja, A.N., Mugo, S.M., Sujee, D.M., Fromm, K.M., Isolation of Cellulose Nanofibers from *Oryza sativa* Residues *via* TEMPO Mediated Oxidation. *J. Nat. Fibers*, *19*, 4, 1310–1322, 2022. https://doi.org/10.1080/15440478.2020.1764454.
64. Magnani, C., Idström, A., Nordstierna, L., Müller, A.J., Dubois, P., Raquez, J.-M., Lo Re, G., Interphase Design of Cellulose Nanocrystals/Poly(hydroxybutyrate- *ran* -valerate) Bionanocomposites for Mechanical and Thermal Properties Tuning. *Biomacromolecules*, *21*, 5, 1892–1901, 2020. https://doi.org/10.1021/acs.biomac.9b01760.
65. Mahmud, R.U. and Rahman, M.Z., Synthesis and characterization of nanocomposites for tissue engineering, in: *Comprehensive Materials Processing*, pp. 241–269, Elsevier, 2024, https://doi.org/10.1016/B978-0-323-96020-5.00183-7.
66. Meng, F., Wang, G., Du, X., Wang, Z., Xu, S., Zhang, Y., Extraction and characterization of cellulose nanofibers and nanocrystals from liquefied banana pseudo-stem residue. *Composites, Part B*, *160*, 341–347, 2019. https://doi.org/10.1016/j.compositesb.2018.08.048.
67. Miao, X., Hua, W., Li, Y., Bian, F., Xiao, T., Extraction of cellulose nanofibrils from pine sawdust by integrated chemical pretreatment. *Heliyon*, *10*, 3, e25355, 2024. https://doi.org/10.1016/j.heliyon.2024.e25355.
68. Mo, Y., Huang, X., Yue, M., Hu, L., Hu, C., Preparation of nanocellulose and application of nanocellulose polyurethane composites. *RSC Adv.*, *14*, 26, 18247–18257, 2024. https://doi.org/10.1039/D4RA01412J.
69. Montero, B., Rico, M., Barral, L., Bouza, R., López, J., Schmidt, A., Bittmann-Hennes, B., Preparation and characterization of bionanocomposite films based on wheat starch and reinforced with cellulose nanocrystals. *Cellulose*, *28*, 12, 7781–7793, 2021. https://doi.org/10.1007/s10570-021-04017-z.
70. Mugo, S.M. and Alberkant, J., Flexible molecularly imprinted electrochemical sensor for cortisol monitoring in sweat. *Anal. Bioanal. Chem.*, *412*, 8, 1825–1833, 2020. https://doi.org/10.1007/s00216-020-02430-0.
71. Murugesan, S., Rajasree, S.R.R., Rajan, R., Extraction and Characterization of Cellulose Nanocrystals from Brown Seaweed Dictyota Bartayresiana, J.V. Lamouroux. *J. Cluster Sci.*, *36*, 1, 12, 2025. https://doi.org/10.1007/s10876-024-02721-x.
72. Mwafy, E.A., Hasanin, M.S., Mostafa, A.M., Cadmium oxide/TEMPO-oxidized cellulose nanocomposites produced by pulsed laser ablation in liquid environment: Synthesis, characterization, and antimicrobial

activity. *Opt. Laser Technol.*, *120*, 105744, 2019. https://doi.org/10.1016/j.optlastec.2019.105744.

73. Nehra, P. and Chauhan, R.P., Eco-friendly nanocellulose and its biomedical applications: Current status and future prospect. *J. Biomater. Sci., Polym. Ed.*, *32*, 1, 112–149, 2021. https://doi.org/10.1080/09205063.2020.1817706.
74. N'Gatta, K.M., Belaid, H., El Hayek, J., Assanvo, E.F., Kajdan, M., Masquelez, N., Boa, D., Cavaillès, V., Bechelany, M., Salameh, C., 3D printing of cellulose nanocrystals based composites to build robust biomimetic scaffolds for bone tissue engineering. *Sci. Rep.*, *12*, 1, 21244, 2022. https://doi.org/10.1038/s41598-022-25652-x.
75. Nkuna, C.N., Mhike, W., Ojijo, V., Shongwe, M.B., John, M.J., Biocomposites and bionanocomposites from poly(lactide) and cellulosic materials – a review. *Cellulose*, *31*, 8, 4709–4732, 2024. https://doi.org/10.1007/s10570-024-05913-w.
76. Norizan, M.N., Shazleen, S.S., Alias, A.H., Sabaruddin, F.A., Asyraf, M.R.M., Zainudin, E.S., Abdullah, N., Samsudin, M.S., Kamarudin, S.H., Norrrahim, M.N.F., Nanocellulose-Based Nanocomposites for Sustainable Applications: A Review. *Nanomaterials*, *12*, 19, 3483, 2022. https://doi.org/10.3390/nano12193483.
77. Ochoa-Yepes, O., Di Giogio, L., Goyanes, S., Mauri, A., Famá, L., Influence of process (extrusion/thermo-compression, casting) and lentil protein content on physicochemical properties of starch films. *Carbohydr. Polym.*, *208*, 221–231, 2019. https://doi.org/10.1016/j.carbpol.2018.12.030.
78. Osei-Bonsu, R., Hoque, M., McMichael, P.S., Foster, E.J., Subcritical water digestion of woody biomass: Extraction of cellulose nanomaterials under acid-lean condition. *Nanoscale Adv.*, *6*, 15, 3923–3933, 2024. https://doi.org/10.1039/D4NA00108G.
79. Pandey, A., Pharmaceutical and biomedical applications of cellulose nanofibers: A review. *Environ. Chem. Lett.*, *19*, 3, 2043–2055, 2021. https://doi.org/10.1007/s10311-021-01182-2.
80. Patil, T.V., Patel, D.K., Dutta, S.D., Ganguly, K., Santra, T.S., Lim, K.-T., Nanocellulose, a versatile platform: From the delivery of active molecules to tissue engineering applications. *Bioact. Mater.*, *9*, 566–589, 2022. https://doi.org/10.1016/j.bioactmat.2021.07.006.
81. Phuong, H.T., Thoa, N.K., Tuyet, P.T.A., Van, Q.N., Hai, Y.D., Cellulose Nanomaterials as a Future, Sustainable and Renewable Material. *Crystals*, *12*, 1, 106, 2022. https://doi.org/10.3390/cryst12010106.
82. Pinteala, T., Sirbu, P.-D., Anghel, N., Rosca, I., Voicu, G., Calin, M., Spiridon, I., Integrating hydroxyapatite and bovine bone mineral into cellulose–collagen matrices for enhanced osteogenesis. *Mater. Adv.*, *5*, 24, 9573–9585, 2024. https://doi.org/10.1039/D4MA00456F.
83. Pitton, M., Fiorati, A., Buscemi, S., Melone, L., Farè, S., Contessi Negrini, N., 3D Bioprinting of Pectin-Cellulose Nanofibers Multicomponent

Bioinks. *Front. Bioeng. Biotechnol.*, *9*, 732689, 2021. https://doi.org/10.3389/fbioe.2021.732689.

84. Poulose, A., Parameswaranpillai, J., George, J.J., Gopi, J.A., Krishnasamy, S., Dominic, C.D.M., Hameed, N., Salim, N.V., Radoor, S., Sienkiewicz, N., Nanocellulose: A Fundamental Material for Science and Technology Applications. *Molecules*, *27*, 22, 8032, 2022. https://doi.org/10.3390/molecules27228032.
85. Pratama, A.W., Piluharto, B., Mahardika, M., Widiastuti, N., Firmanda, A., Norrrahim, M.N.F., Comparative study of oxidized cellulose nanofibrils properties from diverse sources *via* TEMPO-mediated oxidation. *Case Stud. Chem. Environ. Eng.*, *10*, 100823, 2024. https://doi.org/10.1016/j.cscee.2024.100823.
86. Puglia, D., Luzi, F., Tolisano, C., Rallini, M., Priolo, D., Brienza, M., Costantino, F., Torre, L., Del Buono, D., Cellulose Nanocrystals and Lignin Nanoparticles Extraction from Lemna minor L.: Acid Hydrolysis of Bleached and Ionic Liquid-Treated Biomass. *Polymers*, *16*, 10, 1395, 2024. https://doi.org/10.3390/polym16101395.
87. Putra, B.U., Hardiningtyas, S.D., Hastuti, N., Ramadhan, W., Uju, Razi, M.A., Agustini, L., Alginate hydrogel incorporating cellulose nanofiber from solid waste agar industry for hydrophobic antibiotic delivery: Synthesis and characterization. *Mater. Today Commun.*, *38*, 108248, 2024. https://doi.org/10.1016/j.mtcomm.2024.108248.
88. Qi, Y., Guo, Y., Liza, A.A., Yang, G., Sipponen, M.H., Guo, J., Li, H., Nanocellulose: A review on preparation routes and applications in functional materials. *Cellulose*, *30*, 7, 4115–4147, 2023. https://doi.org/10.1007/s10570-023-05169-w.
89. Qiao, Y. and Cao, H., State-of-the-Art Functional Biomaterials in China. *J. Funct. Biomater.*, *15*, 1, 23, 2024. https://doi.org/10.3390/jfb15010023.
90. Rafat, A.A., Verdía Barbará, P., Ullah, A., Kontturi, E., Law, R.V., Hallett, J.P., Efficient extraction of carboxylated nanocellulose from ionoSolv pulps with alkaline H2O2 assisted oxidation. *Cellulose*, *32*, 2, 853–875, 2025. https://doi.org/10.1007/s10570-024-06319-4.
91. Rahmani, S., Khoubi-Arani, Z., Mohammadzadeh-Komuleh, S., Maroufkhani, M., Electrospinning of Cellulose Nanofibers for Advanced Applications, in: *Handbook of Nanocelluloses*, A. Barhoum (Ed.), pp. 1–34, Springer International Publishing, 2021, https://doi.org/10.1007/978-3-030-62976-2_14-1.
92. Raja, I.S., Kim, B., Han, D.-W., Nanofibrous Material-Reinforced Printable Ink for Enhanced Cell Proliferation and Tissue Regeneration. *Bioengineering*, *11*, 4, 363, 2024. https://doi.org/10.3390/bioengineering11040363.
93. Rana, A.K., Gupta, V.K., Hart, P., Thakur, V.K., Cellulose-alginate hydrogels and their nanocomposites for water remediation and biomedical applications. *Environ. Res.*, *243*, 117889, 2024. https://doi.org/10.1016/j.envres.2023.117889.

94. Refate, A., Mohamed, Y., Mohamed, M., Sobhy, M., Samhy, K., Khaled, O., Eidaroos, K., Batikh, H., El-Kashif, E., El-Khatib, S., Mehanny, S., Influence of electrospinning parameters on biopolymers nanofibers, with emphasis on cellulose & chitosan. *Heliyon*, *9*, 6, e17051, 2023. https://doi.org/10.1016/j.heliyon.2023.e17051.
95. Ren, Q., Wu, M., Wang, L., Zheng, W., Hikima, Y., Semba, T., Ohshima, M., Cellulose nanofiber reinforced poly (lactic acid) with enhanced rheology, crystallization and foaming ability. *Carbohydr. Polym.*, *286*, 119320, 2022. https://doi.org/10.1016/j.carbpol.2022.119320.
96. Ren, S., Sun, X., Lei, T., Wu, Q., The Effect of Chemical and High-Pressure Homogenization Treatment Conditions on the Morphology of Cellulose Nanoparticles. *J. Nanomater.*, *2014*, 1, 582913, 2014. https://doi.org/10.1155/2014/582913.
97. Rojas, J., Bedoya, M., Ciro, Y., Current Trends in the Production of Cellulose Nanoparticles and Nanocomposites for Biomedical Applications, in: *Cellulose—Fundamental Aspects and Current Trends*, M. Poletto and H.L. Ornaghi (Eds.), InTech, 2015, https://doi.org/10.5772/61334.
98. Rosa, R.P., Rosace, G., Arrigo, R., Malucelli, G., Preparation and Characterization of 3D-Printed Biobased Composites Containing Micro- or Nanocrystalline Cellulose. *Polymers*, *14*, 9, 1886, 2022. https://doi.org/10.3390/polym14091886.
99. Rubiya, M.H., Melethil, K., James, A., Varghese, S., Thomas, B., Cellulose Nanocrystals (CNCs) Supported Inorganic Nanomaterials for Catalytic Applications, in: *Handbook of Biopolymers*, S. Thomas, A. Ar, C. Jose Chirayil, B. Thomas (Eds.), pp. 907–939, Springer Nature, Singapore, 2023, https://doi.org/10.1007/978-981-19-0710-4_34.
100. Saito, T., Kuramae, R., Wohlert, J., Berglund, L.A., Isogai, A., An Ultrastrong Nanofibrillar Biomaterial: The Strength of Single Cellulose Nanofibrils Revealed *via* Sonication-Induced Fragmentation. *Biomacromolecules*, *14*, 1, 248–253, 2013. https://doi.org/10.1021/bm301674e.
101. Samsalee, N., Meerasri, J., Sothornvit, R., Rice husk nanocellulose: Extraction by high-pressure homogenization, chemical treatments and characterization. *Carbohydr. Polym. Technol. Appl.*, *6*, 100353, 2023. https://doi.org/10.1016/j.carpta.2023.100353.
102. Shahriar, S.M.S., Mondal, J., Hasan, M.N., Revuri, V., Lee, D.Y., Lee, Y.-K., Electrospinning Nanofibers for Therapeutics Delivery. *Nanomaterials*, *9*, 4, 532, 2019. https://doi.org/10.3390/nano9040532.
103. Shanmugam, K., Chandrasekar, N., Balaji, R., Nanocellulose as Sustainable Bio-Nanomaterial for Packaging and Biomedical Applications. *Sci. Social Res.*, *6*, 2, 57–67, 2024. https://doi.org/10.26689/ssr.v6i2.6057.
104. Shojaeiarani, J., Bajwa, D.S., Chanda, S., Cellulose nanocrystal based composites: A review. *Composites, Part C: Open Access*, *5*, 100164, 2021. https://doi.org/10.1016/j.jcomc.2021.100164.

105. Sihag, S.S., Pal, J., Yadav, M., Extraction and Characterization of Nanocellulose from Wheat Straw: Facile Approach. *J. Water Environ. Nanotechnol.*, *7*, 3, 2022. https://doi.org/10.22090/jwent.2022.03.007.
106. Singh, S., Bhardwaj, S., Tiwari, P., Dev, K., Ghosh, K., Maji, P.K., Recent advances in cellulose nanocrystals-based sensors: A review. *Mater. Adv.*, *5*, 7, 2622–2654, 2024. https://doi.org/10.1039/D3MA00601H.
107. Sofiah, A.G.N., Pasupuleti, J., Samykano, M., Kadirgama, K., Koh, S.P., Tiong, S.K., Pandey, A.K., Yaw, C.T., Natarajan, S.K., Harnessing Nature's Ingenuity: A Comprehensive Exploration of Nanocellulose from Production to Cutting-Edge Applications in Engineering and Sciences. *Polymers*, *15*, 14, 3044, 2023. https://doi.org/10.3390/polym15143044.
108. Sun, L., Li, L., Wang, Y., Li, M., Xu, S., Zhang, C., A collagen-based bi-layered composite dressing for accelerated wound healing. *J. Tissue Viability*, *31*, 1, 180–189, 2022. https://doi.org/10.1016/j.jtv.2021.09.003.
109. Tamo, A.K., Nanocellulose-based hydrogels as versatile materials with interesting functional properties for tissue engineering applications. *J. Mater. Chem. B*, *12*, 32, 7692–7759, 2024. https://doi.org/10.1039/D4TB00397G.
110. Tamo, A.K., Doench, I., Morales Helguera, A., Hoenders, D., Walther, A., Madrazo, A.O., Biodegradation of Crystalline Cellulose Nanofibers by Means of Enzyme Immobilized-Alginate Beads and Microparticles. *Polymers*, *12*, 7, 1522, 2020. https://doi.org/10.3390/polym12071522.
111. Tamo, A.K., Doench, I., Walter, L., Montembault, A., Sudre, G., David, L., Morales-Helguera, A., Selig, M., Rolauffs, B., Bernstein, A., Hoenders, D., Walther, A., Osorio-Madrazo, A., Development of Bioinspired Functional Chitosan/Cellulose Nanofiber 3D Hydrogel Constructs by 3D Printing for Application in the Engineering of Mechanically Demanding Tissues. *Polymers*, *13*, 10, 1663, 2021. https://doi.org/10.3390/polym13101663.
112. Tamo, A.K., Tran, T.A., Doench, I., Jahangir, S., Lall, A., David, L., Peniche-Covas, C., Walther, A., Osorio-Madrazo, A., 3D Printing of Cellulase-Laden Cellulose Nanofiber/Chitosan Hydrogel Composites: Towards Tissue Engineering Functional Biomaterials with Enzyme-Mediated Biodegradation. *Materials*, *15*, 17, 6039, 2022. https://doi.org/10.3390/ma15176039.
113. Tayeb, A., Amini, E., Ghasemi, S., Tajvidi, M., Cellulose Nanomaterials—Binding Properties and Applications: A Review. *Molecules*, *23*, 10, 2684, 2018. https://doi.org/10.3390/molecules23102684.
114. Taymaz, E.R. and Uslu, M.E., Innovations in biocompatible materials: Exploring the potential of cellulose nanocrystals from grape pomace. *Chem. Pap.*, *78*, 9, 5445–5455, 2024. https://doi.org/10.1007/s11696-024-03484-5.
115. Tenaye, T., Mohammed, S.A., Jabasingh, S.A., Sustainable synthesis and characterization of Enset cellulose nanocrystals (E-CNp) from Enset ventricosum biomass and its application in the fabrication of Enset cellulose nanocomposite (E-CNc). *Biomass Convers. Biorefin.*, *14*, 4, 5019–5034, 2024. https://doi.org/10.1007/s13399-022-02682-z.

116. Tracey, C.T., Torlopov, M.A., Martakov, I.S., Vdovichenko, Ekaterina., A., Zhukov, M., Krivoshapkin, P.V., Mikhaylov, V.I., Krivoshapkina, E.F., Hybrid cellulose nanocrystal/magnetite glucose biosensors. *Carbohydr. Polym., 247,* 116704, 2020. https://doi.org/10.1016/j.carbpol.2020.116704.
117. Trache, D., Tarchoun, A.F., Derradji, M., Hamidon, T.S., Masruchin, N., Brosse, N., Hussin, M.H., Nanocellulose: From Fundamentals to Advanced Applications. *Front. Chem.*, *8*, 392, 2020. https://doi.org/10.3389/fchem.2020.00392.
118. Triantafyllou, E., Karydis-Messinis, A., Moschovas, D., Kyriakaki, C., Vasilopoulos, K.C., Giannakas, A.E., Karakassides, M.A., Avgeropoulos, A., Zafeiropoulos, N.E., Salmas, C.E., Microwave-Assisted Extraction of Cellulose from Aloe Vera Plant Residue and Preparation of Cellulose Nanocrystal–Poly(vinyl alcohol) Hydrogels. *Molecules*, *29*, 24, 6012, 2024. https://doi.org/10.3390/molecules29246012.
119. Trivedi, A.K., Gupta, M.K., Singh, H., PLA based biocomposites for sustainable products: A review. *Adv. Ind. Eng. Polym. Res.*, *6*, 4, 382–395, 2023. https://doi.org/10.1016/j.aiepr.2023.02.002.
120. U.S. National Science Foundation, *NSF announces 4 new Engineering Research Centers focused on biotechnology, manufacturing, robotics and sustainability*, 2024, August 21. [Online post]. https://www.nsf.gov/news/nsf-announces-4-new-engineering-research-centers-focused.
121. Wang, C., Bai, J., Tian, P., Xie, R., Duan, Z., Lv, Q., Tao, Y., The Application Status of Nanoscale Cellulose-Based Hydrogels in Tissue Engineering and Regenerative Biomedicine. *Front. Bioeng. Biotechnol.*, *9*, 732513, 2021a. https://doi.org/10.3389/fbioe.2021.732513.
122. Wang, D.-C., Yu, H.-Y., Qi, D., Wu, Y., Chen, L., Li, Z., Confined Chemical Transitions for Direct Extraction of Conductive Cellulose Nanofibers with Graphitized Carbon Shell at Low Temperature and Pressure. *J. Am. Chem. Soc., 143*, 30, 11620–11630, 2021b. https://doi.org/10.1021/jacs.1c04710.
123. Wang, H., Du, H., Liu, K., Liu, H., Xu, T., Zhang, S., Chen, X., Zhang, R., Li, H., Xie, H., Zhang, X., Si, C., Sustainable preparation of bifunctional cellulose nanocrystals *via* mixed H2SO4/formic acid hydrolysis. *Carbohydr. Polym., 266*, 118107, 2021c. https://doi.org/10.1016/j.carbpol.2021.118107.
124. Wang, H., Zuo, M., Ding, N., Yan, G., Zeng, X., Tang, X., Sun, Y., Lei, T., Lin, L., Preparation of Nanocellulose with High-Pressure Homogenization from Pretreated Biomass with Cooking with Active Oxygen and Solid Alkali. *ACS Sustainable Chem. Eng.*, *7*, 10, 9378–9386, 2019. https://doi.org/10.1021/acssuschemeng.9b00582.
125. Wang, Q., Liu, S., Liu, J., Sun, J., Zhang, Z., Zhu, Q., Sustainable cellulose nanomaterials for environmental remediation - Achieving clean air, water, and energy: A review. *Carbohydr. Polym.*, *285*, 119251, 2022. https://doi.org/10.1016/j.carbpol.2022.119251.
126. Wang, W., Xue, Y., Wang, X., Zheng, X., Quan, J., Wang, Z., Zhang, J., Yao, J., Xue, B., Lignin-enhanced wet strength and UV-Blocking properties

of nanocellulose-based composite films obtained by casting/hot-pressing methods. *Eur. Polym. J.*, *213*, 113109, 2024. https://doi.org/10.1016/j.eurpolymj.2024.113109.

127. Wisnom, M.R., Pimenta, S., Shaffer, M.S.P., Robinson, P., Potter, K.D., Hamerton, I., Czél, G., Jalalvand, M., Fotouhi, M., Anthony, D.B., Yu, H., Longana, M.L., Wu, X., Bismarck, A., High performance ductile and pseudo-ductile polymer matrix composites: A review. *Composites, Part A*, *181*, 108029, 2024. https://doi.org/10.1016/j.compositesa.2024.108029.
128. Wright, Z.M., Arnold, A.M., Holt, B.D., Eckhart, K.E., Sydlik, S.A., Functional Graphenic Materials, Graphene Oxide, and Graphene as Scaffolds for Bone Regeneration. *Regener. Eng. Transl. Med.*, *5*, 2, 190–209, 2019. https://doi.org/10.1007/s40883-018-0081-z.
129. Wu, H., Wang, C., Mao, H., Wen, Y., Zheng, C., Zhang, W., Yi, S., Value-added bio-products extracted from straw utilized as carriers for pesticide delivery: Innovative repurposing of agricultural waste. *Ind. Crops Prod.*, *223*, 120294, 2025. https://doi.org/10.1016/j.indcrop.2024.120294.
130. Xiang, J., Mlambo, R., Shaw, I., Seid, Y., Shah, H., He, Y., Kpegah, J.K.S.K., Tan, S., Zhou, W., He, B., Cryopreservation of bioflavonoid-rich plant sources and bioflavonoid-microcapsules: Emerging technologies for preserving bioactivity and enhancing nutraceutical applications. *Front. Nutr.*, *10*, 1232129, 2023. https://doi.org/10.3389/fnut.2023.1232129.
131. Xu, H., Sanchez-Salvador, J.L., Blanco, A., Balea, A., Negro, C., Recycling of TEMPO-mediated oxidation medium and its effect on nanocellulose properties. *Carbohydr. Polym.*, *319*, 121168, 2023. https://doi.org/10.1016/j.carbpol.2023.121168.
132. Yang, G., Zhang, Z., Liu, K., Ji, X., Fatehi, P., Chen, J., A cellulose nanofibril-reinforced hydrogel with robust mechanical, self-healing, pH-responsive and antibacterial characteristics for wound dressing applications. *J. Nanobiotechnol.*, *20*, 1, 312, 2022. https://doi.org/10.1186/s12951-022-01523-5.
133. Yao, C., Li, F., Chen, T., Tang, Y., Green preparation of cellulose nanofibers *via* high-pressure homogenization and their film-forming properties. *Ind. Crops Prod.*, *206*, 117575, 2023. https://doi.org/10.1016/j.indcrop.2023.117575.
134. Yasin, S., Hussain, M., Zheng, Q., Song, Y., Effects of ionic liquid on cellulosic nanofiller filled natural rubber bionanocomposites. *J. Colloid Interface Sci.*, *591*, 409–417, 2021. https://doi.org/10.1016/j.jcis.2021.02.029.
135. Yu, W., Wang, C., Yi, Y., Wang, H., Yang, Y., Zeng, L., Tan, Z., Direct pretreatment of raw ramie fibers using an acidic deep eutectic solvent to produce cellulose nanofibrils in high purity. *Cellulose*, *28*, 1, 175–188, 2021. https://doi.org/10.1007/s10570-020-03538-3.
136. Yue, C., Ding, C., Yang, N., Luo, Y., Su, J., Cao, L., Cheng, B., Strong and tough collagen/cellulose nanofibril composite films *via* the synergistic effect of hydrogen and metal–ligand bonds. *Eur. Polym. J.*, *180*, 111628, 2022. https://doi.org/10.1016/j.eurpolymj.2022.111628.

137. Yue, C., Xu, M., Zhong, L., Tang, S., Cai, G., Zhang, R., Cheng, B., pH-responsive release features of collagen/carboxylated cellulose nanofiber composite aerogels through the incorporation of cyclodextrin/5-fluorouracil inclusion complexes. *Eur. Polym. J.*, *207*, 112807, 2024. https://doi.org/10.1016/j.eurpolymj.2024.112807.
138. Yusefi, M., Soon, M.L.-K., Teow, S.-Y., Monchouguy, E.I., Neerooa, B.N.H.M., Izadiyan, Z., Jahangirian, H., Rafiee-Moghaddam, R., Webster, T.J., Shameli, K., Fabrication of cellulose nanocrystals as potential anticancer drug delivery systems for colorectal cancer treatment. *Int. J. Biol. Macromol.*, *199*, 372–385, 2022. https://doi.org/10.1016/j.ijbiomac.2021.12.189.
139. Zaczek-Moczydłowska, M.A., Joszko, K., Kavoosi, M., Markowska, A., Likus, W., Ghavami, S., Łos, M.J., Biomimetic Natural Biomaterial Nanocomposite Scaffolds: A Rising Prospect for Bone Replacement. *Int. J. Mol. Sci.*, *25*, 24, 13467, 2024. https://doi.org/10.3390/ijms252413467.
140. Zhang, S., Gatsi, B., Yao, X., Jin, Y., Amhal, H., Cellulose nanofiber-reinforced antimicrobial and antioxidant multifunctional hydrogel with self-healing, adhesion for enhanced wound healing. *Carbohydr. Polym.*, *352*, 123189, 2025. https://doi.org/10.1016/j.carbpol.2024.123189.
141. Zhou, Z., Liu, D., Zhao, X., Conversion of lignocellulose to biofuels and chemicals *via* sugar platform: An updated review on chemistry and mechanisms of acid hydrolysis of lignocellulose. *Renewable Sustainable Energy Rev.*, *146*, 111169, 2021. https://doi.org/10.1016/j.rser.2021.111169.

5

Interfacial Modification of Polyvinyl Chloride (PVC)-Based Biocomposites and Bionanocomposites

Majed M. Alghamdi* and Adel A. El-Zahhar

Department of Chemistry, College of Science, King Khalid University, Abha, Saudi Arabia

Abstract

This chapter provides an overview of strategies for modifying polyvinyl chloride (PVC) by incorporating natural materials and nanostructured fillers to enhance its properties and sustainability for diverse applications. It highlights the importance of physicochemical modifications in overcoming the hydrophilicity challenges of biomaterials and ensuring compatibility with hydrophobic PVC matrices, resulting in the development of reinforced PVC biocomposites and bionanocomposites with superior physical and mechanical properties. Additionally, the chapter discusses how these modifications can reduce plastic waste and improve recycling capabilities, highlighting the role of natural materials in minimizing environmental impact.

Keywords: Polyvinyl chloride, PVC, natural fillers, biocomposites, bionanocomposites, physicochemical modification, interfacial compatibility, sustainable polymers

5.1 Introduction

Polyvinyl chloride (PVC) is a thermoplastic material that is extensively utilized across multiple sectors due to its affordability, excellent chemical stability, mechanical properties, and electrical insulation [1–3]. It has been produced on an industrial scale since 1920s, and nowadays, it can be

**Corresponding author*: MMALGHAMDI@kku.edu.sa

Shiv Kumari Panda and Visakh P. M (eds.) Polyvinyl Chloride-Based Biocomposites and Bionanocomposites, (125–162) © 2026 Scrivener Publishing LLC

considered one of the most commonly used vinyl polymers, coming as the third largest usable thermoplastic polymer after polyethylene and polypropylene [2]. Despite its many advantages, PVC presents certain challenges, including limited thermal and photo stability. It begins to degrade at temperatures around 100 °C and is also compromised by UV light and oxygen exposure, resulting in further degradation. To overcome these limitations and enhance the performance of PVC in various applications, it is essential to modify it with a range of additives. This can be done by incorporating various additives or modified substances into the PVC matrix. These include stabilizers, plasticizers, fillers, nanoparticles, colorants, and other polymers, which are crucial for improving the processing and the overall performance of PVC without altering its molecular structure.

Modifying PVC with various fillers is crucial for tailoring its properties to meet a wide range of application needs [4]. Mechanical reinforcements using fibers, both natural and synthetic, are commonly employed in polymer composites. While traditional synthetic fibers like aramid, carbon, and glass have been widely used, environmental concerns have driven the shift towards more sustainable natural fibers. Synthetic fibers are known for their superior mechanical properties and thermal stability. However, their inclusion can lead to an increase in the overall weight of the composite, which diminishes the lightweight benefit usually associated with an all-natural fiber composite. Extensive research supports the replacement of synthetic fibers with natural ones, which are derived from plant and animal sources. This variety in natural fiber sources highlights the broad diversity and potential for customization in composite materials. For example, natural fiber-reinforced polymer composites have seen extensive use in construction components as well as in automotive applications [5–11]. Mercedes-Benz pioneered this approach in the early 90s, becoming the first carmaker to adopt sustainable materials by building door panels using jute fibers [12].

Thermal modifiers like organotin, calcium–zinc, and magnesium hydroxide are essential for improving PVC's thermal stability and fire resistance, which are critical for safety in electrical installations and building materials [13–15]. Electrical and magnetic enhancers, including conductive elements like carbon nanotubes, graphene, and various metal nanoparticles, significantly boost PVC's electrical and magnetic properties, making it ideal for advanced electronic applications and effective as shielding materials [16–21]. Moreover, the integration of photocatalytic and antimicrobial agents like TiO_2 and zinc oxide not only endows PVC with environmental cleaning capabilities but also imparts antibacterial and antifungal properties, expanding its use in healthcare and sanitary products [22–25]. Additionally, embedding materials such as PbO, Bi_2O_3, and tungsten enhances PVC's ability to shield against harmful radiation, making it suitable for protective gear and medical applications [26–29]. These advancements in mechanical, thermal, electrical, and biological properties underscore the critical role of innovative materials engineering in adapting PVC to meet the diverse demands of modern applications.

Although a variety of fillers have been incorporated into PVC composites to enhance mechanical and thermal properties, extensive research continues to explore the potential of biomaterials in developing high-performance engineering products. Biocomposites, which are not only cost-effective and biodegradable but also characterized by their low density and availability, offer significant adaptability for various applications [30–33]. This shift towards biomaterials is also driven by the growing demand for sustainable and environmentally friendly alternatives, with a particular emphasis on utilizing plant fibers. The selection of the bio-fillers and manufacturing procedure to achieve the required properties is the challenge. As shown in Figure 5.1, approximately 88% of PVC modifications utilize plant fibers, highlighting their predominant role compared to animal fibers in this field.

Plant fibers, derived from various parts of plants, play a crucial role in the development of biocomposite materials, particularly in enhancing the performance of PVC composites. Figure 5.2 illustrates the classification of natural fibers based on plant parts, highlighting the diversity within plant fibers.

Moreover, plant fibers are known for their unique mechanical properties, making them valuable as reinforcement materials in polymer composites

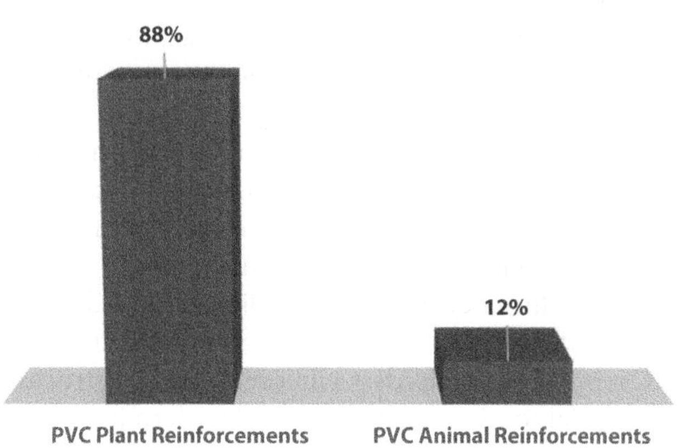

Figure 5.1 Comparison of the use of plant and animal fibers for PVC biocomposites [11].

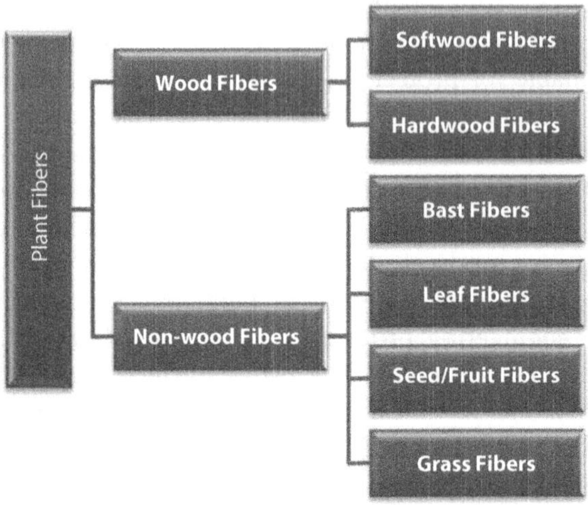

Figure 5.2 Diagram illustrating the classification of natural fibers based on the plant parts.

like PVC. One of their key advantages is the combination of high tensile strength and low density, offering a lightweight yet strong material for reinforcement. For example, fibers such as flax and hemp exhibit tensile strengths ranging from 345 MPa to 1500 MPa, which surpasses many conventional materials used in composites. Additionally, these fibers have relatively high Young's modulus values, contributing to the stiffness of the composites. Figure 5.3 illustrates the range of Young's modulus and tensile strength for different natural fibers, in comparison to rigid and flexible PVC. Therefore, commercial wood–plastic composites (WPCs), which consist of wood particles derived from renewable resources, a thermoplastic polymer, and various additives, are widely used to enhance durability, mechanical strength, and manufacturing efficiency, while also reducing water absorption [34].

The composition of plant fibers is vital in determining their mechanical properties and compatibility in reinforcement applications. The interplay of its components not only dictates the mechanical performance of the PVC composite but also impacts its durability, environmental resistance, and overall efficiency in various applications. Understanding these compositional differences is essential for optimizing fiber selection for specific PVC reinforcement needs. Cellulose, a primary structural component, plays a crucial role in plant fibers, largely determining their utility by contributing to the tensile strength and stiffness of composites. However, its effectiveness is significantly influenced by the surrounding lignin and

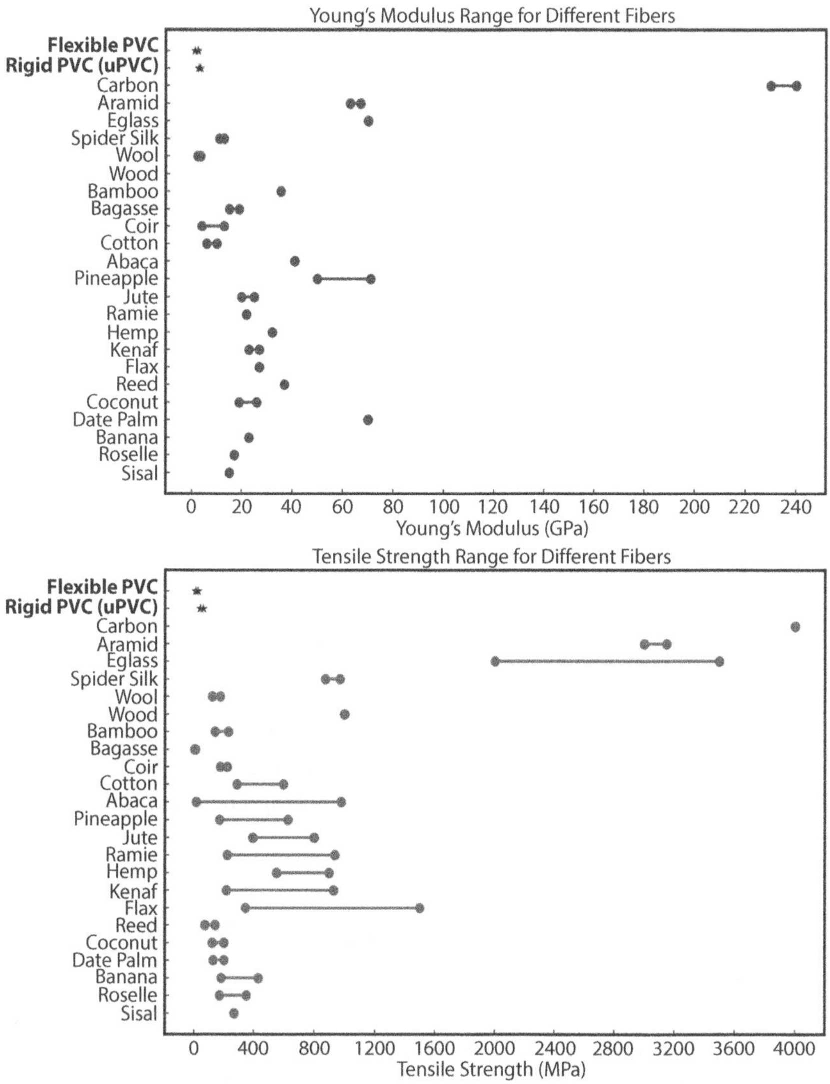

Figure 5.3 Comparison of Young's modulus and tensile strength ranges for various natural fibers and PVC.

hemicellulose. These factors critically affect the mechanical and thermal properties of the composites. The overall performance of the composite is further enhanced by electrostatic forces, interdiffusion, and mechanical interlocking at the fiber–matrix interface. Figure 5.4 provides a detailed comparison of the compositional differences among various plant fibers.

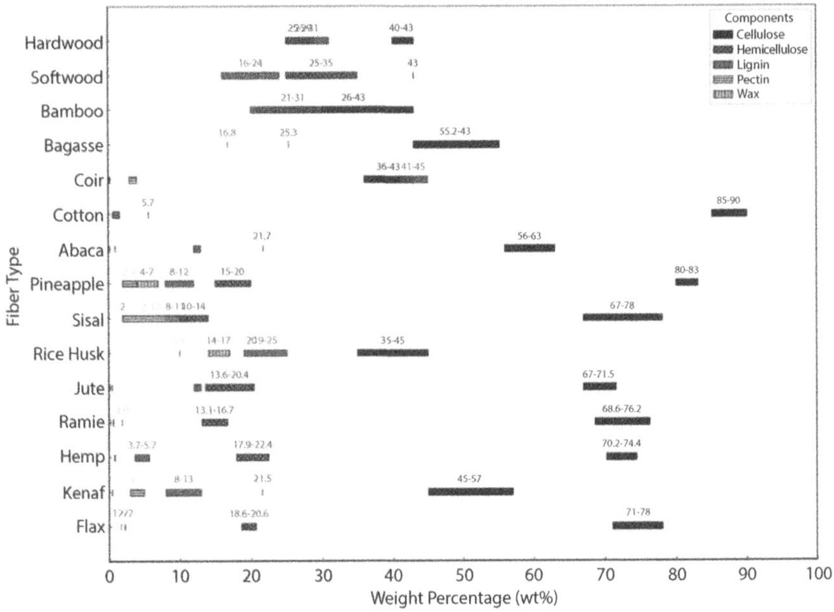

Figure 5.4 Comparison of fiber components across different fiber [11].

The upcoming sections will explore PVC biocomposites and PVC bionanocomposites using various biomaterial modifications, with a particular focus on the synergistic effects of combining chemical and physical approaches to optimize performance.

5.2 Sustainable Development of PVC-Reinforced Biocomposites

It is important to note that simply adding wood fibers to a non-biodegradable matrix does not produce biodegradable materials, as the matrix remains chemically unaltered and non-degradable [31, 35]. Given that WPCs are typically used for durable applications, biodegradability may not be a desired property. Instead, improving sustainability through recycling is a more beneficial approach, as it reduces energy consumption and CO_2 emissions. Reducing plastic usage and enhancing recycling capabilities are key objectives in the development of reinforced bionanocomposites with the required physical and mechanical properties. The incorporation of biomaterials into polymers leads to the creation of highly recyclable composites that significantly reduce plastic waste and mitigate environmental impact [36]. For instance, Augier *et al.* demonstrated that PVC–vegetable fiber composites not only improve the stability and mechanical properties of PVC but also allow the composite to be recycled up to five times without substantial degradation in performance [37]. This recycling capability highlights the potential for internal waste to be effectively reused within the production cycle, reinforcing the sustainability of the composite from an industrial perspective.

5.3 PVC Enhancement through Biocomposites

The application of biocomposites necessitates satisfying several key requirements such as mechanical properties and thermal stability. Extensive research has explored the incorporation of plant fibers to boost the performance and properties of PVC [3, 7, 34]. Various natural materials, including natural fibers, have been extensively evaluated for their effectiveness in PVC composites, as detailed in Table 5.1. For instance, bamboo fiber/PVC composites have demonstrated enhanced mechanical properties including tensile strength, flexural stress, strain, and modulus, alongside improved thermal stability. The introduction of bamboo fibers led to a 10% increase in Young's modulus, peaking at 1250 MPa with an 8% fiber content before diminishing with further increases [38]. Additionally, low-cost olive pits flour (OPF) has been utilized to create PVC biocomposites, resulting in increased tensile modulus, mechanical hardness, and thermal stability [39]. Various biopolymers such as chitosan, starch, and lignin have also been studied as fillers. Animal-derived materials, including chicken feathers, eggshells, and cow horn leather, have further contributed to enhancing PVC properties [40–42].

To further enhance PVC properties, biocomposite design focuses on optimizing interfacial interactions through strategies like employing compatibilizers, modifying fiber surfaces, and reducing fibers to nanoscale dimensions. The effectiveness of these composites largely depends on the robust interfacial bonding between biomaterials and the PVC polymer matrix. Weak adhesion between hydrophilic biomaterials and the hydrophobic polymer matrix can lead to insufficient reinforcement and compromised properties [32, 72]. Thus, surface modifications or the inclusion of compatibilizers becomes crucial for enhancing the performance of PVC biocomposites. Moreover, the interplay between crystal size, crystallinity, and degradation highlights the importance of precise material selection and processing [73, 74].

Table 5.1 Bio-fillers used in PVC composites.

Filler	Reference
Wood flour	[43–45]
Sorghum straw	[46]
Rice husk	[10, 25, 47–49]
Cotton stalk	[50–52]
Eucalyptus	[53]
Citrus maxima	[9]
Areca sheath	[54]
Bagasse	[55, 56]
Corn stalk	[57]
Kenaf core powder and fibers	[58, 59]
Waste rice husk ash	[48]
Raspberry pomace filler	[60]
Flour of decayed wood	[61]
Modified eggshell	[62]
Chitosan	[63, 64]
Fly ash	[65, 66]
Date palm fibers	[67, 68]
Bamboo	[8, 38, 69]
Coconut fiber	[8, 70]
Olive pits flour	[39]
Banana fiber	[71]

Achieving the desired level of interaction at the interface requires careful selection of preparation methods that include both physical and chemical modifications. These modifications are essential for enhancing compatibility and effectiveness, as well as increasing the cellulose content in fibers, which significantly contributes to the overall performance of natural fiber polymer composites. The following section will explore various enhancements of PVC biocomposites, examining how these modifications can further boost their effectiveness.

5.4 PVC Biocomposites Based on Chemically Modified Biomaterials

The incompatibility between hydrophilic biomaterials and hydrophobic thermoplastic matrices often leads to weak interfacial bonding, which compromises the reinforcement capabilities of the composite. This issue stems from the high hydrophilicity of the biomaterials, resulting in poor interfacial properties at the polymer–biomaterial interface. To improve this, fibers are frequently chemically modified to reduce their hydrophilicity, thus enhancing their compatibility and adherence within the polymer matrix [7, 75]. The impacts of different chemical modification processes on the physicochemical properties of the composites have been extensively studied and reported. To demonstrate the effect of chemical modification, comparative analysis of the mechanical properties of treated and untreated rice husk PVC biocomposites is presented in Figure 5.5. PVC/modified rice husk composites exhibited apparent improved mechanical properties compared to unmodified [17]. This treatment also contributed to enhanced thermal stability and reduced water absorption.

Various chemical modification techniques such as alkaline treatment (mercerization), acid treatment, acetylation, benzoylation, oxidative modifications, and silane treatment have been utilized to enhance the properties of fiber–polymer matrices. These methods have successfully improved adherence, mechanical stability, and overall compatibility for polymer composite reinforcement [76–78]. Modifications may include chemical incorporation of aliphatic or cyclic structures with reactive functionalities of the fibers, significantly altering the biomaterials' fine structure and improving their mechanical and thermal stability. In the context of PVC, benzoylation, which substitutes hydroxyl groups of cellulose for benzoyl groups, enhances fiber-matrix adhesion, thus increasing tensile strength, toughness, and thermal stability while reducing water absorption. For example, benzoylation treatment of olive husk flour and kenaf core powder resulted in a higher Young's modulus and decreased moisture absorption in the PVC composites [59, 79]. Acetylation treatment enhances fiber hydrophobicity, leading to stronger interfacial bonding and improved mechanical properties. Composites with acetylated fibers also exhibit significantly

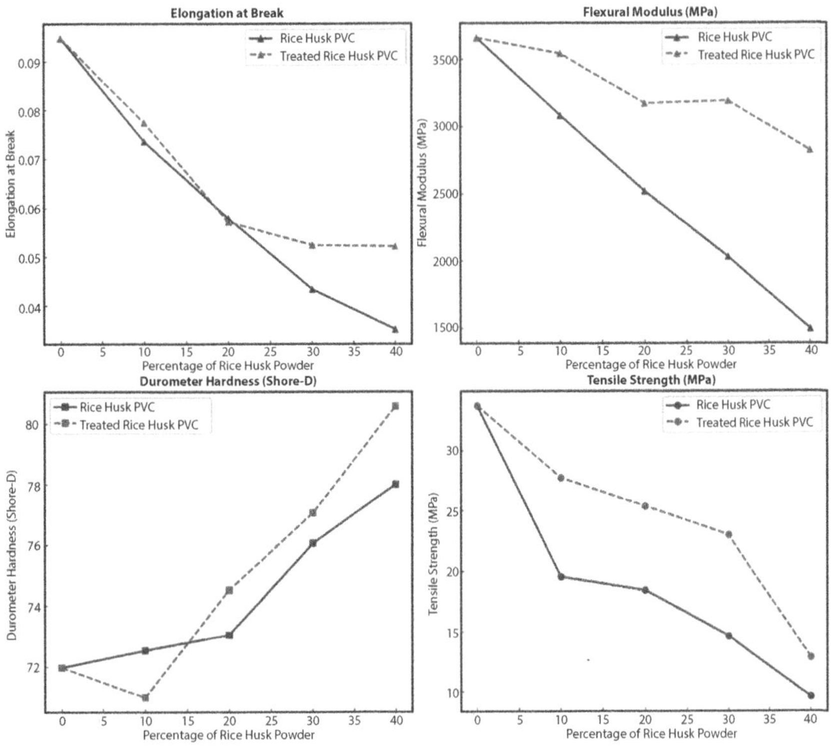

Figure 5.5 Comparative analysis of the mechanical properties of treated and untreated rice husk PVC biocomposites as indicated [17].

lower water absorption rates. Moreover, mechanical tests indicated that the addition of the natural plasticizer obtained through esterification of rice fatty acid resulted in an increase in the elongation at break [80].

Acid treatment can also be employed to improve interfacial interaction, as demonstrated by treating palm fibers in an acetic acid solution, significantly enhancing the PVC mechanical properties [81]. Sorghum straw fiber was pretreated with a mixture of stearic and palmitic acids to boost its water resistance [82]. Similarly, alkali treatment has been widely employed to improve the interfacial bonding between natural fibers and the polymer matrix by effectively removing lignin, pectin, and hemicellulose from the fibers, thus increasing their surface area [83]. For example, treating cotton stalk and palm fibers with a NaOH solution resulted in significant improvements in the mechanical and physical properties of their PVC composites, including increased tensile strength, Young's modulus, impact strength, water resistance, and heat distortion temperature [51, 84].

Similarly, using a 3% NaOH solution to treat waste paper for the fabrication of PVC/paper biocomposites notably enhanced both mechanical and thermal stability [85]. These composites displayed improved mechanical properties, with tensile and flexural strengths reaching up to 184 MPa and 67 MPa, respectively. These treated composites also exhibited reduced water absorption, achieving a low rate of 9.3% with a PVC-to-treated paper ratio of 7:3, featuring the positive effect of reducing moisture uptake.

Similarly, PVC/rice straw composites have been studied extensively to assess the effects of various factors such as pretreatment, PVC content, and temperature on their properties [86]. Results (Figure 5.6) show that rice straw treated with 5% and 10% NaOH at 80°C resulted in composites with higher tensile strength, while those treated with water at 130°C exhibited the lowest strength [86]. Figure 5.7 illustrates the effect of treatment on the swellability percentage, highlighting the effectiveness of alkali treatment in enhancing the water resistance of PVC composites compared to the untreated control. Alkali treatment not only improves fiber adherence, but it also provides a rougher surface morphology and decreases fiber diameter, thereby increasing the composite's aspect ratio and enhancing interfacial binding. These enhancements collectively improve the mechanical properties of the composites, demonstrating the critical role of both

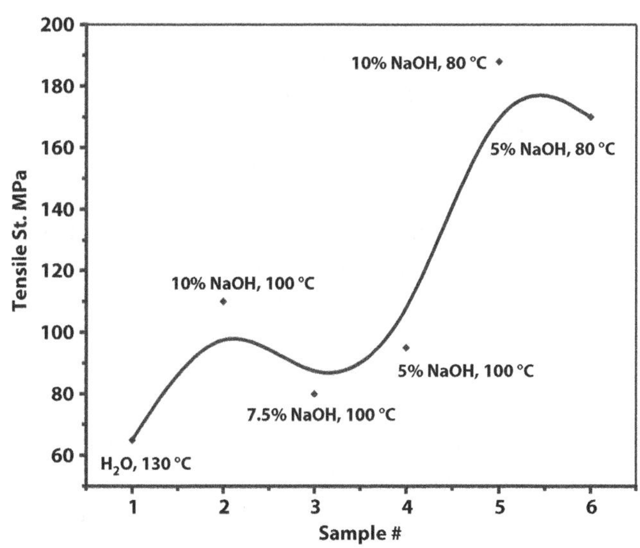

Figure 5.6 Effect of treatment solution and temperature on tensile strength of PVC/rice straw composites.

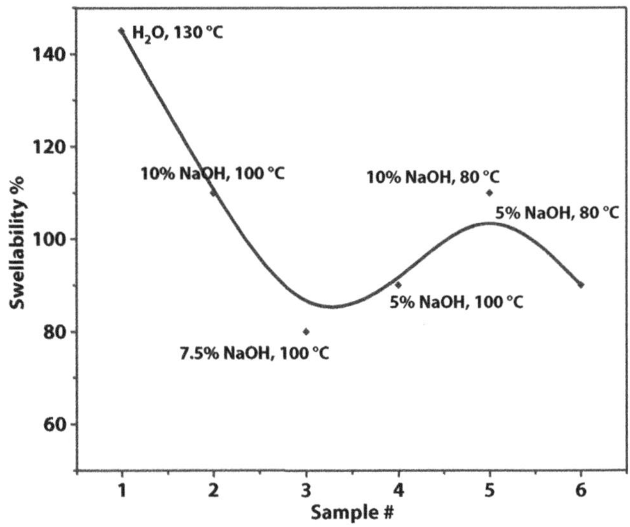

Figure 5.7 Effect of treatment solution and temperature on swelling behavior of PVC/rice straw composites.

chemical modifications and process conditions in optimizing PVC composite performance.

Oxidative modifications have been also studied for improving bonding and adherence between plant fibers biomaterial and polymers as ozonation and carboxylate format on the cellulosic fibrous and composite surfaces [86–88]. For example, dichromate in oxalic acid and nitric acid solution were studied as oxidizing agent for wood fibrous surfaces to enhance their interaction with polymer matrix [86]. The use of epoxidized agents further highlights the broad applicability of chemical modifications in optimizing the interfacial bonding and overall performance of PVC-based biocomposites. The role of epoxidized agents as compatibilizers lies in enhancing interfacial interaction through hydrogen bonding with the hydroxyl groups of fibers and cross-linking with PVC (Figure 5.8). Beyond improving mechanical properties, epoxidized groups also significantly improve the thermal stability of PVC by adsorbing hydrogen chloride released during thermal decomposition. In PVC–rice husk biocomposites, epoxidized natural rubber (ENR) has been employed as a plasticizer and a compatibilizer, demonstrating considerable miscibility with PVC and enhancing its mechanical properties due to the polar nature of both polymers [89]. The study further explores the effects of rice husk content, particle size, and ENR content, identifying an optimal formulation of 90:10:20

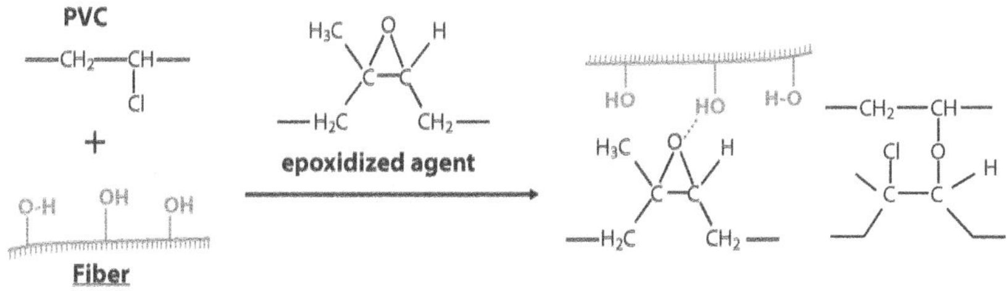

Figure 5.8 Schematic explanation for the formation of the PVC–fiber composite in the presence of an epoxidized compatibilizer, illustrating the formation of hydrogen bonding between the cellulosic hydroxyl groups and epoxy groups, as well as the formation of a cross-linked network between PVC and the compatibilizer.

wt.% for PVC:ENR:rice husk, which results in superior composite properties. Similarly, epoxidized soybean oil (ESBO) has been utilized as an alternative bioplasticizer, showing significant improvements in the properties of PVC composites. Replacing 20% of dioctyl phthalate (DOP) with ESBO increased elongation at break, tensile strength, and Young's modulus by 48%, 24%, and 4.5%, respectively [90]. Furthermore, a study by Skórczewska *et al.* demonstrated the potential of modifying plasticized PVC with cassia oil to create a material with antimicrobial properties and enhanced mechanical performance [91]. The PVC–cassia oil composite exhibited strong antimicrobial activity, while mechanically, the cassia oil acted as a plasticizer, reducing the modulus of elasticity and glass transition temperature and increasing elongation at break.

Chemical treatment is not the only method to enhance the interfacial interactions properties of biomaterials; incorporating a compatibilizer reagent can also significantly improve interfacial interactions, resulting in more robust composites [92]. To improve interfacial interaction between olive husk flour (OHF) and PVC, two chemical modifications were applied: the use of maleic anhydride grafted onto PVC as a compatibilizer reagent and benzoylation of OHF. The resulting PVC–OHF biocomposites demonstrated substantially improved thermal and mechanical stability with varied OHF content ratios. Composites using chemically treated OHF showed enhanced tensile strength compared to those with untreated OHF, which suffered from poor dispersion and agglomeration, leading to non-uniform stress and reduced binding forces within the composite [93]. Furthermore, the mechanical, thermal, and morphological properties of PVC–alfa composites have shown remarkable improvement in thermal stability with the addition of maleic anhydride-grafted PVC, illustrating the significant role of compatibilizers in enhancing composite performance [94].

Significant advancements in PVC composites have been achieved with the use of silane-based coupling agents, which are extensively utilized to enhance interfacial cross-linking and strengthen polymer–fiber bonding. For instance, PVC/rice husk ash composites treated with 3-aminopropyltrimethoxysilane (KH550) exhibited notable improvements, with a 44% increase in impact strength and a 38% reduction in water absorption compared to untreated composites [48]. Moreover, the application of KH550 also enhanced the compatibility and thermal stability of silane-modified fly ash within PVC composites, increasing thermal stability up to 187°C and significantly improving tensile strength and Young's modulus compared to untreated composites [65]. In another study by Shen *et al.*, the impact of KH550 and methylenediphenyl-4,4′-diisocyanate (MDI) as coupling agents on PVC/wood composites was examined, revealing improvements in the modulus of rupture and elasticity. This study highlights the broad efficacy of coupling agents in enhancing the properties of diverse PVC-based composites [95]. Particularly, MDI showed superior mechanical properties, with a modulus of rupture at 91.9 MPa, a modulus of elasticity at 10.9 GPa, and minimal thickness swelling of 2.4%. These improvements are attributed to the formation of Si–O–C bonds, which significantly reduced the content of hydrophilic hydroxyl groups on the wood surface. Additionally, stable urethane bonds were formed between the isocyanate groups of MDI and the hydroxyl groups of the wood, contributing further to the enhanced durability and strength of the composites. Furthermore, the interfacial bond strength of PVC composites was found to be the highest when compared to those of PP, HDPE, and LDPE, both with and without the use of KH550 and MDI coupling agents [7]. This significant bond underlines the effectiveness of these coupling agents in enhancing the mechanical properties of PVC composites, highlighting the importance of carefully selecting both the coupling agents and the thermoplastic matrices to achieve optimal performance [95].

Coupling agents also play a crucial role in controlling water uptake, which directly impacts mechanical performance. In composites lacking coupling agents, higher water absorption is often observed, leading to a decline in material properties. Natural fiber–polymer composites, particularly those with higher fiber content, tend to absorb more water due to the presence of cellulose, lignin, and hemicellulose, all of which contain hydroxyl groups that contribute to hydrophilicity. However, this hydrophilicity can be mitigated through chemical treatments or the application of coupling agents, such as maleic anhydride. These agents interact with the hydroxyl groups, preventing hydrogen bonding between the fibers and water, thereby enhancing the material's hydrophobic characteristics. This effect is particularly evident in PVC–alfa fiber composites, which were tested with and without coupling agents through hydrothermal aging [94]. The composites were immersed in distilled water at 100°C for 120 hours, revealing that those without coupling agents exhibited significantly higher water uptake, leading to decreased thermal and mechanical stability due to fiber cracking and weakened interfacial bonding. Similarly, in sisal/PP composites, tensile properties declined with increased water absorption, prolonged immersion, and higher fiber loading [76]. The moisture absorption of fibers is further influenced by their non-crystalline content and void content, emphasizing the necessity of using coupling agents to enhance the durability and performance of these composites.

Modification can also be carried out using lignin, an amorphous aromatic natural biopolymer that imparts rigidity to plant fibers by encapsulating cellulose strands within its matrix. Due to its renewable, biodegradable nature, lightweight, reinforcing properties, and antimicrobial characteristics, lignin has increasingly become a key material in the development of PVC composites. Its distinctive chemical structure allows lignin to function as a filler, stabilizer, compatibilizer, cross-linking agent, and reinforcement in composites. Notably, it has the potential to serve as a compatibilizer between hydrophilic natural fiber reinforcements and hydrophobic matrix polymers such as PVC, thereby enhancing interfacial bonding and improving overall composite performance. Despite the challenges posed by its extensive cross-linking and strong intermolecular interactions, which make it difficult to form miscible blends, specific modifications have successfully integrated lignin into advanced polymeric composites [78]. These modifications have led to composites with enhanced properties such as fire retardancy, improved dispersion, surfactant functionality, and adhesion, especially in PVC-based composites [96–100]. However, for effective interaction between the polymer matrix and lignin, maintaining a low lignin content is crucial to avoid particle agglomeration [13]. This highlights the need for careful management of lignin incorporation to maximize the performance of the resulting composites. Additionally, hybrid materials and modified lignin have been extensively applied in PVC composite preparation. For instance, lignin–silica and lignin–magnesium hydroxide hybrid materials have been employed to create PVC composites, showing good dispersion of modified lignin within the polymer matrix and significantly improving mechanical properties [14, 15, 101–103]. Halloysite–lignin (HL) hybrid materials have also been prepared and incorporated into the PVC matrix through rigid processing, resulting in composites with enhanced thermal and mechanical stability [13]. Notably, the PVC–HL composite with 2.5 wt.% of HL markedly increased impact strength and doubled the elongation at break.

Combining various methods can significantly enhance PVC composite properties. For instance, chemically treated date palm fibers (DPFs), derived from palm leaflets and rachis, were used as reinforcing filler fibers for composites of PVC and high-density polyethylene (HDPE) with a ratio of 20:80 [67]. These DPFs were subjected to multiple chemical treatments, including sodium hydroxide, KH550, a combined alkali-silane process, and bleaching with $NaClO_2$ and H_2O_2+HNO_3. The resulting composites demonstrated improved properties, such as enhanced surface morphology, increased thermal and mechanical stability, and reduced water absorption behavior, making them suitable for use as reinforcing fibers in various applications. Hybridization of fibers in composites is another approach for enhancing mechanical properties while maintaining sustainability and cost-effectiveness. For instance, Adediran *et al.* developed a hybrid PVC composite by combining bamboo fiber and coconut shell particulates [8]. This combination within a PVC matrix not only effectively improves mechanical properties but also ensures sustainability. The resulting composite demonstrates significant improvements in yield strength, modulus of elasticity, flexural strength, and modulus of rupture. Understanding and addressing these factors is critical for advancing the development of sustainable, high-performance biocomposites, highlighting the synergistic effects of integrating various methods to optimize performance.

Moving from plant-based biomaterials, we next consider the use of animal-derived fibers in the formulation of PVC composites. Seashell waste (SS), which is abundant in coastal areas and primarily composed of over 90% calcium carbonate with a small amount of organic matter, exhibits unique morphological forms that make it an effective material for composite reinforcement [104]. In conjunction with reduced graphene oxide (RGO), known for its high surface area and antibacterial activity, SS has been utilized to enhance PVC composites [105]. Analysis of the SS content ratio indicated that optimal reinforcement was achieved with 15–20 wt.% of SS in the polymer matrix, ensuring uniform RGO dispersion and improved interfacial interaction between SS and PVC. This led to significantly enhanced properties through an environmentally friendly and cost-effective process. Another innovative approach involves the use of green reduced graphene oxide (GRGO) to enhance the mechanical and thermal properties of PVC composites. GRGO, reduced using vitamin C (ascorbic acid), which also imparts antibacterial properties, was chosen as a filler [106]. The PVC/GRGO composites were produced through a colloidal blending method, with varying concentrations of GRGO in the PVC matrix. Thermal analysis *via* TGA and DSC revealed that composites with a 1% GRGO loading exhibited significantly enhanced thermal stability. Additionally, the mechanical properties showed remarkable improvement; for instance, the composite containing just 0.1% GRGO demonstrated a 42% increase in tensile strength and a 98% increase in hardness.

Building on the discussion of animal-derived fibers such as SS for PVC reinforcement, chitosan (CS) emerges as another pivotal biomaterial in this category, and it is considered as the second most abundant polysaccharide on the planet, after cellulose. Extracted from the chitin of crustacean shells, chitosan is valued for its biocompatibility, biodegradability, and natural antimicrobial properties, which make it an excellent enhancer for PVC composites, and most widely used in biomedical applications [107]. Its chemical affinity for synthetic polymers allows it to significantly improve the mechanical and barrier properties of these composites. Research has shown that incorporating chitosan into PVC formulations can enhance PVC properties; a mechanical characterization shows an increase in the elastic modulus by 150% and elongation at break drops by 170% [108]. CS has also been used as a coupling agent for wood flour/PVC composite to improve interfacial adhesion, addressing key performance and safety concerns [86, 109].

5.5 PVC Biocomposites Based on Physically Modified Biomaterials

Physical modification processes are widely used to enhance the mechanical interlocking and interface properties of biocomposites. Unlike chemical modifications, these physical treatments do not alter the chemical structure of biomaterials but instead modify surface characteristics, thereby influencing the interfacial bonding between fibers and the polymer matrix [7, 26]. These modifications include techniques such as stretching [110], thermal treatment [111], calendaring [110], impregnation [4, 112], plasma treatment, gamma irradiation [113], ultrasonic irradiation, and surface roughening [110–112, 114]. These treatments play a crucial role in improving the interaction between biomaterials and the polymer matrix, complementing any chemical modifications applied. Physical modifications primarily aim to separate fiber bundles into single filaments and to enhance the potential applications of the fiber composites. Interestingly, processes like mercerization and steam explosion, while involving the use of chemicals such as sodium hydroxide (NaOH), are often discussed alongside physical modifications due to their significant impact on the physical properties of fibers. Mercerization, for example, involves treating fibers with NaOH, which increases surface area and strength, thus enhancing the fiber physical properties and its compatibility with the polymer matrix [4]. Similarly, steam explosion also utilizes NaOH to impregnate fibers, which are then subjected to high-pressure steam to separate lignocellulosic materials into cellulose, lignin, and hemicellulose. This process not only alters the surface characteristics of the fibers but also produces shorter fibers with irregular surfaces, which can be beneficial or detrimental depending on the application [26, 115]. As a result, the compatibility between the modified fiber and the polymer matrix is enhanced, increasing interfacial interaction through various binding forces, such as hydrogen bonding, van der Waals forces, and ionic interactions. These forces lead to stronger adherence, better dispersion, and more efficient reinforcement. For instance, surface roughness, determined by physical treatment, directly impacts the wettability of the fiber in the polymer matrix and other properties critical to interfacial interactions. Thermomechanical processing is another technique used to modify fibers, producing single, separated filaments. During this process, fibers undergo thermal treatment near the lignin glass transition temperature, allowing lignin to be released without altering the other fiber constituents. This method produces fibers with greater mechanical strength than the original fibers [26]. Additionally, techniques such as plasma treatments and ultrasonic irradiation are employed to ensure effective fiber dispersion within the polymer matrix, further enhancing the overall properties of the composite.

5.6 PVC Bionanocomposites

Bionanocomposites are a novel class of multiphase nanostructured materials that include at least one biologically derived phase and nanoparticles with nanometer-scale dimensions. The integration of nanoparticles into natural fibers has gained significant attention due to their ability to enhance interfacial properties, even at low concentrations (0.1–1.0%), leading to notable improvements in strength and Young's modulus [116]. In recent years, interest in bionanocomposites has surged due to their superior properties and versatility. Research has particularly focused on incorporating various nano-additives, such as metal oxides, graphene oxide, and carbon nanotubes, into PVC-based biocomposites. These additives enhance the materials' performance, contributing to better recycling potential and enhanced antimicrobial and antioxidant properties [117]. While chemical modifications have successfully enhanced chemical interactions, resulting in better interfacial bonding, mechanical, thermal, and antimicrobial properties, the addition of inorganic nanoparticles further boosts the material's overall performance.

Although integrating these nanomaterials into PVC biocomposites has shown substantial enhancements across a wide range of properties, effective dispersion and optimal particle size are crucial for maximizing the performance for various applications [118–120]. Nanomaterials in these composites, characterized by their high surface area-to-volume ratio and surface functionality (especially the abundance of –OH groups), tend to aggregate, which can negatively affect properties. To counter this, surface modification of nanoparticles is essential to block these functionalities, prevent aggregation, and ensure uniform dispersion. Proper dispersion of nanoparticles significantly increases the matrix/filler interfacial area, enhancing molecular mobility, relaxation behavior, and, consequently, the thermal and mechanical properties of the material. For example, modifying ZnO nanoparticles with citric acid (CA) has been demonstrated to significantly improve their dispersion within PVC matrix films [24]. The resulting PVC/ZnO-CA nanocomposites, prepared at various hybrid nanoparticle concentrations, exhibited notable enhancements in both mechanical and thermal properties. Figure 5.9 illustrates the nanoparticle modification process using carboxylic acid, effectively showing how this method optimizes dispersion.

Building on this knowledge, further reinforcement of PVC nanocomposite films with ZnO nanoparticles, modified by CA and vitamin C, was

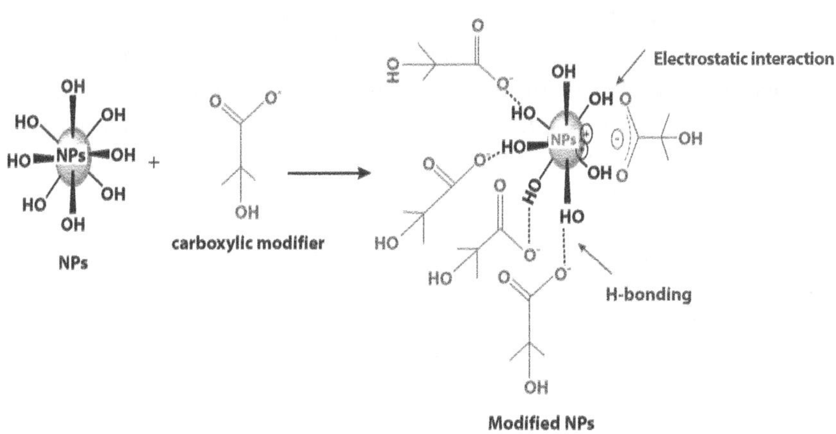

Figure 5.9 Schematic explanation for NPs modification *via* carboxylic coupling agent.

found to substantially increase flame retardancy [121]. Moreover, the mechanical properties, especially elongation at maximum stress, were markedly improved. These findings emphasize the effectiveness of ZnO as an additive in enhancing both the safety and durability of PVC-based materials. The inclusion of hybrid additive of TiO_2 within PVC also highly affected the composite mechanical properties [122]. Lignin/TiO_2 within PVC matrix provides excellent UV aging resistance along with enhanced mechanical properties [123]. Amina *et al.* investigated the fabrication of polymeric bionanocomposite films by incorporating 5–10 wt.% naturally extracted *Moringa oleifera* oil into PVC, alongside silver nanoparticles for antimicrobial activity. Their study demonstrated that the silver nanoparticles were well-dispersed within the PVC–*Moringa oleifera* oil matrix, significantly enhancing the composite's thermal stability up to 250°C and effectively inhibiting the growth of various bacteria [124].

The ZnO–chitosan hybrid material was also synthesized and incorporated into a PVC matrix to serve as a hybrid reinforcement [22]. The surface characterization of the resulting composite revealed improved dispersion and strong adherence, attributed to the presence of multiple functional groups that facilitated the formation of strong hydrogen bonds and van der Waals forces. The inclusion of the ZnO–chitosan hybrid nano-additive (at 1 wt.%) within the PVC matrix significantly enhanced the thermal stability of the prepared bionanocomposites, compared to pure PVC. This improvement indicates a high level of compatibility between the ZnO–chitosan hybrid and the PVC matrix. The glass transition temperature (Tg) was also analyzed and found to increase in the PVC/ZnO–chitosan. Similarly, this rise in Tg is attributed to the stronger intermolecular bonding within the composite, where the hybrid nano-additive effectively restricted the free movement of the polymer chains, thereby enhancing thermal stability. Regarding mechanical properties, these are strongly influenced by the characteristics of the incorporated nanoparticles, including their structure, shape, distribution within the polymer matrix, and interaction with the polymer itself. The mechanical parameters, such as tensile strength, Young's modulus, and elongation at maximum stress, were measured and compared between PVC and PVC/ZnO–chitosan composites. The results showed significant improvements in tensile strength, Young's modulus, or the composite films compared to neat PVC. This enhancement is largely due to the presence of chitosan and its strong bonding with both ZnO and PVC. Furthermore, the elongation at maximum stress was found to improve with the incorporation of 2 wt.% of the hybrid nanoparticles, further demonstrating the superior mechanical properties of the PVC/ZnO–chitosan composite compared to pure PVC.

Nanocellulose particles (CNs) are derived from extensive chemical modifications of plant fibers, involving both alkaline and acidic treatments. This process removes amorphous components, leaving only the crystalline cellulose with exposed OH groups, which enables a wide range of applications. The impact of cellulose particle size and distribution on composite properties is substantial. Smaller cellulose particles tend to provide stronger bonding with the polymer and ensure more uniform dispersion, which affected both thermal and mechanical properties [125].

Thermal analysis of the PVC/CNs revealed considerable improvements in thermal stability, including decomposition temperature, combustion, and degradation rates, compared to pure PVC (Table 5.2). Although the decomposition temperature showed a slight reduction with the incorporation of cellulose, indicating an effect on the thermal degradation process, the total weight loss at 500°C decreased as cellulose content increased, suggesting that cellulose contributes to a lower overall weight loss compared to pure PVC.

Beyond thermal properties, the presence of CNs in PVC composites significantly impacts mechanical stability. The effect of cellulose particle size was investigated using PVC-g-maleic anhydride as a coupling agent and CNs with diameters up to 170 nm, produced *via* a top–down milling process [125]. Generally, tensile strength increased with larger cellulose particle sizes, but beyond a certain threshold, further size increases led to a decline in tensile strength (Figure 5.10). Although the tensile strength of

Table 5.2 Thermal degradation of PVC and PVC/nanocellulose biocomposites at different cellulose content.

Composite	Temp. °C	Weight loss	Total weight loss at 500 °C
PVC	278	53.38%	90.2%
	444	82.85%	
PVC/cellulose 10%	270	46%	83.8%
	458	75%	
PVC/cellulose 20%	270	46%	85.2%
	458	75%	
PVC/cellulose 30%	270	46%	85.5%
	458	75%	

PVC/CNs composites was consistently lower than that of pure PVC, the influence of particle size was significant. Moreover, the data in Figure 5.10 illustrates that, unlike tensile strength, the percentage elongation at break, a key indicator of composite toughness, decreased as cellulose particle size increased. Smaller particles promoted better dispersion within the polymer matrix, enhancing the composite's toughness. However, it is crucial to note that nanoparticle aggregation can detrimentally affect the mechanical properties of the composite, underscoring the importance of optimizing particle size for the desired balance of properties.

Although nanocellulose excels in boosting the mechanical stability and toughness of the composites, there are other materials like nanoclays that offer additional benefits, particularly in enhancing thermal stability, performance, and flame retardancy [119, 120, 126]. While nanocellulose and other plant-based fibers are derived from renewable biological sources, nanoclays also originate from natural sources and bring their unique benefits to composite materials. Therefore, clay-reinforced PVC composites have emerged as another important category of advanced materials, providing unique advantages that complement those offered by nanocellulose. Nanoclay can be incorporated within the polymer matrix using different procedures such as *in situ* polymerization, melt intercalation, or solution

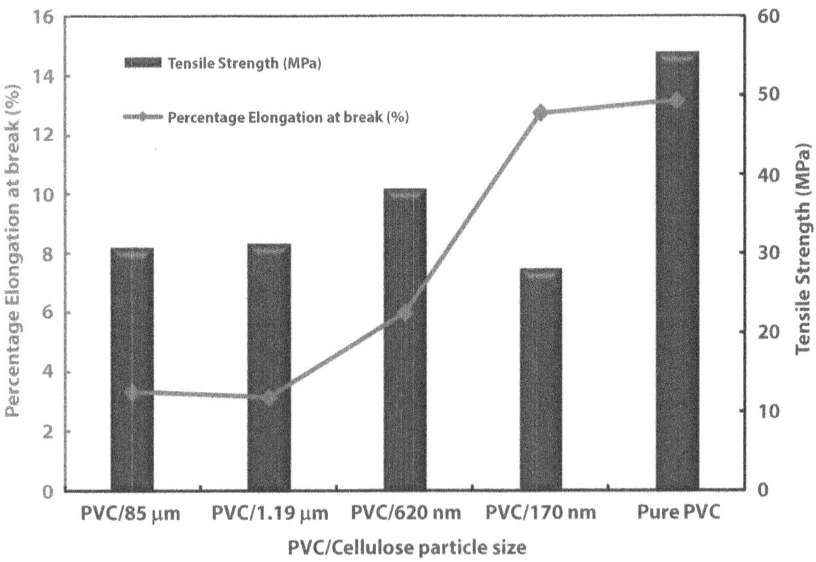

Figure 5.10 Tensile strength and percentage elongation at break of PVC/cellulose composites as a function of cellulose particle size.

blending. The desired properties of the polymer clay composite are related to the obtained particle dispersion which act as a challenge because of the clay high surface area and high surface energy, which facilitate particle aggregation. Different factors were identified to control obtaining uniform dispersion intercalated or exfoliated polymer–clay composite structures such as modifying agent type, preparation process, and temperature [127].

The inclusion of nanoclays such as bentonite, montmorillonite, and kaolinite in the PVC matrix further enhances mechanical properties, thermal stability, and other functional attributes, making these composites suitable for a wide range of applications. It was reported that uniform dispersion of MMT layers within the polymer matrix produces composites with improved physicochemical, mechanical, thermal, and fire-retarding properties, compared to pure polymers [127]. Moreover, the integration of nanoclays works synergistically with other fillers, like nanocellulose, to produce PVC composites with superior performance characteristics, addressing a broader spectrum of material demands in industries ranging from biomedical to automotive and construction [128]. In the study by Dutta *et al.*, polymer nanocomposites were prepared using rice husk (RH), microcrystalline cellulose (MCC), and montmorillonite (MMT) to enhance the properties of polyvinyl chloride (PVC) [129]. The study explored the effects of varying MCC and MMT content on the mechanical, thermal, and flame retardancy properties of the composites. The composite containing 2 phr MCC and 3 phr MMT exhibited the best performance, with significant improvements in tensile strength, flexural strength, and hardness, compared to other formulations. Specifically, the composite containing 2 phr MCC and 3 phr MMT showed a tensile strength of 11.33 MPa and flexural strength of 38.16 MPa. This enhancement is attributed to the synergistic effect of MCC and MMT, which reduced the mobility of the polymer chains, resulting in better interfacial adhesion and overall composite stability, consistent with other studies using cellulose nanofiber-montmorillonite hybrid filler on starch composite [130]. However, further increasing MCC content led to a decline in mechanical properties, likely due to the agglomeration of MCC, which decreased interfacial adhesion and increased the flexibility of the PVC matrix, reducing both tensile and flexural properties. Similarly, hybrid PVC bionanocomposites were developed by incorporating both calcinated halloysite and wood flour into the PVC matrix [131]. The simultaneous inclusion of these two fillers, each derived from different natural sources, significantly enhanced the thermal stability of the composites, by approximately 116%, compared to those containing only wood flour. This synergistic effect is also evident in the mechanical properties, where the addition of 5 wt.% halloysite alongside wood flour resulted in a 14.4% increase in stiffness and a 22.5% improvement in tensile strength. Moreover, the presence of wood flour in the PVC/halloysite nanocomposites improved the overall homogeneity by effectively breaking up halloysite agglomerates, leading to a more uniform composite structure. Yang *et al.* investigated the use of nano-MMT to enhance the fire safety and mechanical properties of wood flour/PVC biocomposites [132]. In their study, MMT@ZnO nanosheets were synthesized and incorporated into the wood flour/PVC matrix. The inclusion of these nanosheets significantly improved both flame retardancy and mechanical performance. Specifically, the composite containing 6 wt.% MMT@ZnO nanosheets demonstrated a 30.1% reduction in total heat release rate and a 32.2% decrease in total smoke production compared to the standard WF/PVC composite. Additionally, the tensile strength and flexural strength of this bionanocomposite increased by 24.6% and 36.8%, respectively, highlighting the effectiveness of this hybrid composite in enhancing the overall performance of the material.

Also, nanoclay can be modified into organoclay to provide better interaction with polymers. The impact of incorporating organoclays nanoparticles into polyvinyl chloride grafted with styrene and maleic anhydride (PVC-g-(St-MA)) was investigated. The nanocomposites, with nanoclay content ranging from 0.2 wt.% to 1 wt.%, exhibited significantly enhanced mechanical properties compared to the cross-linked PVC-g-(St-MA) samples [133]. Also, to enhance the interfacial adhesion in PVC/wood composites, Yim *et al.* introduce organoclay as a nanofiller (PVC/wood/clay), which significantly improved the overall performance of the composite [134].

Furthermore, the incorporation of different carbonaceous (carbon fiber, carbon nanotube, graphene fillers) within polymer matrix has been widely studied for property modification [18, 21, 135]. The polymeric composites containing these nanocarbon additives highly improve the composite properties [136]. Carbon nanotubes (CNTs) and nanographite, while synthetic, are derived from carbon, an element abundantly found in nature and a fundamental building block of all life forms. These carbon-based nanomaterials can be conceptually linked to biomaterials, as they share the same elemental composition as many natural fibers and have the potential to interact synergistically with bio-based polymers. CNTs have 50–100 times higher tensile strength than steel, and if compared with carbon fibers, nanotubes show high stiffness and low brittleness. CNTs are highly flexible, and their electrical properties are affected by size chiral behavior. Carbon fibers (CFs) have excellent mechanical properties, thermal properties, and suitable strength–density ratio, and therefore used as a reinforcement material in composite-applied aviation, automotive, biomedical, and electric technologies. CFs are also characterized with non-polar, inert, and smooth surface, while they have low wettability and low mechanical strength if not suitably modified [137, 138]. By combining CNTs with natural polymers or modifying them for compatibility with biocomposites, it is possible to create advanced materials that leverage the strengths of both synthetic and natural sources, paving the way for innovative applications that align with sustainable practices. CNT effect on PVC characteristics has been studied and discussed. Farsheh *et al.* reported the effectiveness of MWCNT reinforcing of PVC/wood-flour biocomposites [139]. The experimental findings revealed that the addition of MWCNTs significantly reduced water absorption and thickness swelling compared to wood-plastic composites without MWCNTs. Additionally, the tensile strength and modulus of the bionanocomposites were enhanced by up to 20% and 23%, respectively. In another study, aminated PVC composites were modified and incorporated with MWCNTs with 2.5% carboxyl groups and various percentages of nanographite [16]. The study found that the Tg of the aminated PVC increased with higher concentrations of oxidized MWCNTs, indicating enhanced thermal stability. Conductivity measurements were also conducted and revealed a significant increase in electrical conductivity due to the addition of the nanographite. Furthermore, ball-milled MWCNTs were evaluated as thermal stabilizers in PVC composites and proved to be the most effective in enhancing the thermal stability of PVC, offering superior resistance to both discoloration and hydrogen chloride (HCl) degradation [140]. However, this is not always true. In another study, Skórczewska *et al.* developed PVC composites modified with a hybrid carbon filler composed of CFs and MWCNTs. The hybrid filler was synthesized using a solvent-based method, with poly(vinyl acetate) as an adhesive. The findings revealed that, while there was a significant increase in the elastic modulus, the mechanical properties of the PVC/CF/CNT composites showed no substantial improvement overall. However, slight enhancements in thermal stability and glass Tg were observed [141]. Dutta *et al.* also explored the impact of incorporating a graphene oxide (GO) nanomaterial into waste rice husk/PVC biocomposites using green additives and a compatibilizer [142]. Furthermore, the addition of GO nanomaterial significantly enhanced the thermal, mechanical, chemical, and flame resistance properties of the composite. Specifically, the composite reinforced with 0.5 phr GO exhibited improvements in tensile and flexural by 34% and 37%, respectively.

In the development of PVC/SiO_2 nanocomposites, silica nanoparticles were selected for their natural origin and reinforcing properties.

To improve the dispersion and compatibility of these nanoparticles within the PVC matrix, surface modifications were applied. For instance, Zhu *et al.* grafted SiO_2 with polymethyl methacrylate (PMMA), which significantly enhanced the interfacial adhesion between the polymer matrix and the PMMA-g-SiO_2 hybrid nanoparticles [143]. This modification resulted in a higher Tg compared to PVC composites containing untreated nanosilica. Furthermore, these hybrid nanoparticles improved mechanical properties, including tensile strength and rigidity, particularly at low silica content levels below 3 wt.%. Additionally, incorporating 6 wt.% micro-silica (MS) into PVC/sorghum straw composites significantly enhanced their wear and corrosion resistance, toughness, strength, and heat resistance [46].

5.7 Conclusion

In conclusion, this chapter provides a comprehensive overview of the advancements in PVC to enhance its properties for a diverse range of applications. Despite its inherent limitations, such as low thermal and mechanical stability, PVC's performance can be significantly improved through various modifications, including the incorporation of natural fibers and nanostructured fillers. The chapter highlights the critical role of natural fibers, both unmodified and modified, in boosting the performance of PVC composites. These fibers not only enhance mechanical and thermal properties but also contribute to overall sustainability, addressing the limitations of traditional PVC and expanding its practical uses. Effective modification techniques, whether chemical or physical, are essential for optimizing fiber dispersion, ensuring homogeneity, and achieving desirable surface morphologies. These modifications are crucial for improving interfacial bonding and adhesion between the PVC matrix and fibers, which is key to achieving the desired reinforcement and enhanced composite properties. Furthermore, the chapter examines the impact of incorporating nanostructured fillers into PVC matrices. These fillers, often derived from natural sources, improve interfacial properties, adhesion, dispersion, and water absorption behavior, contributing to the development of high-performance PVC/bionanocomposites. By combining natural fibers with nanostructured materials, significant advancements have been made in creating sustainable, high-performance composites. These insights not only advance the understanding of PVC modification but also guide future research and development in creating innovative, sustainable materials for diverse industrial applications.

AI disclosure statement: The author acknowledges the use of AI-based tools for minor language editing or grammatical correction. However, all intellectual contributions, analyses, and conclusions presented in this chapter are entirely the author's own.

References

1. Iulianelli, G.C.V., Maciel, P.d.M.C., Tavares, M.I.B., Preparation and Characterization of PVC/Natural Filler Composites. *Macromol. Symp.*, 299–300, 227–233, 2011.
2. Braun, D., Poly(vinyl chloride) on the way from the 19th century to the 21st century. *J. Polym. Sci., Part A: Polym. Chem.*, 42, 578–586, 2004.
3. Wirawan, R., Zainudin, E.S., Sapuan, S., Mechanical Properties of Natural Fibre Reinforced PVC Composites: A Review. *Sains Malaysiana*, 38, 531–535, 2009.
4. Pandey, J.K., Ahn, S.H., Lee, C.S., Mohanty, A.K., Misra, M., Recent Advances in the Application of Natural Fiber Based Composites. *Macromol. Mater. Eng.*, 295, 975–989, 2010.
5. Bolin, C.A. and Smith, S., Life cycle assessment of ACQ-treated lumber with comparison to wood plastic composite decking. *J. Cleaner Prod.*, 19, 620–629, 2011.
6. Sun, G., Ibach, R.E., Faillace, M., Gnatowski, M., Glaeser, J.A., Haight, J., Laboratory and exterior decay of wood–plastic composite boards: Voids analysis and computed tomography. *Wood Mater. Sci. Eng.*, 12, 263–278, 2017.
7. Khalil, H.A., Tehrani, M., Davoudpour, Y., Bhat, A., Jawaid, M., Hassan, A., Natural fiber reinforced poly(vinyl chloride) composites: A review. *J. Reinf. Plast. Compos.*, 32, 330–356, 2013.
8. Adediran, A.A., Akinwande, A.A., Balogun, O.A., Olasoju, O.S., Adesina, O.S., Experimental evaluation of bamboo fiber/particulate coconut shell hybrid PVC composite. *Sci. Rep.*, 11, 5465, 2021.
9. Joglekar, J.J., Munde, Y.S., Jadhav, A.L., Bhutada, D.S., Radhakrishnan, S., Kulkarni, M.B., Studies on effective utilization of Citrus Maxima fibers based PVC composites. *Mater. Today Proc.*, 42, 578–583, 2021.
10. Jiang, L., Fu, J., Liu, L., Du, P., Wear and thermal behavior of basalt fiber reinforced rice husk/polyvinyl chloride composites. *J. Appl. Polym. Sci.*, 138, 50094, 2021.
11. Farrokhabadi, A. and Arabqomi, H., Natural Fiber Reinforced PVC Composites and Nanocomposites, in: *Poly(Vinyl Chloride) Based Composites and Nanocomposites*, A. H and T. Sabu (Eds.), pp. 87–116, Springer International Publishing, Cham, 2024.
12. Marsh, G., Next step for automotive materials. *Mater. Today*, 6, 36–43, 2003.

13. Tomaszewska, J., Wieczorek, M., Skórczewska, K., Klapiszewska, I., Lewandowski, K., Klapiszewski, L., Preparation, Characterization and Tailoring Properties of Poly(Vinyl Chloride) Composites with the Addition of Functional Halloysite–Lignin Hybrid Materials. *Materials*, 15, 8102, 2022.
14. Bula, K., Klapiszewski, L., Jesionowski, T., A novel functional silica/lignin hybrid material as a potential bio-based polypropylene filler. *Polym. Compos.*, 36, 913–922, 2015.
15. Klapiszewski, L., Tomaszewska, J., Skórczewska, K., Jesionowski, T., Preparation and Characterization of Eco-Friendly Mg(OH)2/Lignin Hybrid Material and Its Use as a Functional Filler for Poly(Vinyl Chloride). *Polymers*, 9, 258, 2017.
16. Haruna, H., Pekdemir, M.E., Tukur, A., Coşkun, M., Characterization, thermal and electrical properties of aminated PVC / oxidized MWCNT composites doped with nanographite. *J. Therm. Anal. Calorim.*, 139, 3887–3895, 2020.
17. Chand, N. and Das Jhod, B., Mechanical, Electrical, And Thermal Properties Of Maleic Anhydride Modified Rice Husk Filled PVC Composites. *Bioresources*, 3, 1228–1242, 2008.
18. Mousavi, S.R., Estaji, S., Paydayesh, A., Arjmand, M., Jafari, S.H., Nouranian, S., Khonakdar, H.A., A review of recent progress in improving the fracture toughness of epoxy-based composites using carbonaceous nanofillers. *Polym. Compos.*, 43, 1871–1886, 2022.
19. Qin, J., Wang, C., Yao, Z., Ma, Z., Gao, Q., Wang, Y., Wang, Q., Wei, H., Growing carbon nanotubes on continuous carbon fibers to produce composites with improved interfacial properties: A step towards commercial production and application. *Compos. Sci. Technol.*, 211, 108870, 2021.
20. Petrény, R. and Mészáros, L., Moisture dependent tensile and creep behaviour of multi-wall carbon nanotube and carbon fibre reinforced, injection moulded polyamide 6 matrix multi-scale composites. *J. Mater. Res. Technol.*, 16, 689–699, 2022.
21. Sayam, A., Rahman, A.N.M.M., Rahman, M.S., Smriti, S.A., Ahmed, F., Rabbi, M.F., Hossain, M., Faruque, M.O., A review on carbon fiber-reinforced hierarchical composites: mechanical performance, manufacturing process, structural applications and allied challenges. *Carbon Lett.*, 32, 1173–1205, 2022.
22. Hajibeygi, M., Maleki, M., Shabanian, M., Ducos, F., Vahabi, H., New polyvinyl chloride (PVC) nanocomposite consisting of aromatic polyamide and chitosan modified ZnO nanoparticles with enhanced thermal stability, low heat release rate and improved mechanical properties. *Appl. Surf. Sci.*, 439, 1163–1179, 2018.
23. Tsai, J.-L. and Wu, M.-D., Organoclay Effect on Mechanical Responses of Glass/Epoxy Nanocomposites. *J. Compos. Mater.*, 42, 553–568, 2008.

24. Mallakpour, S. and Javadpour, M., Design and characterization of novel poly(vinyl chloride) nanocomposite films with zinc oxide immobilized with biocompatible citric acid. *Colloid. Polym. Sci.*, 293, 2565–2573, 2015.
25. Saidi, M.A.A., Ahmad, M., Arjmandi, R., Hassan, A., Rahmat, A.R., 10 - The Effect of Titanate Coupling Agent on Water Absorption and Mechanical Properties of Rice Husk Filled Poly(vinyl Chloride) Composites, in: *Natural Fibre Reinforced Vinyl Ester and Vinyl Polymer Composites*, S.M. Sapuan, H. Ismail, E.S. Zainudin (Eds.), pp. 197–210, Woodhead Publishing, Cambridge, UK, 2018.
26. Mukhopadhyay, S. and Fangueiro, R., Physical Modification of Natural Fibers and Thermoplastic Films for Composites — A Review. *J. Thermoplast. Compos. Mater.*, 22, 135–162, 2009.
27. Nuñez-Briones, A.G., Benavides, R., Mendoza-Mendoza, E., Martínez-Pardo, M.E., Carrasco-Abrego, H., Kotzian, C., Saucedo-Zendejo, F.R., García-Cerda, L.A., Preparation of PVC/Bi2O3 composites and their evaluation as low energy X-Ray radiation shielding. *Radiat. Phys. Chem.*, 179, 109198, 2021.
28. Elsad, R.A., Mahmoud, K.A., Rammah, Y.S., Abouhaswa, A.S., Fabrication, structural, optical, and dielectric properties of PVC-PbO nanocomposites, as well as their gamma-ray shielding capability. *Radiat. Phys. Chem.*, 189, 109753, 2021.
29. Abdolahzadeh, T., Morshedian, J., Ahmadi, S., Ay, M.R., Mohammadi, O., Introducing a novel Polyvinyl chloride/Tungsten composites for shielding against gamma and X-ray radiations. *Iran. J. Nucl. Med.*, 29, 58–64, 2021.
30. Paul, S.A., Joseph, K., Mathew, G.D.G., Pothen, L.A., Thomas, S., Influence of polarity parameters on the mechanical properties of composites from polypropylene fiber and short banana fiber. *Composites, Part A*, 41, 1380–1387, 2010.
31. Ramesh, M., Kenaf (Hibiscus cannabinus L.) fibre based bio-materials: A review on processing and properties. *Prog. Mater Sci.*, 78-79, 1–92, 2016.
32. Sliwa, F., El Bounia, N.-e., Charrier, F., Marin, G., Malet, F., Mechanical and interfacial properties of wood and bio-based thermoplastic composite. *Compos. Sci. Technol.*, 72, 1733–1740, 2012.
33. Nair, K.C.M. and Thomas, S., Effect of interface modification on the mechanical properties of polystyrene-sisal fiber composites. *Polym. Compos.*, 24, 332–343, 2003.
34. Ashik, K. and Sharma, R., A Review on Mechanical Properties of Natural Fiber Reinforced Hybrid Polymer Composites. *J. Miner. Mater. Charact. Eng.*, 03, 420–426, 2015.
35. Chen, M.-J., Meister, J.J., Gunnells, D.W., Gardner, D.J., A process for coupling wood to thermoplastic using graft copolymers. *Adv. Polym. Technol.*, 14, 97–109, 1995.
36. Burgstaller, C. and Renner, K., Recycling of Wood–Plastic Composites—A Reprocessing Study. *Macromol.*, 3, 754–765, 2023.

37. Augier, L., Sperone, G., Vaca-Garcia, C., Borredon, M.-E., Influence of the wood fibre filler on the internal recycling of poly(vinyl chloride)-based composites. *Polym. Degrad. Stab.*, 92, 1169–1176, 2007.
38. Kabir, H., Gafur, M., Ahmed, F., Begum, F., Qadir, M., Investigation of Physical and Mechanical Properties of Bamboo Fiber and PVC Foam Sheet Composites. *Universal J. Mater. Sci.*, 2, 119–124, 2014.
39. Abdellah Ali, S.F., Althobaiti, I.O., El-Rafey, E., Gad, E.S., Wooden Polymer Composites of Poly(vinyl chloride). *Olive Pits Flour Precipitated Bio-Calcium Carbonate ACS Omega*, 6, 23924–23933, 2021.
40. Lakhdar, A., Moumen, A., Laabid, Z., Mansouri, K., Mechanical Behavior of Flexible Recycled PVC Bio-Loaded By 10% Chicken Feathers. *Int. J. Tech. Phys. Probl. Eng.*, 13, 14–19, 2021.
41. Skórczewska, K., Lewandowski, K., Szewczykowski, P., Wilczewski, S., Szulc, J., Stopa, P., Nowakowska, P., Waste Eggshells as a Natural Filler for the Poly(Vinyl Chloride) Composites. *Polymers*, 14, 4372, 2022.
42. Madera-Santana, T.J., Torres, A.C., Lucero, A.M., Extrusion and mechanical characterization of PVC-leather fiber composites. *Polym. Compos.*, 19, 431–439, 1998.
43. Bendjaouahdou, C. and Aidaoui, K., Synthesis and characterization of polyvinyl chloride/wood flour/organoclay ternary composites. *Polym. Polym. Compos.*, 29, S949–S958, 2021.
44. Jiang, H., Pascal Kamdem, D., Bezubic, B., Ruede, P., Mechanical properties of poly(vinyl chloride)/wood flour/glass fiber hybrid composites. *J. Vinyl Add. Tech.*, 9, 138–145, 2003.
45. Zong, G., Hao, J., Hao, X., Fang, Y., Song, Y., Wang, H., OU, R., Wang, Q., Enhancing the flame retardancy and mechanical properties of veneered wood flour/polyvinyl chloride composites. *Polym. Compos.*, 41, 848–857, 2020.
46. Jiang, L., He, C., Fu, J., Xu, D., Enhancement of wear and corrosion resistance of polyvinyl chloride/sorghum straw-based composites in cyclic sea water and acid rain conditions. *Constr. Build. Mater.*, 223, 133–141, 2019.
47. Wang, L. and He, C., Effects of rice husk fibers on the properties of mixed-particle-size fiber-reinforced polyvinyl chloride composites under soil accelerated aging conditions. *J. Eng. Fibers Fabr.*, 14, 1558925019879288, 2019.
48. Petchwattana, N. and Sanetuntikul, J., Static and Dynamic Mechanical Properties of Poly (vinyl chloride) and Waste Rice Husk Ash Composites Compatibilized with γ-aminopropyltrimethoxysilane. *Silicon*, 10, 287–292, 2018.
49. Dutta, N. and Maji, T.K., Valorization of waste rice husk by preparing nanocomposite with polyvinyl chloride and montmorillonite clay. *J. Thermoplast. Compos. Mater.*, 34, 801–816, 2021.
50. Kazi, M.-K., Eljack, F., Mahdi, E., Predictive ANN models for varying filler content for cotton fiber/PVC composites based on experimental load displacement curves. *Compos. Struct.*, 254, 112885, 2020.

51. Li, Y.-Y., Wang, B., Wang, B., Ma, M.-G., The enhancement performances of cotton stalk fiber/PVC composites by sequential two steps modification. *J. Appl. Polym. Sci.*, 135, 46090, 2018.
52. Mahdi, E. and Dean, A., The Effect of Filler Content on the Tensile Behavior of Polypropylene/Cotton Fiber and poly(vinyl chloride)/Cotton Fiber Composites. *Materials*, 13, 753, 2020.
53. Zhang, K., Cui, Y., Yan, W., Thermal and three-body abrasion behaviors of alkali-treated eucalyptus fiber reinforced polyvinyl chloride composites. *BioResources*, 14, 1229–1240, 2019.
54. Nayak, S., Mohanty, J.R., Samal, P.R., Nanda, B.K., Polyvinyl Chloride Reinforced with Areca Sheath Fiber Composites—An Experimental Study. *J. Nat. Fibers*, 17, 781–792, 2020.
55. Youssef, A.M., Abd El-Aziz, M.E., Abouzeid, R.E., A morphological and mechanical analysis of composites from modified bagasse fibers and recycled polyvinyl chloride. *Polym. Compos.*, 43, 2878–2887, 2022.
56. Zheng, Y.-T., Cao, D.-R., Wang, D.-S., Chen, J.-J., Study on the interface modification of bagasse fibre and the mechanical properties of its composite with PVC. *Composites, Part A*, 38, 20–25, 2007.
57. Shen, T., Li, M., Zhang, B., Zhong, L., Lin, X., Yang, P., Li, M., Zhuang, W., Zhu, C., Ying, H., Enhanced Mechanical Properties of Polyvinyl Chloride-Based Wood–Plastic Composites With Pretreated Corn Stalk. *Front. Bioeng. Biotechnol.*, 9, 2022.
58. Edeerozey, A.M.M., Akil, H.M., Azhar, A.B., Ariffin, M.I.Z., Chemical modification of kenaf fibers. *Mater. Lett.*, 61, 2023–2025, 2007.
59. Abdul Majid, R., Ismail, H., Mat Taib, R., Processing, Tensile, and Thermal Studies of Poly(Vinyl Chloride)/Epoxidized Natural Rubber/Kenaf Core Powder Composites with Benzoyl Chloride Treatment. *Polym. Plast. Technol. Eng.*, 57, 1507–1517, 2018.
60. Mirowski, J., Oliwa, R., Oleksy, M., Tomaszewska, J., Ryszkowska, J., Budzik, G., Poly(vinyl chloride) Composites with Raspberry Pomace Filler. *Polymers*, 13, 1079, 2021.
61. Ge, S., Zuo, S., Zhang, M., Luo, Y., Yang, R., Wu, Y., Zhang, Y., Li, J., Xia, C., Utilization of decayed wood for polyvinyl chloride/wood flour composites. *J. Mater. Res. Technol.*, 12, 862–869, 2021.
62. Khaleghi, M., Experimental and computational study of thermal behavior of PVC composites based on modified eggshell biofiller for UPVC product. *J. Polym. Res.*, 29, 2, 2021.
63. Huo, Z., Wu, H., Song, Q., Zhou, Z., Wang, T., Xie, J., Qu, H., Synthesis of zinc hydroxystannate/reduced graphene oxide composites using chitosan to improve poly(vinyl chloride) performance. *Carbohydr. Polym.*, 256, 117575, 2021.
64. Song, Q., Wu, H., Liu, H., Wang, T., Meng, W., Qu, H., Chitosan-regulated inorganic oxyacid salt flame retardants: preparation and application in PVC composites. *J. Therm. Anal. Calorim.*, 146, 1629–1639, 2021.

65. Gohatre, O.K., Biswal, M., Mohanty, S., Nayak, S.K., Effect of silane treated fly ash on physico-mechanical, morphological, and thermal properties of recycled poly(vinyl chloride) composites. *J. Appl. Polym. Sci.*, 138, 50387, 2021.
66. Gohatre, O.K., Biswal, M., Mohanty, S., Nayak, S.K., Study on thermal, mechanical and morphological properties of recycled poly(vinyl chloride)/fly ash composites. *Polym. Int.*, 69, 552–563, 2020.
67. Maou, S., Meghezzi, A., Grohens, Y., Meftah, Y., Kervoelen, A., Magueresse, A., Effect of various chemical modifications of date palm fibers (DPFs) on the thermo-physical properties of polyvinyl chloride (PVC)–high-density polyethylene (HDPE) composites. *Ind. Crops Prod.*, 171, 113974, 2021.
68. Awad, S., Hamouda, T., Midani, M., Zhou, Y., Katsou, E., Fan, M., Date palm fibre geometry and its effect on the physical and mechanical properties of recycled polyvinyl chloride composite. *Ind. Crops Prod.*, 174, 114172, 2021.
69. Wang, H., Chang, R., Sheng, K.-c., Adl, M., Qian, X.-q., Impact Response of Bamboo-Plastic Composites with the Properties of Bamboo and Polyvinylchloride (PVC). *J. Bionic Eng.*, 5, 28–33, 2008.
70. Ferreira, R.L., Furtado, C.R.G., Visconte, L.L.Y., Leblanc, J.L., Optimized Preparation Techniques for PVC–Green Coconut Fiber Composites. *Int. J. Polym. Mater. Polym. Biomater.*, 55, 1055–1064, 2006.
71. Balzer, P., Vicente, L.L., Briesemeister, R., Becker, D., Soldi, V., Rodolfo Jr., A., Feltran, M., Study of the mechanical properties of PVC/banana fiber composites. *Polimeros*, 17, 1–4, 2007.
72. Fiber, E., L'effet de l'interface dans les composites de fibres naturelles et de matières plastiques. *Rev. Roum. Chim.*, 52, 409–414, 2007.
73. Kanesawa, Y., Tanahashi, N., Doi, Y., Saito, T., Enzymatic degradation of microbial poly(3-hydroxyalkanoates). *Polym. Degrad. Stab.*, 45, 179–185, 1994.
74. Bahari, K., Mitomo, H., Enjoji, T., Yoshii, F., Makuuchi, K., Degradability of poly(3-hydroxybutyrate) and its copolymer grafted with styrene by radiation. *Polym. Degrad. Stab.*, 61, 245–252, 1998.
75. Abu Bakar, A. and Baharulrazi, N., Mechanical Properties of Benzoylated Oil Palm Empty Fruit Bunch Short Fiber Reinforced Poly(vinyl chloride) Composites. *Polym. Plast. Technol. Eng.*, 47, 1072–1079, 2008.
76. Joseph, P.V., Rabello, M.S., Mattoso, L.H.C., Joseph, K., Thomas, S., Environmental effects on the degradation behaviour of sisal fibre reinforced polypropylene composites. *Compos. Sci. Technol.*, 62, 1357–1372, 2002.
77. Ray, D. and Sarkar, B.K., Characterization of alkali-treated jute fibers for physical and mechanical properties. *J. Appl. Polym. Sci.*, 80, 1013–1020, 2001.
78. Mishra, S.B., Mishra, A.K., Kaushik, N.K., Khan, M.A., Study of performance properties of lignin-based polyblends with polyvinyl chloride. *J. Mater. Process. Technol.*, 183, 273–276, 2007.

79. Canché-Escamilla, G., Cauich-Cupul, J.I., Mendizábal, E., Puig, J.E., Vázquez-Torres, H., Herrera-Franco, P.J., Mechanical properties of acrylate-grafted henequen cellulose fibers and their application in composites. *Composites, Part A*, 30, 349–359, 1999.
80. Altenhofen da Silva, M., Adeodato Vieira, M.G., Gomes Maçumoto, A.C., Beppu, M.M., Polyvinylchloride (PVC) and natural rubber films plasticized with a natural polymeric plasticizer obtained through polyesterification of rice fatty acid. *Polym. Test.*, 30, 478–484, 2011.
81. Hamida, B., Mazouzi, D.E., Guerira, B., Effect of Chemicals Treatments on the Morphological, Mechanical, Thermal and Water Uptake Properties of Polyvinyl Chloride/Palm Fibers Composites, Revue Des Composites Et Des Materiaux Avances. *J. Compos. Adv. Mater.*, 29, 1–8, 2019.
82. Jiang, L.P., He, C.X., Li, X.L., Fu, J.J., Wear properties of wood-plastic composites pretreated with a stearic acid-palmitic acid mixture before exposure to degradative water conditions. *Bioresources*, 13, 3817–3831, 2018.
83. Ouarhim, W., Zari, N., Bouhfid, R., Qaiss, A.e.k., 3 - Mechanical performance of natural fibers–based thermosetting composites, in: *Mechanical and Physical Testing of Biocomposites, Fibre-Reinforced Composites and Hybrid Composites*, M. Jawaid, M. Thariq, N. Saba (Eds.), pp. 43–60, Woodhead Publishing, Cambridge, UK, 2019.
84. Hamida, B., Mazouzi, D.E., Guerira, B., Effect of Chemicals Treatments on the Morphological, Mechanical, Thermal and Water Uptake Properties of Polyvinyl Chloride/ Palm Fibers Composites. *Rev. Des. Compos. Des. Matériaux Avancés J. Compos. Adv. Mater.*, 29, 1–8, 2019.
85. Ye, H., Yang, Y., Jiang, J., Sun, H., Zhang, L., Ge, S., Zhang, Y., Zhou, Y., Liew, R.K., Zhang, Z., Preparation of high-performance waste polyvinyl chloride/ paper biocomposites by hot pressing, 2023.
86. Shah, B.L., Matuana, L.M., Heiden, P.A., Novel coupling agents for PVC/ wood-flour composites. *J. Vinyl Add. Tech.*, 11, 160–165, 2005.
87. Taib, M.N.A.M., Antov, P., Savov, V., Fatriasari, W., Madyaratri, E.W., Wirawan, R., Osvaldová, L.M., Hua, L.S., Ghani, M.A.A., Edrus, S.S.A.O.A., Chen, L.W., Trache, D., Hussin, M.H., Current progress of biopolymer-based flame retardant. *Polym. Degrad. Stab.*, 205, 110153, 2022.
88. Muessig, J., Influence of fiber fineness on the properties of natural fiber composites, in: *Proceedings of the 4th International Wood and Natural Fiber Composites Symposium*, Kassel, Germany, pp. 10–11, 2002.
89. Bui, Q., Colin, J., Nguyen, T., Mao, N., Nguyen, T., Perré, P., Mechanical and thermal properties of a biocomposite based on polyvinylchloride/epoxidized natural rubber blend reinforced with rice husk microfiller. *J. Thermoplast. Compos. Mater.*, 34, 1180–1192, 2021.
90. Nguyen, T.D., Nguyen, C.T., Tran, V.T.T., Nguyen, G.V., Le, H.V., Tran, L.D., Bach, G.L., Thai, H., Enhancement of the Thermomechanical Properties of a Fly Ash- and Carbon Black-Filled Polyvinyl Chloride Composite by Using

Epoxidized Soybean Oil as a Secondary Bioplasticizer. *Int. J. Polym. Sci.*, 2018, 8428736, 2018.

91. Skórczewska, K., Szulc, J., Lewandowski, K., Ligocka, A., Wilczewski, S., Modification of Poly(vinyl chloride) with Bio-Based Cassia Oil to Improve Thermo-Mechanical and Antimicrobial Properties. *Materials*, 16, 2698, 2023.
92. Aouat, H., Hammiche, D., Boukerrou, A., Djidjelli, H., Grohens, Y., Pillin, I., Effects of interface modification on composites based on olive husk flour. *Mater. Today Proc.*, 36, 94–100, 2021.
93. Amar, B., Salem, K., Hocine, D., Chadia, I., Juan, M.J., Study and characterization of composites materials based on polypropylene loaded with olive husk flour. *J. Appl. Polym. Sci.*, 122, 1382–1394, 2011.
94. Hammiche, D., Boukerrou, A., Djidjelli, H., Corre, Y.-M., Grohens, Y., Pillin, I., Hydrothermal ageing of alfa fiber reinforced polyvinylchloride composites. *Constr. Build. Mater.*, 47, 293–300, 2013.
95. Shen, Z., Ye, Z., Li, K., Qi, C., Effects of Coupling Agent and Thermoplastic on the Interfacial Bond Strength and the Mechanical Properties of Oriented Wood Strand–Thermoplastic Composites. *Polymers*, 13, 4260, 2021.
96. Thakur, V.K., Thakur, M.K., Raghavan, P., Kessler, M.R., Progress in Green Polymer Composites from Lignin for Multifunctional Applications: A Review. *ACS Sustainable Chem. Eng.*, 2, 1072–1092, 2014.
97. Ridho, M.R., Agustiany, E.A., Rahmi Dn, M., Madyaratri, E.W., Ghozali, M., Restu, W.K., Falah, F., Rahandi Lubis, M.A., Syamani, F.A., Nurhamiyah, Y., Hidayati, S., Sohail, A., Karungamye, P., Nawawi, D.S., Iswanto, A.H., Othman, N., Mohamad Aini, N.A., Hussin, M.H., Sahakaro, K., Hayeemasae, N., Ali, M.Q., Fatriasari, W., Lignin as Green Filler in Polymer Composites: Development Methods. *Characteristics Potential Applications Adv. Mater. Sci. Eng.*, 2022, 1363481, 2022.
98. Yong, M., Zhang, Y., Sun, S., Liu, W., Properties of polyvinyl chloride (PVC) ultrafiltration membrane improved by lignin: Hydrophilicity and antifouling. *J. Membr. Sci.*, 575, 50–59, 2019.
99. Solihat, N.N., Hidayat, A.F., Taib, M.N.A.M., Hussin, M.H., Lee, S.H., Ghani, M.A.A., Edrus, S.S.O.A., Vahabi, H., Fatriasari, W., Recent Developments in Flame-Retardant Lignin-Based Biocomposite: Manufacturing, and characterization. *J. Polym. Environ.*, 30, 4517–4537, 2022.
100. Agustiany, E.A., Rasyidur Ridho, M., M. Rahmi, D.N., Madyaratri, E.W., Falah, F., Lubis, M.A.R., Solihat, N.N., Syamani, F.A., Karungamye, P., Sohail, A., Nawawi, D.S., Prianto, A.H., Iswanto, A.H., Ghozali, M., Restu, W.K., Juliana, I., Antov, P., Kristak, L., Fatriasari, W., Fudholi, A., Recent developments in lignin modification and its application in lignin-based green composites: A review. *Polym. Compos.*, 43, 4848–4865, 2022.

101. Klapiszewski, L., Pawlak, F., Tomaszewska, J., Jesionowski, T., Preparation and Characterization of Novel PVC/Silica–Lignin Composites. *Polymers*, 7, 1767–1788, 2015.
102. Tomaszewska, J., Klapiszewski, L., Skórczewska, K., Szalaty, T., Jesionowski, T., Advanced organic-inorganic hybrid fillers as functional additives for poly(vinyl chloride). *Polimery*, 62, 19–26, 2017.
103. Klapiszewski, L., Nowacka, M., Milczarek, G., Jesionowski, T., Physicochemical and electrokinetic properties of silica/lignin biocomposites. *Carbohydr. Polym.*, 94, 345–355, 2013.
104. Huang, P., Lv, L., Liao, W., Lu, C., Xu, Z., Microstructural Properties of Cement Paste and Mortar Modified by Low Cost Nanoplatelets Sourced from Natural Materials. *Materials*, 11, 783, 2018.
105. Mindivan, F. and Göktaş, M., Production and characterization of graphene/PVC biocomposite from seashell wastes. *Mater. Today Proc.*, 27, 3119–3123, 2020.
106. Mindivan, F. and Göktaş, M., Preparation of new PVC composite using green reduced graphene oxide and its effects in thermal and mechanical properties. *Polym. Bull.*, 77, 1929–1949, 2020.
107. Ranjan, N., Chitosan with PVC polymer for biomedical applications: A bibliometric analysis. *Mater. Today Proc.*, 81, 894–898, 2023.
108. Taurino, R., Sciancalepore, C., Collini, L., Bondi, M., Bondioli, F., Functionalization of PVC by chitosan addition: Compound stability and tensile properties. *Composites, Part B*, 149, 240–247, 2018.
109. Xu, K., Li, K., Zhong, T., Guan, L., Xie, C., Li, S., Effects of chitosan as biopolymer coupling agent on the thermal and rheological properties of polyvinyl chloride/wood flour composites. *Composites, Part B*, 58, 392–399, 2014.
110. Thomas, S., Pothan, L., Cherian, B., Natural Fibre Reinforced Polymer Composites: From Macro to Nanoscale, International Journal of Materials & Product Technology. *Int. J. Mater. Prod. Technol.*, 36, 2009.
111. Bogoeva-Gaceva, G., Avella, M., Malinconico, M., Buzarovska, A., Grozdanov, A., Gentile, G., Errico, M.E., Natural fiber eco-composites. *Polym. Compos.*, 28, 98–107, 2007.
112. Oksman, K., High Quality Flax Fibre Composites Manufactured by the Resin Transfer Moulding Process. *J. Reinf. Plast. Compos.*, 20, 621–627, 2001.
113. Al Naim, A., Alnaim, N., Ibrahim, S.S., Metwally, S.M., Effect of gamma irradiation on the mechanical properties of PVC/ZnO polymer nanocomposite. *J. Radiat. Res. Appl. Sci.*, 10, 165–171, 2017.
114. Oksman, K., Wallström, L., Berglund, L., Toledo Filho, R., Morphology and mechanical properties of unidirectional sisal- epoxy composites. *J. Appl. Polym. Sci.*, 84, 2358–2365, 2002.
115. Eichhorn, S.J., Sirichaisit, J., Young, R.J., Deformation mechanisms in cellulose fibres, paper and wood. *J. Mater. Sci.*, 36, 3129–3135, 2001.

116. Ptak, M., Kaczyński, P., Wilhelm, J., Margarido, J.M.T., Marques, P.A.A.P., Pinto, S.C., Alves de Sousa, R.J., Fernandes, F.A.O., Graphene-Enriched Agglomerated Cork Material and Its Behaviour Under Quasi-Static and Dynamic Loading. *Materials*, 12, 151, 2019.
117. Azeredo, H.M.C.d., Nanocomposites for food packaging applications. *Food Res. Int.*, 42, 1240–1253, 2009.
118. Khaleghi, M., Didehban, K., Shabanian, M., Effect of new melamine-terephthaldehyde resin modified graphene oxide on thermal and mechanical properties of PVC. *Polym. Test.*, 63, 382–391, 2017.
119. Alghamdi, M.M., El-Zahhar, A.A., Asiri, B.M., Incorporation of magnetite nanoparticles in poly(vinyl chloride) microfiltration membrane for improving antifouling property and desalination performance. *Desalin. Water Treat.*, 165, 54–62, 2019.
120. El-Zahhar, A.A., Alghamdi, M.M., Asiri, B.M., Poly (vinyl chloride)-MMT composite membranes with enhanced properties and separation performance. *Desalin. Water Treat.*, 155, 381–389, 2019.
121. Mallakpour, S. and Javadpour, M., Comprehensive study on reinforcement of poly(vinyl chloride) nanocomposite films with ZnO nanoparticles modified by citric acid and vitamin C. *Int. J. Polym. Anal. Charact.*, 23, 415–429, 2018.
122. Yang, T.-C., Noguchi, T., Isshiki, M., Wu, J.-H., Effect of titanium dioxide on chemical and molecular changes in PVC sidings during QUV accelerated weathering. *Polym. Degrad. Stab.*, 104, 33–39, 2014.
123. Wang, H., Wang, Y., Fu, F., Qian, Y., Xiao, Y., Yang, D., Qiu, X., Controlled preparation of lignin/titanium dioxide hybrid composite particles with excellent UV aging resistance and its high value application. *Int. J. Biol. Macromol.*, 150, 371–379, 2020.
124. Amina, M., Al Musayeib, N.M., Alarfaj, N.A., El-Tohamy, M.F., Orabi, H.E., Bukhari, S.I., Mahmoud, A.Z., Exploiting the Potential of Moringa oleifera Oil/Polyvinyl Chloride Polymeric Bionanocomposite Film Enriched with Silver Nanoparticles for Antimicrobial Activity. *Int. J. Polym. Sci.*, 2019, 5678149, 2019.
125. Ghozali, M. and Haryono, A., Effect of Size of Cellulose Particle as Filler in the PVC Biocomposites on Their Thermal and Mechanical Properties. *Mater. Sci. Forum*, 737, 67–73, 2013.
126. Benderly, D., Osorio, F., IJdo, W.L., PVC nanocomposites—Nanoclay chemistry and performance. *J. Vinyl Add. Tech.*, 14, 155–162, 2008.
127. Fu, H., Yan, C., Zhou, W., Huang, H., Preparation and characterization of a novel organic montmorillonite/fluorinated waterborne polyurethane nanocomposites: Effect of OMMT and HFBMA. *Compos. Sci. Technol.*, 85, 65–72, 2013.
128. Abdollahi Boraei, S.B., Bakhshandeh, B., Mohammadzadeh, F., Haghighi, D.M., Mohammadpour, Z., Clay-reinforced PVC composites and nanocomposites. *Heliyon*, 10, e29196, 2024.

129. Dutta, N. and Maji, T.K., Synergic effect of montmorillonite and microcrystalline cellulose on the physicochemical properties of rice husk/PVC composite. *SN Appl. Sci.*, 2, 439, 2020.
130. Son, D., Lee, J., Kim, S.K., Hong, J., Jung, H., Shim, J.K., Kang, D., Effect of cellulose nanofiber-montmorillonite hybrid filler on the melt blending of thermoplastic starch composites. *Int. J. Biol. Macromol.*, 254, 127236, 2024.
131. Wieczorek, M. and Tomaszewska, J., Poly(vinyl chloride) composites reinforced with wood flour and calcinated halloysite. *Wood Mater. Sci. Eng.*, 16, 2024.
132. Yang, R.F., Fang, Y.Q., Zhu, J.H., Ren, G.L., Lin, B.H., Wang, F.Q., Ou, R.X., Wang, Q.W., Song, Y.M., Fabrication ZnO functionalized nano-montmorillonite for enhancing the fire-safety and mechanical properties of wood flour/poly (vinyl chloride) composites. *Constr. Build. Mater.*, 420, 10, 2024.
133. Izadpanah, R., Rezaei, M., Talebi, S., Preparation of cross-linked PVC-g-(St-MA) copolymer/organoclay nanocomposites and investigation of chemical characteristics, morphology, thermal, and mechanical properties. *J. Thermoplast. Compos. Mater.*, 35, 211–229, 2022.
134. Yim, H. and Kim, D.S., Physical properties of PVC/aminosilane-treated wood flour/organoclay composites. *Polym. Adv. Technol.*, 23, 1441–1445, 2012.
135. Ismail, A.H.M., Al-Oqla, F.M., Risby, M.S., Sapuan, S.M., On the enhancement of the fatigue fracture performance of polymer matrix composites by reinforcement with carbon nanotubes: a systematic review. *Carbon Lett.*, 32, 727–740, 2022.
136. Lemartinel, A., Castro, M., Fouché, O., De-Luca, J.-C., Feller, J.-F., A Review of Nanocarbon-Based Solutions for the Structural Health Monitoring of Composite Parts Used in Renewable Energies. *J. Compos. Sci.*, 6, 32, 2022.
137. Zheng, H., Zhang, W., Li, B., Zhu, J., Wang, C., Song, G., Wu, G., Yang, X., Huang, Y., Ma, L., Recent advances of interphases in carbon fiber-reinforced polymer composites: A review. *Composites, Part B*, 233, 109639, 2022.
138. Zakaria, M.R., Md Akil, H., Abdul Kudus, M.H., Ullah, F., Javed, F., Nosbi, N., Hybrid carbon fiber-carbon nanotubes reinforced polymer composites: A review. *Composites, Part B*, 176, 107313, 2019.
139. Farsheh, A.T., Talaeipour, M., Hemmasi, A.H., Khademieslam, H., Ghasemi, I., Investigation On The Mechanical And Morphological Properties Of Foamed Nanocomposites Based On Wood Flour/Pvc/Multi-Walled Carbon Nanotube. *Bioresources*, 6, 841–852, 2011.
140. Shieh, Y.-T., Hsieh, K.-C., Cheng, C.-C., Carbon nanotubes stabilize poly(vinyl chloride) against thermal degradation. *Polym. Degrad. Stab.*, 144, 221–230, 2017.

141. Skórczewska, K., Lewandowski, K., Wilczewski, S., Novel Composites of Poly(vinyl chloride) with Carbon Fibre/Carbon Nanotube Hybrid Filler. *Materials*, 15, 5625, 2022.
142. Dutta, N. and Maji, T.K., Development of waste rice husk/PVC/GO nanocomposite using TA-CaO adduct and ESO as green additives. *J. Thermoplast. Compos. Mater.*, 36, 2212–2232, 2023.
143. Zhu, A., Cai, A., Zhang, J., Jia, H., Wang, J., PMMA-grafted-silica/PVC nanocomposites: Mechanical performance and barrier properties. *J. Appl. Polym. Sci.*, 108, 2189–2196, 2008.

6

Sustainable PVC Biocomposites and Bionanocomposites: A Path Towards Green Innovation

Shiv Kumari Panda

Department of Chemistry, U.N. Autonomous College of Science & Technology, Adaspur, Cuttack, Odisha, India

Abstract

Because of ecological sustainability issues, this current period has perceived remarkable accomplishments in the field of biocomposite and bionanocomposite developments. In the latest discoveries of biocomposites, PVC-based polymeric materials are gaining rapid attention due to its high applicability in numerous sectors. Thus, this chapter can showcase the recent progresses in using various reinforcing biofillers and bionanofillers from several sources and types for preparing PVC biomaterials. Polyvinyl chloride is a unique, mostly useful, durable, thermoplastic material having good water absorption, chemical resistance, ultraviolet, and fire resistance. But it is nonbiodegradable in nature. So, its modification with some bio-based material and natural nanomaterial for the formation of corresponding hybrid biocomposites and bionanocomposites can positively modify its nature from nonbiodegradable to biodegradable in an innovative way.

Keywords: PVC, biofillers, reinforcement, bionanofillers, biocomposites, bionanocomposites

6.1 Introduction

Non-biodegradable, synthetic, thermoplastic polymer such as polyvinyl chloride is ornamented with various physical characteristics like stability, versatility, and inexpensiveness, and it is the third most globally available synthetic polymer next to polypropylene and polyethylene. It is highly

Email: shivkumaripanda@gmail.com

Shiv Kumari Panda and Visakh P. M (eds.) Polyvinyl Chloride-Based Biocomposites and Bionanocomposites, (163–184)

valued for its good mechanical strength, thermal characteristics, good weather and chemical resistance, flexibility, and inexpensiveness. For that reason, it has wide varieties of applications starting from several automobile industries to biomedical fields. It is also highly used in the field of cables, house sliding, pipes, and fibers. It possesses film-forming capacity, and because of this, it is used in various rigid and flexible molded plastics.

But virgin PVC exhibits some bottlenecks, and the vital problems associated with this is its less toughness and minimum thermal-related stability while being introduced to excess temperatures through the process of molding and during application in comparison to other contemporary valuable polymers like polypropylene, polyethylene, and polystyrene [1, 2]. Actually, PVC is the final consequence of polymerization of many vinyl chloride monomers that is considered as the main building block of PVC. So, the pure PVC polymer consists of many repeating functionalities of $-CH_2-CHCl-$, where each particular monomer provides a chlorine atom to the polymeric backbone. Only this chlorine content imparts exceptional features to PVC by differentiating it from other synthetic polymers. The existences of unstable structural moieties like adjoining carbon-chlorine and carbon-hydrogen linkages or some imperfections present in the polymeric backbone are the main reasons behind this less thermal stability of PVC [3]. So, it is necessary to renovate this PVC to its corresponding conjugated polyene bonds by means of de-hydrochlorination processes due to this more chlorine content and less thermal stability [4]. These are the main reasons of overall instability as well as nonbiodegradability of PVC.

However, now-a-days, there are various concerns that rise by considering the environmental impacts as well as the performance of PVC. So, the synthesis of biocomposite and bionanocomposite by introducing natural fibers, biofillers, and bionanofillers is the solution for this. According to current trends, many research works have been published related to the petroleum-based polymers that resulted to biobased composites, nanocomposites, and their applications [5–11]. Similarly, a number of new research works were concentrated on the successful synthesis of PVC biocomposites and bionanocomposites with improved efficiency. Because of this reason, many reinforcing fillers and additives have been used to enhance various properties of PVC like thermal degradation temperature, fire retardancy and combustion, barrier properties, mechanical stability and toughness, hardness, stiffness, dimensional stability, and mold contraction [1, 12]. Integration of biodegradable nanofillers in PVC resin matrix and synthesis of PVC bionanocomposites has been a suitably valuable technique for increasing the overall characteristics as well as thermal stability of PVC [13]. Again interestingly, different reinforcing inorganic biofiller additives like zinc oxide, graphene oxide, antimony, and zinc borate can show their efficiency during the production process for the improvement of the physical as well as chemical characteristics of PVC like thermal and mechanical stability and fire resistance [14–16].

After all this, preparation of PVC-based biocomposites follows the 12 well-known principles of green chemistry with special emphasis on the role and impact of various bio-based reinforcing materials as a biodegradable matrix component on this biocomposite performance. This chapter showcases the forthcoming outlooks in the field of biocomposites from petroleum-based polymer like PVC. The 12 principles of green chemistry give priority for preventing waste or reintroduce the same as the initial starting material for new innovative and revolutionary creation, which is far more sustainable than cleaning it up afterward. In line with this philosophy, it is crucial to modify synthetic materials like PVC to their corresponding biodegradable form by using several biopolymers to improve their environmental impact.

6.2 Fundamentals of PVC-Based Composites

6.2.1 Molecular Structure of PVC

Molecular arrangement of PVC shows a very continuous and linear arrangement of several carbon atoms as backbone chain that contain hydrogen and chlorine moieties at alternating position. As we know, chlorine has higher electronegativity than hydrogen, so this is the main reason behind the unique properties in this widely used polymer. In case of PVC, the vinyl chloride repeating units can be interlinked through free radical polymerization technique in the presence of an initiator like benzoyl peroxide to produce a long-chain polymer with 500 to 2000 repeating entities. Again, the presence and specific arrangement of chlorine atoms in the backbone chain decides the tacticity of the chain. The importance of this tacticity is that it decides the structure and characteristics of the chain. Tacticity is of three types given in Figure 6.1 like atactic, syndiotactic and isotactic. Among these three types of PVC, syndiotactic types have the highest degree of crystallinity, and here, the chlorine functionalities are present in alternate positions of the polymer backbone in an ordered pattern. Whereas in case of isotactic PVC, the chlorine atoms are present in the similar side of the carbon backbone chain and impart a partly crystalline structure. Finally, in the atactic PVC, the building functional blocks

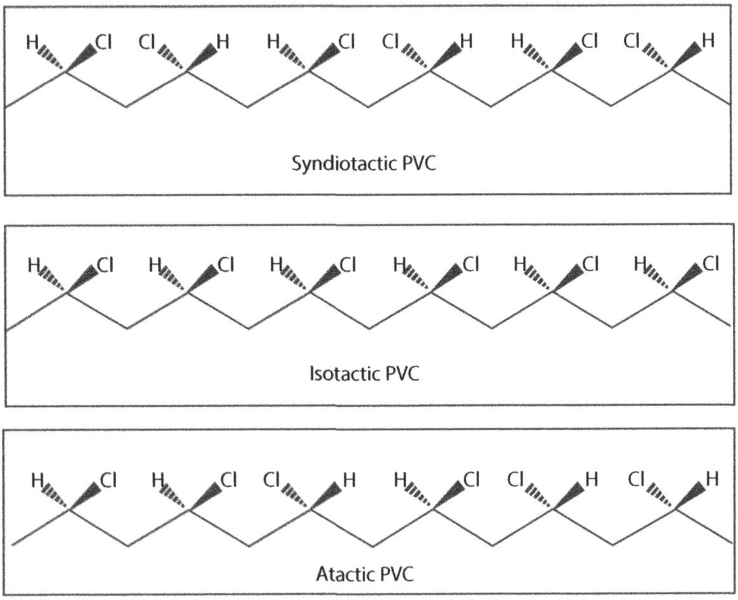

Figure 6.1 Types of PVC through the arrangement of chlorine atoms.

are arranged in a random manner and results in an amorphous structure. Among all these structures, atactic polymers are mostly used due to the unsystematic arrangement of chlorine groups which makes the polymeric chain more flexible with restricted crystallinity and amorphous arrangement. So, a commonly used PVC polymer is atactic and considered as amorphous in nature that limits the mechanical and thermal properties.

But after all, PVC has many weaknesses like various environmental concerns, stiffness and brittleness, restricted recyclability, toxic and corrosive chemical discharge, thermal instability and temperature intolerance, high flammability, absence of biodegradability, and low mechanical properties that confine its application in a number of ways in numerous sectors. Simply, it shows lack of flexibility which is the main requirement for many applications. So in whole, to avoid such type of bottleneck, this PVC has to be transformed to its corresponding composite, biocomposites, and bionanocomposite that can solve the problem and enhance its candidature for application in several industries. It can be modified by introducing additives, fillers, and plasticizers in the main polymer. Actually, on this basis also, it is categorized into two types of broad classes as UPVC (rigid PVC or unplasticized PVC) that comprises little, i.e., less than 10% or no plasticizers and FPVC (flexible PVC or plasticized PVC) that contains about

30–70% amount of plasticizers along with stabilizers and other additives [17]. Also, there are less significant groups of PVC available that take account of chlorinated PVC that has higher chlorine content with molecularly oriented PVC as well as modified PVC [17]. Moreover, combining these petroleum-based polymers like PVC with bio-based resources can drive the research for the development of commonly used economical manufactured goods with diverse applications [18].

6.2.2 PVC-Based Biocomposites

Generally, petroleum-based polymeric materials which are technically modified by renewable resources by direct or indirect way through combination are called as eco-friendly materials or biomaterials or green materials that are gaining wide-reaching importance in the present day [19]. Nonbiodegradable, synthetic, petroleum-based PVC polymeric resin can be modified through various ways. So, for this reason, only most of the polymers are introduced to some reinforcing filler and suitable additive that can process well while providing the necessary physical and environmental performance as well as applicability [20]. Fillers can develop the performance of the polymeric envelope by permitting it to compete against other resins as well as it can create new revolution in different areas of polymer industry with high applicability. So, these fillers are used extensively for many unique and novel invention and discoveries in polymeric field. They can accelerate the overall performance of the polymeric matrix, increase rigidity and strength, and expand the application areas. But biofillers are an exceptional candidate in nature and mostly acceptable by polymeric community as they can enhance the properties as well as the biodegradability of the resulted biocomposites [21]. Various biopolymers, such as cellulose, starch, chitin, chitosan, and lignin, are being extensively used as a biofiller in the PVC matrix [22]. Poly(vinyl chloride) is a non-biodegradable and synthetic polymer which can go through limitless physical and chemical modifications by using extremely simple procedures, according to specific necessities associated to its proposed purpose of use [23, 24]. In 2021, the global poly(vinyl chloride) capacity was 60.27 MPTA [25], and the demand for this raw material is constantly increasing. By considering the environmental issues, the shortcoming of synthetic plastics arises in the form of their non-biodegradability in a natural environment for many years [26]. So intermixing of these artificial polymers with appropriate natural biopolymers may lead to convert their properties from non-biodegradable to biodegradable configurations [27]. However, commonly, only the biopolymer part undergoes disintegration

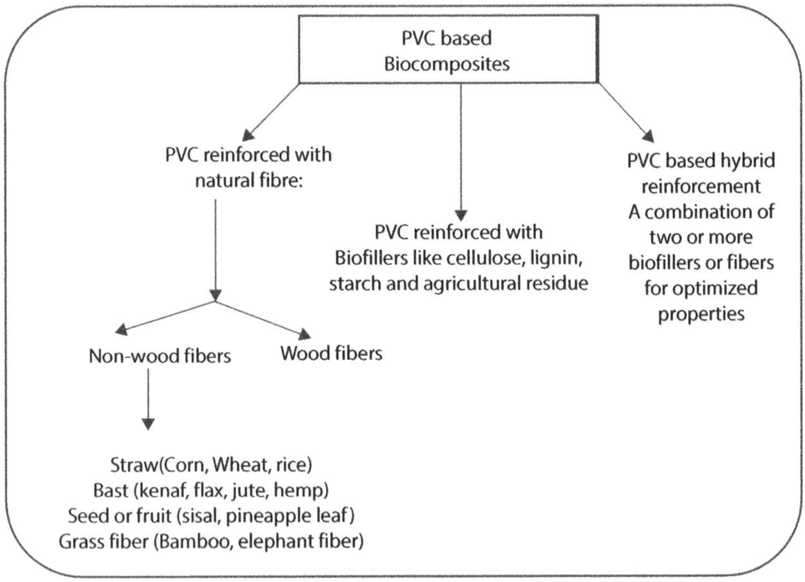

Figure 6.2 Various types of PVC-based biocomposites.

through the breakdown process activated by microorganisms, whereas the synthetic constituents only become decomposed into small fragments and dissipated [28–31]. So, now more focus is coming over the preparation of biocomposites and bionanocomposites by using these PVCs with some fibers or fillers. PVC may be successfully reinforced with natural fibers or a number of biofillers or through various hybrid reinforcements and called as PVC biocomposites. Again, all forms of PVC based biocomposites possess unique characteristics and overall properties. This chapter discovers a comprehensive innovative idea regarding the types and properties of biocomposites and bionanocomposites reinforced with PVC and their uses in various fields with emphasizing their prospective as ecologically approachable replacements to conventional PVC materials. Figure 6.2 shows various types of PVC-based biocomposites that are generally used in various sectors on the basis of various reinforcements.

6.2.2.1 Natural Fiber–Reinforced PVC Biocomposites

Introduction of natural fibers such as jute, bamboo, rice straw, palm, rice husk, hemp, flax, bagasse, and various types of agricultural residues not only improves the biological effect of PVC-based biocomposites but also develop some unique specific characteristics on them [32–35]. These economical natural fibers are commonly available from the waste part of various agricultural industries. As natural fibers possess varieties of chemical compositions, physical and mechanical characteristics according to their inherited structure, inconsistency of characteristics can be anticipated from the composites resulted from them [36]. Different researchers have successfully testified the characteristics of PVC reinforced with biofiber-based biocomposites. According to latest research, there are nine different variants of natural fibers that have been used for reinforcing filler and materials. They are wood [37–39], bamboo, pine [40], rice straw [41], sisal [42], banana [43], sugarcane bagasse [44], and coconut [45].

Biodegradable fibers may be categorized into two groups: wood-based fibers and non-wood–derived fibers. Again, non-wood–based fibers can be separated into various types [46] like straw, bast, seed or fruit, and grass fibers. Interestingly, the straw type includes rice straw; bast includes jute like natural fiber; seed/fruit covers sisal and pineapple leaf; finally, grass includes bamboo fibers. The detailed examples of various non-wood fibers are given in Figure 6.2. Wood flour–based natural fibers are often used as bioadditives or reinforcing filler for preparation of natural fiber biocomposites. The characteristic values of different natural fibers like cellulose content of the fiber, the cellulosic degree of polymerization, and the reinforcement type of the fibers vary from one to another according to their structure. Generally, cellulose content of a fiber is directly proportional to the degree of polymerization and angle or orientation of the reinforced fiber as well as it affects the tensile strength and modulus in positive approach. The dissimilarities in their characteristic may directly affect the overall properties as well as mechanical properties both among different types of fibers as well as along the length of an individual fiber [47].

Basically, the natural fibers are loaded with some positive characteristics like low density, superior specific strength, and specific modulus than even synthetic glass fibers also. Hence, natural fibers are the good option to introduce in various petroleum-based synthetic polymers for enhanced overall properties [48]. The famous techniques like melt compounding and compression molding are very appropriate to prepare poly(vinyl chloride) (PVC) and natural fiber-based biocomposites [49].

But the challenge arises to prepare the natural fibers well-matched with the PVC polymer matrix to synthesize a successful biocomposite. It is only because there arises a major shortcoming for the application when they are used as reinforcements or bioadditives for various synthetic polymers. It is very common as natural fibers show high moisture absorption due to the presence of hydroxyl moieties and other polar groups on their surface. It leads to the poor wettability of the polymers and weak interfacial bonding among the hydrophilic natural fibers and hydrophobic polymers, as the matrix [50]. It is necessary to solve the problems related to the compatibility among natural fibers and the manmade polymer matrices through suitable treatments in order to prepare biocomposites with better enhanced mechanical properties [50].

In addition to this, nowadays, it is of more realistic importance to develop the impact strength for PVC/natural fiber–based biocomposites only due to the comparative lower impact resistance of the virgin PVC-based matrix. Contemporary investigation in this field is mainly concentrated on the surface treatments like addition of various coupling agents and additives and using impact modifiers to develop the biocomposite properties. The addition of this coupling agents for example like poly [methylene poly (phenyl isocyanate)] (PMPPIC) and other various types has been studied for PVC/natural fiber composites to improve their overall properties [51, 52]. Furthermore, the surface treatments by acrylic acid, maleic anhydride, alkali, and various other anhydrides and acids can also be performed for natural fibers to enhance its compatibility with PVC to prepare the biocomposites [53, 54]. There is a rule of mixture for determining the approach of the arrangement of fibers in PVC-based polymers which shows that the use of unidirectional continuous fibers can result in a good compatibility among polymer matrix and fibers. It is only because the effective transference of stress in a biocomposite can be determined largely by fiber length with orientation, stress concentration at the fiber ends, interfacial shear strength, and lastly compatibility in between the polymer matrix and the used fiber [55].

So finally, the usage of bio-based fibers as reinforcing material for PVC can successfully enhance the overall mechanical and physical properties of the resulted biocomposites. Wood-based fibers show very good reinforcement effect with overall good properties for PVC biocomposites. Because of minimum density as well as maximum specific characteristics, these PVC biocomposites have proved its candidature for application in automotive as well as furniture industry.

6.2.2.2 Biofiller-Reinforced PVC Biocomposites

Fillers are like natural, mineral, or manmade type of additive materials which are used for synthetic or petroleum-based polymeric compounds like PVC to enhance its mechanical and thermal characteristics, as well as help to achieve some particular functionalities by making the resulting biocomposites strong, durable, and inexpensive with improved texture. But reinforcing biofillers are used broadly as a sustainable and environmentally friendly substitute for PVC-based biocomposites over the conventional mineral fillers. These biofillers are resulted from different natural sources and decorated with various exceptional characteristics. This chapter deals with the introduction of reinforcing biofillers for PVC polymer to prepare the corresponding biocomposites by considering the better carbon footprint. Actually, the exact selection of biofiller can be affected by the required characteristics of the resulted PVC that can successfully prove its candidature in the resulted biocomposites decorated with cost-effectiveness, accessibility of the filler, and the specific processing techniques employed. Here, Figure 6.3 shows some various types of biofillers used in PVC-based biocomposites. General breakdown of various types of biofillers shows that these are of broadly three types based on their origin from where they are produced. They may be waste biofillers from agriculture or animal origin and mineral-based biofillers.

If we consider the agricultural origin–based biofillers, we can say that recently, waste rice husk ash is applied as a good reinforcement in polymers only because of its easy availability, inexpensiveness, and improved mechanical characteristics [56, 57]. Agricultural origin–based waste biofillers like waste rice husk ash is an abundant agrarian waste product which can also be successfully used in PVC matrix to enhance the strength and thermal stability of PVC-based biocomposites. This rice husk can be used as a filler to develop the thermal stability as well as to diminish the water absorption of PVC composites [58]. Generally, the vital parameters affecting whole properties of this biofiller and PVC polymer biocomposites are the type of filler, moisture and filler content, and interfacial connection between polymer matrix and the biofillers [59–61]. Without any

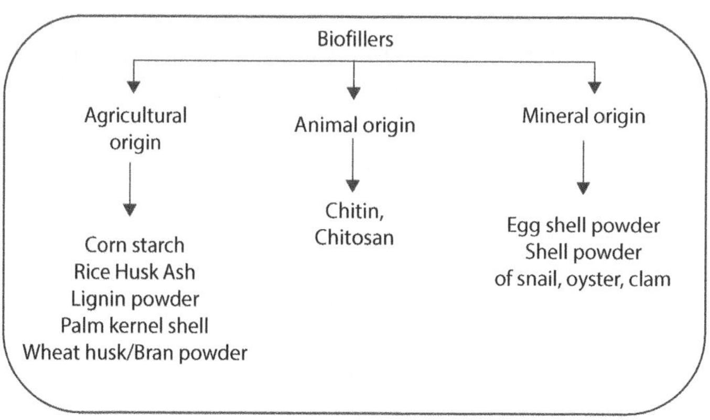

Figure 6.3 Types of biofillers for PVC-based biocomposites.

compatibilizer, this rice husk ash was generating some drawbacks and large amounts of voids in high density–based biocomposites that indicate the reduced interfacial force of attraction among the polymer matrix and biofiller [62]. This implied the reduced interfacial bonding among the used polymer resin matrix and the reinforcing biofillers. So at the present time, to recover the interfacial connection and mechanical performance of the PVC and rice husk ash in its corresponding biocomposites, γ-aminopropyl trimethoxysilane coupling agent was used as an adhesion promoter [63]. This PVA/rice husk ash–based biocomposite was synthesized by a reactive extrusion process, and the introduction of this specific coupling agent at even minimum concentrations of 1 wt% can enhance mechanical characteristics like tensile strength, impact strength, tensile modulus, and storage modulus and reduce water absorption capacity. So, the introduction of some specific suitable coupling agent in this discussed biocomposite can evidently reduce water absorption tendency as it is the major issue related to it and reduces the major properties of the biocomposites as well as its applicability.

Again, plant-based cellulose, corn starch, and lignin powder can be used as effective biofillers for this synthetic PVC polymer for the preparation of corresponding biocomposites. Actually, cellulose is one of the gifted biomaterials for the production of biodegradable composite among various reinforcing biofillers that is only because of its extraordinary structural strength. The combination of cellulose like green polymer with petroleum-based non-biodegradable PVC polymers can efficiently minimize plastic waste by biodegradation. Biocomposite materials are very pertinent and demanding for the current scenario for its cost-effectiveness and superior properties. So when this cellulose biofiller was uniformly dispersed into the polymer matrix, then the resultant biocomposite shows remarkable reinforcing effect with positive increment in thermal and mechanical properties, even at low concentration [64]. Agricultural cellulosic biofillers produced from upcycled olive pits, walnut shell powder, and other plant material–based sources can help to enhance the scratch resistance and texturing in PVC applications in various sectors [65].

Lignin is considered as the second most plentily available natural aromatic polymer originate in plant cell walls. It is an agricultural origin–based biofiller created from wood or paper pulp or bioethanol industries. It is a by-product from paper factory but can perform as a proficient UV stabilizer that can be used for increasing the thermal features of the PVC polymer [66]. But pristine lignin biopolymer is hydrophilic and cannot intermix completely with hydrophobic PVC. For that reason, some research for surface modifications and introduction of compatibilizer are commonly desirable. So, sometimes, modifications like alkylation and acetylation of this lignin can give better reinforcing effect to the PVC polymer.

Animal origin–based waste biofillers like chitin and chitosan-reinforced PVC biocomposites are successfully introduced to improve the overall thermal, physical, chemical, and mechanical properties with positive contribution to more environmentally friendly technologies [67]. But these prospects related to this chitin and chitosan for various unique applications can be extended through the preparation of novel biocomposite materials with this PVC polymer matrix being associated with biopolymers. Among the biopolymers, chitin and chitosan stand out given their natural occurrence, maximum availability, biodegradability, antimicrobial action, biocompatibility, adsorption capacity, and easiness of modification [67]. Interestingly, among the various applications of these chitosan PVC biocomposites, they have also many applications in various medical sectors due to the inherent antimicrobial effect of chitosan. This particular PVC and chitosan-based hybrid biocomposites can be successfully used for common blood bag application in clinical area [68, 69].

After all, mineral origin–based biofillers like waste eggshell powder, shell powder of snail and oyster, and bio-calcium carbonate can be used as an exceptional reinforcing biomaterial for PVC polymers [70, 71]. Such biomaterials are very rich in calcium carbonate that can help to boost rigidity, dimensional stability, thermal stability, whiteness in texture, and hardness to the resulted PVC hybrid biocomposites. Biomineral-based fillers like bio-calcium carbonate were productively used as fillers in PVC [72]. The synthesized biocomposites have possessed superior thermal, mechanical, and physical stability as well as better stiffness properties in comparison to neat PVC. It was found that the properties like tensile modulus, flexural strength, tensile strength, and hardness are increased by the combination of this mineral biofiller into the artificial PVC matrix. The consistent distribution of this bio-calcium carbonate biofiller in the PVC polymer hybrid biocomposite is the most important aspect accountable for the enhancement of the mechanical properties as well as overall characteristics [71]. The most innovative and unique advantage of using egg shell–based calcium carbonate is to transform PVC, a significant development in the thermal stability of this material [60]. The benefit of this proposed biocomposite of PVC and egg shell–based mineral biofiller can improve the overall required properties successfully with high applicability.

6.2.2.3 *PVC-Based Hybrid Biocomposites*

These are a special type of hybrid composite material that can successfully incorporate two or more types of reinforcing fillers and fibers or even two different polymer matrices. Here, this approach can help to gain the benefits of each participating material in a positive way like enhanced mechanical strength, biodegradability, thermal stability, durability, and applicability. There are various literatures of scientific studies available that successfully describe a huge number of research on these PVC-based hybrid biocomposites. Here, we can discuss the various types of PVC-based hybrid biocomposites derived from only the reinforcement of natural filler and fiber resources. For example, PVC with jute fiber and wood flour–based hybrid biocomposites can describe a good combination of performance and applicability [73]. Here, in this particular hybrid biocomposite, the chemical resistance, durability, and processability properties of PVC combines with the high mechanical property of jute fiber as primary reinforcement and dimensional stability and enhanced stiffness of wood flour as secondary reinforcement through the process of extrusion or compression molding. But there some problems that arise because of the interfacial adhesion generated among the biopolymer and synthetic polymer matrix like PVC, for which maleic anhydride coupling agents are used to improve the filler–fiber–polymer matrix interfacial adhesion. This final resulted hybrid biocomposite has a good application in various industries [73]. PVC can successfully transfer to its corresponding hybrid biocomposite in the presence of tough, durable, lignin-based coir fiber, and cellulose-based rice husk [74]. This resulted biocomposite has high mechanical and thermal properties. But here, the surface modification of the biofiber and reinforcing biofiller was performed to develop the filler–polymer matrix connection and maintain the cost and water absorption characteristics. It has great application in automotive, packaging industry, as flooring tiles, etc. In the biocomposites of PVC with reinforcing fiber and filler, the only challenge is the good combination among the biofiber and PVC. But this successful combination of two different types of lignocellulosic biofillers and biofibers can make PVC stiffer and greener with improved biodegradability.

6.2.3 PVC-Based Hybrid Bionanocomposites

In the current scenario, numerous investigations were based on the synthesis of PVC-based hybrid bionanocomposites ornamented with unique efficiency and high applicability. The primary aim of most of the researches is based on the improvement of the overall stability of PVC in contrast to de-hydrochlorination [75, 76]. Because of this reason, different types of nanofillers and additives are broadly utilized to successfully enhance the whole properties of the PVC like mechanical and dimensional stability, thermal degradation temperature, fire retardancy and hardness, stiffness, and mold contraction [12]. The introduction of more than one bio-based nanofillers in PVC polymeric resin matrix and synthesis of PVC hybrid bionanocomposites is the most beneficial technique positively intended for improving the thermal steadiness of PVC [13]. An overall idea about various bionanofillers for PVC-based bionanocomposites is given in Figure 6.4.

Generally, bionanofillers are used in various significant forms like cellulose and starch nanocrystals, nanocellulose, starch, clay, lignin and chitosan nanoparticles, clay and organically modified clays, nanosilica and nanotitanium dioxide, and finally nanofibers, i.e., fibers in nanosize. These nanofillers can develop the overall mechanical, thermal, physical, and barrier properties of the resulted bionanocomposites through biodegradability and applicability. For illustration, the successful incorporation of only 3–5wt% of reinforcing cellulose nanocrystals in PVC polymer matrix can significantly improve the thermal stability, Young's modulus by up to 45%, and tensile strength up to 40%, by setting the onset thermal degradation by approximately 30°C [77]. Similarly, bionanofillers like chitosan nanoparticles with very less concentration help for improving mechanical properties as well as transfer potent antimicrobial activity, with approximately 99% of *E. coli* and *S. aureus* bacterial reduction against some particular bacterial

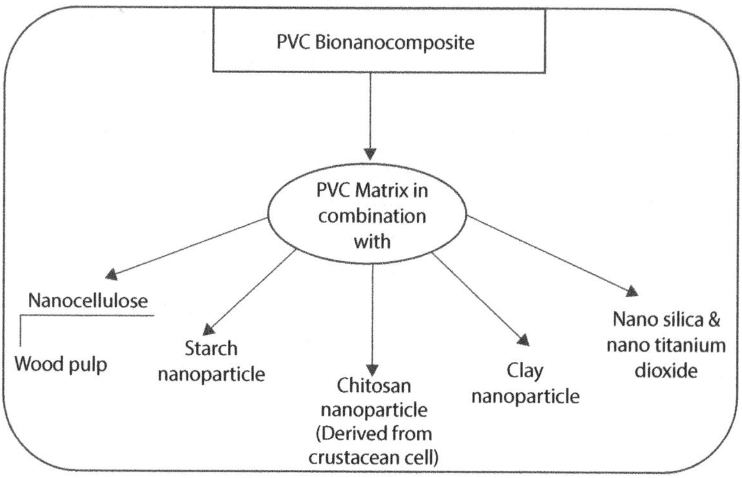

Figure 6.4 Types of bionanofillers for PVC bionanocomposites.

infection [78]. Sometimes, biopolymers like chitosan also grafted with other polymers show some better properties than the original one. When this chitosan was modified by ZnO inorganic nanofillers, the flame retardancy and optical and catalytic characteristic was observed in the resulted PVC-based bionanocomposites that can enhance the applicability of the bionanocomposites [79].

So, PVC bionanocomposites show various multifunctional enhancements that make it extremely a smart bionanomaterial for several applications as medicinal devices, food wrapping, and sustainable manufacturing industries. These positive transformations of PVC to its corresponding bionanocomposites can make it a more potential and suitable material for various new applications.

6.3 Conclusion

These revolutionary transformations of nonbiodegradable, petroleum-based PVC polymer to its corresponding bio-friendly structure through the preparation of biocomposites and bionanocomposites can allow to successfully improve various characteristics like thermal, mechanical, and biological characteristics and contribute to more environmentally friendly technologies. The usage of natural fibers, biofillers, and bionanofillers for efficient reinforcement to PVC-based plastic industry can provide an exceptional renewable resource for the source of economic improvement for agriculture in rural areas. But to implement these natural fibers for PVC and to achieve a good interfacial bonding among the reinforcing filler and polymer matrix, it is necessary to introduce some suitable modifier agents. So, these PVC-based biocomposites and bionanocomposites can prove their candidature successfully in various sectors like construction [80], electronics [81], automotive [82, 83], packaging [84, 85], optoelectronic [86–89], healthcare [90], and biodegradability by solving the limitations like thermal stability and brittleness previously associated with it. PVC-based biocomposites and bionanocomposites can create a beautiful bridge among performance and conservational responsibility. Finally, in this chapter, petroleum-based PVC synthetic polymer has proved itself as a successful biomaterial by undergoing a revolutionary change towards a path of green and high-performance material to challenge the contemporary global issues.

AI disclosure statement: The author confirms that no artificial intelligence (AI)-generated content or AI-assisted writing tools were used in the preparation of this chapter (including summaries and the abstract). All text, analysis, and conclusions are the result of the author's original work.

References

1. Yu, J., Sun, L., Ma, C., Qiao, Y., Yao, H., Thermal degradation of PVC: A review. *Waste Manage. (Oxford)*, 48, 300–314, 2016.
2. Titow, W.V., *PVC Plastics: Properties, Processing, and Applications*, Springer, Netherlands, 2012.
3. Sanchez-Jimenez, P.E., Perejon, A., Criado, J.M., Dianez, M.J., Perez-Maqueda, L.A., Kinetic model for thermal dehydrochlorination of poly (vinyl chloride). *Polymer*, 51, 3998–4007, 2010.
4. Gilbert, M., Ho, K.C., Hitt, D.J., Vrsaljko, D., Assessment of PVC stabilisation using hydrotalcites - Raman spectroscopy and other techniques. *Polym. Degrad. Stab.*, 98, 1537–1547, 2013.
5. Panda, S.K., Synthesis and Overall Migration Study of Chitosan-Encapsulated ZnO-Based ESO Bionanocomposite with Synergistic Antimicrobial Activity for Packaging Purpose. *Chem. Select*, 7, 2022. doi.org/10.1002/slct.202201433.
6. Mohanty, P., Behera, D., Panda, S.K., Bastia, T.K., Hybrid Composite Laminates from UPE/ESOA Blend Reinforced with Chitosan and Bamboo Fiber: A Study of Mechanical and Thermal Properties. *Hybrid Compos. Laminates UPE/ESOA Blend Reinf. With Chitosan Bamboo Fiber A Study Mech. Therm. Prop.*, 32, 6, 1321–1328, 2020.
7. Mohanty, P., Bastia, T.K., Behera, D., Panda, S.K., Chitosan Grafted Carbon Nanotubes Reinforced Vinyl Ester/UPE Blend Based Partially Bio Nanocomposite. *Asian J. Chem.*, 31, 9, 1943–1948, 2019.
8. Panda, S.K., Behera, D., Rath, P., Bastia, T.K., Enhanced properties of UPE/ESOA partially bio-nanocomposites reinforced with chitosan functionalized graphene nanoplatelets: An innovative approach. *Bull. Mater. Science*, 41, 4, 102, 2018.
9. Panda, S.K., Behera, D., Rath, P., Bastia, T.K., Advances in physical performances of UPE/ESOA partially bio-nanocomposites reinforced with graphene nanoplatelets. *Int. J. Res. Innovation Appl. Sci.*, 2, 8, 2017.
10. Panda, S.K. and Behera, D., Materials Today: Proceeding, Elsevier, 2017, ISSN:2214-7853.
11. Panda, S.K., Mohanty, P., Behera, D., Bastia, T.K., ESOA modified unsaturated polyester hybrid networks: A new perspective. *J. Appl. Polym. Sciences (Wiley)*, 133, 44345, 2016. DOI: 10.1002/APP.44345.

12. Bao, Y.z., Zhi-ming, H., Shen-xing, L., Zhi-xue, W., Thermal stability, smoke emission and mechanical properties of poly (vinyl chloride)/hydrotalcite nanocomposites. *Polym. Degrad. Stab.*, 93, 448–455, 2008.
13. Rabiee, H., Vatanpour, V., Farahani, M.H.D.A., Zarrabi, H., Improvement in flux and antifouling properties of PVC ultrafiltration membranes by incorporation of zinc oxide (ZnO) nanoparticles. *Sep. Purif. Technol.*, 156, 299-310, 2015.
14. Zarrinkhameh, M., Zendehnam, A., Hosseini, S.M., Fabrication of polyvinylchloride based nanocomposite thin film filled with zinc oxide nanoparticles: Morphological, thermal and optical characteristics. *J. Ind. Eng. Chem.*, 30, 295–301, 2015.
15. Basfar, A.A., Effect of various combinations of flame-retardant fillers on flammability of radiation cross-linked poly (vinyl chloride) (PVC). *Polym. Degrad. Stab.*, 82, 333–340, 2003.
16. Khaleghi, M., Didehban, K., Shabanian, M., Effect of new melamine-terephthaldehyde resin modified graphene oxide on thermal and mechanical properties of PVC. *Polym. Test.*, 63, 382–391, 2017.
17. Amobonye Emmanuel, A., Bhagwat, P., Singh, S., Pillai, S., Biodegradability of Conventional Plastics: Opportunities, Challenges, and Misconceptions, Chapter 10 - Biodegradability of Polyvinyl chloride, pp. 201–220, 2023. https://doi.org/10.1016/B978-0-323-89858-4.00017-8.
18. John, M.J. and Anandjiwala, R.D., Recent developments in chemical modification and characterization of natural fiber-reinforced composites. *Polym. Compos.*, 29, 2, 187–207, 2008.
19. Selke, S.E. and Wichman, I., Wood fiber/polyolefin composites. *Composites Part A*, 35, 321–326, 2004.
20. Muhammad, G. and Haryono, A., Effect of Size of Cellulose Particle as Filler in the PVC Biocomposites on Their Thermal and Mechanical Properties. *Mater. Sci. Forum*, 737, 67–73, 2013. doi:10.4028/www.scientific.net/MSF.737.67.
21. Wiebking, H., *Filler in PVC: A Review of the Basic*, Specialty Minerals Inc, 1998.
22. Sionkowska, A., Current research on the blends of natural and synthetic polymers as newbiomaterials: Review. *Prog. Polym. Sci.*, 36, 1254–1276, 2011.
23. Sterzy nski, Tomaszewska, T., Andrzejewski, J., Skorczewska, K., Evaluation of glass transition temperature of PVC/POSS nanocomposites. *Compos. Sci. Technol.*, 117, 398–403, 2015.
24. Wypych, G., *PVC Degradation and Stabilization*, 3rd, ChemTec Publishing, Toronto, ON, Canada, 2015, ISBN 9781927885000.
25. Polyvinyl Chloride Industry Installed Capacity and Capital Expenditure (CapEx) Forecast by Region and Countries Including Details of All Active Plants, Planned and Announced Projects, pp. 2022–2026. Available online: https://www.globaldata.com/store/report/polyvinyl-chloride-market-analysis.

26. Kaczmarek, H., Bajer, K., Podgórski, A., Properties of Poly(vinyl chloride) Modified by Cellulose. *Polym. J.*, 37, 340–349, 2005. https://doi.org/10.1295/polymj.37.340.
27. Scott, G. and Gilead, D., *Degradable Polymers. Principles and Applications*, Chapman and Hall, London, U.K., 1995.
28. Guillet, J., Environmental aspects of photodegradable plastics. *Macromol. Symp.*, 123, 1, 209–224, 1997.
29. Hamid, S.H. (Ed.), *Handbook of Polymer Degradation*, 2nd, Marcel Dekker, Inc., New York, N.Y, 2000.
30. Domb, A.J., Kost, J., Wiseman, D.M. (Eds.), *Handbook of Biodegradable Polymers*, Harwood Academic Publishers, Amsterdam, 1997.
31. Mucha, M. and Mucha, J., Heterogeneous polymers degradation – microplastic. *Polimery*, 67, 7–8, 291–404, 2022.
32. Jiang, H. and Kamdem, D.P., Development of poly (vinyl chloride)/wood composites. A literature review. *J. Vinyl Addit. Technol.*, 10, 2, 59–69, 2004.
33. Joo, Y.L. and Cho, M.H., Utilization of Rheology Control to Develop Wood-Grain Patterned PVC/Wood Flour Composites. *Int. Polym. Process*, 14, 1, 1999.
34. Ayora, M., Rios, R., Quijano, J., Marquez, A., Evaluation by torque-rheometer of suspensions of semi-rigid and flexible natural fibers in a matrix of poly (vinyl chloride). *Polym. Compos.*, 18, 4, 549–560, 1997.
35. Sombatsompop, N., Chaochanchaikul, K., Phromchirasuk, C., Thongsang, S., Effect of wood sawdust content on rheological and structural changes, and thermo-mechanical properties of PVC/sawdust composites. *Polym. Int.*, 52, 12, 1847–1855, 2003.
36. Singha, A.S. and Thakur, V.K., Mechanical properties of natural fibre reinforced polymer composites. *Bull. Mater. Sci.*, 31, 791–799, 2008.
37. Mengeloglu, F., Mauana, L., King, J., Effects of impact modifiers on the properties of rigid PVC/wood-fibre composites. *J. Vinyl Add. Tech.*, 6, 3, 143–157, 2000.
38. Zhao, Y., Wang, K., Zhu, F., Xue, P., Jia, M., Properties of poly (vinyl chloride)/wood flour/montmorillonite composites: effects of coupling agents and layered silicate. *Polym. Degrad. Stab.*, 2874–2883, 2006.
39. Jomboh, K.J., Garkida, A.D., Alemaka, E.M., Yakubu, M.K., AlkaliV, E.C., Eze, W.U., Lawal, N., Properties and applications of natural, synthetic and hybrid fiber reinforced polymer composite: A review. *AIMS Mater. Sci.*, 11, 4, 774–801, 2024.
40. Ge, X., Li, X., Meng, Y., Tensile properties, morphology, and thermal behavior of PVC composites containing pine flour and bamboo flour. *J. Appl. Polym. Sci.*, 93, 1804–1811, 2004.
41. Kamel, S., Preparation and properties of composites made from rice straw and poly (vinyl chloride) (PVC). *Polym. Adv. Technol.*, 15, 612–616, 2004.

42. Djidjelli, H., Boukerrou, A., Founas, R., Rabouhi, A., Kaci, M., Farenc, J., Preparation and characterization of poly (vinyl chloride)/virgin and treated sisal fiber composites. *J. Appl. Polym. Sci.*, 103, 3630–3636, 2007.
43. Zainudin, E., Sapuan, S., Abdan, K., Mohamed, M., Maleque, M., Mechanical properties of banana-pseudostem fiber reinforced plasticized PVC composities, pp. 132–133, 2007.
44. Zheng, Y.-T., Cao, D.-R., Wang, D.-S., Chen, J.-J., Study on the interface modification of bagasse fiber and the mechanical properties of its composite with PVC. *Composites Part A*, 38, 20–25, 2007.
45. Leblanc, J., Furtado, C., Leite, M., L.L.Y, V., Souza, d. & A.M.F., Effect of the fiber content and plasticizer type on the rheological and mechanical properties of poly (vinyl chloride)/green coconut fiber composites. *J. Appl. Polym. Sci.*, 106, 6, 3653–3665, 2007.
46. Mohanty, A., Misra, M., Drzal, L., Selke, S., Harte, B., Hinrichsen, G., Natural fibers, biopolymers, and biocomposites: an introduction, in: *Natural Fibers, Biopolymers and Biocomposites*, vol. 1–36, Taylor & Francis inc, New York, 2005.
47. Jayaraman, K., Manufacturing sisal-polypropylene composites with minimum fiber degradation. *Compos. Sci. Technol.*, 63, 367–374, 2003.
48. Drzal, L., Mohanty, D., Burgueno, R., Misra, M., Biobased structural composite materials for housing and infrastructure applications: opportunities and challenges. *Proc. NSF Hous. Res. Agenda Workshop*, 2, 129–140, 2004.
49. Saheb, D.N., *et al.*, Review: Advances in Polymer Technology. *Nat. Fiber Polym. Composites A*, 18, 4, 351–363, 1999.
50. Brouwer, W., Natural fibre composites in structural components: Alternative application for sisal?, in: *Seminar, Commond Fund for Commodities-Alternative Applications for Sisal and Henequen*, Food and Agriculture Organization of the UN (FAO) and the Common Funds for Commodities (CFC), Rome, 2000.
51. Kayode, F.A., Surface Treatments of Natural Fibres—A Review: Part 1. *Open J. Polym. Chem.*, 5, 3, 41–46, 2015.
52. Maldas, D. and Kokta, B.V., Improving adhesion of wood fiber with polystyrene by chemical treatment of fiber with coupling agent and the influence on the mechanical properties of the composites. *J. Adhes. Sci. Technol.*, 3, 529–539, 1989.
53. Kokta, B.V., Maldas, D., Daneault, C., Beland, P., Composites of polyvinyl chloride-wood fibers. I. effect of isocyanate as a bonding agent. *Polym. Plast. Technol. Eng.*, 29, 1&2, 87–118, 1990.
54. Vilay, V., Mariatti, M., Taib, R., Todo, M., Effect of fiber surface treatment and fiber loading on the properties of bagasse fiber–reinforced unsaturated polyester composites. *Compos. Sci. Technol.*, 68, 633–638, 2008.
55. Kalaprasad, G., Joseph, K., Thomas, S., Pavithran, C., Theoretical modelling of tensile properties of short sisal fibre-reinforced lo-density polyethylene composites. *J. Mater. Sci.*, 32, 4261–42, 1997.

56. Sae-Oui, P., Rakdee, C., Thanmathorn, P., Use of rice husk ash as filler in natural rubber vulcanizates: In comparison with other commercial fillers. *J. Appl. Polym. Sci.*, 83, 11, 2485–2493, 2002.
57. Ismail, H., Mega, L., Abdul Khalil, H.P.S., Effect of a silane coupling agent on the properties of white rice husk ash–polypropylene/natural rubber composites. *Polym. Int.*, 50, 5, 606–611, 2001.
58. Recent Progress of Rice Husk Reinforced Polymer Composites: A Review. by Mohamed Azlan Suhot Mohamed Azlan Suhot,Mohamad Zaki Hassan,Sa'ardin Abdul Aziz and Mohd Yusof Md Daud. *Polymers*, 13, 15, 2391, 2021. https://doi.org/10.3390/polym13152391.
59. Petchwattana, N., Covavisaruch, S., Chanakul, S., Mechanical properties, thermal degradation and natural weathering of high density polyethylene/rice hull composites compatibilized with maleic anhydride grafted polyethylene. *J. Polym. Res.*, 19, 1–9, 9921, 2012.
60. Sombatsompop, N. and Chaochanchaikul, K., Average mixing torque, tensile and impact properties, and thermal stability of poly (vinyl chloride)/sawdust composites with different silane coupling agents. *J. Appl. Polym. Sci.*, 96, 1, 213–221, 2005.
61. Petchwattana, N. and Covavisaruch, S., Influences of particle sizes and contents of chemical blowing agents on foaming wood plastic composites prepared from poly (vinyl chloride) and rice hull. *Mater. Des.*, 32, 5, 2844–2850, 2011.
62. Petchwattana, N. and Covavisaruch, S., Effects of rice hull particle size and content on the mechanical properties and visual appearance of wood plastic composites prepared from poly (vinyl chloride). *J. Bionic Eng.*, 10, 1, 110–117, 2013.
63. Petchwattana, N. and Sanetuntikul, J., Static and Dynamic Mechanical Properties of Poly (vinyl chloride) and Waste Rice Husk Ash Composites Compatibilized with γ-aminopropyltrimethoxysilane. *Silicon*, 10, 287–292, 2018. https://doi.org/10.1007/s12633-016-9440-x.
64. Favier, V., Chanzy, H., Cavaille, J.Y., Polymer Nanocomposites Reinforced by Cellulose Whiskers. *Macromolec.*, 28, 6365–6367, 1996.
65. Shameem, A.S., Schuster, Y.P.S., Monis, K., Manufacturing of Biocomposites for Domestic Applications Using Bio-Based Filler Materials. *J. Compos. Sci.*, 6, 3, 78, 2022. https://doi.org/10.3390/jcs6030078.
66. Duc, H.P., Kuma, B., Kim, J., Environment Friendly, High Performance Lignin Derived and Polyvinyl Alcohol Blended Resin for All Green Natural Fiber Reinforced Composite. *Int. J. Precis. Eng. Manuf. Green Technol.*, 12, 607–619, 2025. https://doi.org/10.1007/s40684-024-00650-8.
67. Marcelo, B.S. and Rosa, D.S., Chitin and Chitosan Based PVC Composites and Nanocomposites, in: *Poly (Vinyl Chloride) Based Composites and Nanocomposites*, pp. 117–128, DOI:10.1007/978-3-031-45375-5_5.

68. Zahra, N.M., Siswanto and Widiyanti, P., The Role of Chitosan on Polyvinyl Chloride (PVC)-Glycerol Biocomposites for Blood Bag Application. *J. Biomimetics Biomater. Biomed. Eng.*, 37, 94–106, 2018.
69. Furqan, M.S., Mohammed, A.A., Salih, W.M., Improving the performance of polyvinyl chloride with added natural extracts for medical storage. *Eng. Res. Express*, 6, 045409, 2024.
70. Katarzyna, S., Krzysztof, L., Sławomir, W., Joanna, S., Paulina, R., Utilization of Ground Eggshell as a Biofiller of Plasticized PVC-Based Materials Fabricated Using Melt Blending. *Polymers*, 17, 4, 434, 2025. https://doi.org/10.3390/polym17040434.
71. Salah, F.A., Ibrahim, O.A., El-Rafey, E., Ehab, S.G., Wooden Polymer Composites of Poly (vinyl chloride), Olive Pits Flour, and Precipitated Bio-Calcium Carbonate. *ACS Omega*, 6, 37, 23924–23933, 2021. doi: 10.1021/acsomega.1c02932.
72. Schiller, M., *PVC Additives Performance, Chemistry, Developments, and Sustainability*, Hanser Publications, Cincinnati, OH, USA, 2013, ISBN 978-1-56990-543-2.
73. Mohanty, A.K., Misra, M., Drzal, L.T., Chapter: 1; Natural Fibers, Biopolymers, and Biocomposites: An Introduction, in: *Natural fibers, biopolymers, and biocomposites*, pp. 1–35, CRC Press, 2005.
74. Bledzki, A.K. and Gassan, J., Composites reinforced with cellulose based fibers. *Prog. Polym. Sci.*, 24, 2, 221–274, 1999.
75. Fang, Y., Wang, Q., Guo, C., Song, Y., Cooper, P.A., Effect of zinc borate and wood flour on thermal degradation and fire retardancy of polyvinyl chloride (PVC) composites. *J. Anal. Appl. Pyrolysis*, 100, 230–236, 2013.
76. Bashta, B., Brostow, W., Granowski, G., Hnatchuk, N., Modification of Poly (Vinyl Chloride) + Epoxy Systems for Improved Thermal and Aging Stability. *Macromol. Symp.*, 365, 239–245, 2016.
77. Mokhena, T.C. and John, M.J., Cellulose nanomaterials: new generation materials for solving global issues. *Cellulose*, 27, 3, 1149–1194, 2020.
78. Marcelo, B.O. and Rosa, D.S., Chitin and Chitosan Based PVC Composites and Nanocomposites, in: *Engineering materials*, pp. 117–128, Springer Nature, 2023.
79. Hajibeygi, M., Maleki, M., Shabanian, M., Ducos, F., Vahabi, H., New polyvinyl chloride (PVC) nanocomposite consisting of aromatic polyamide and chitosan modified ZnO nanoparticles with enhanced thermal stability, low heat release rate and improved mechanical properties, 2018.
80. Quoc Pham, L., Uspenskaya, M.V., Olekhnovich, R., Bernal, R.O., A review on Electrospun PVC Nanofibers: fabrication, properties, and application. *Fibers*, 9, 2, 12, 2021.
81. Mohamed, A.T., Thermal experimental verification on effects of nanoparticles for enhancing electric and dielectric performance of polyvinyl chloride. *Measurement*, 89, 28–33, 2016.

82. Zattini, G., Ballardini, S., Benelli, T., Mazzocchetti, L., Giorgini, L., Safer plasticized polyvinyl chloride synthetic leathers for the automotive industry: Evaluation of alternatives to antimony compounds as flame retardants. *Polym. Eng. Sci.*, 59, 12, 2488–2497, 2019.
83. Kim, J., Jaewoong, L., Chaehyun, J., Chankyu, K., Development of low cost carbon fibers based on chlorinated polyvinyl chloride(CPVC) for automotive applications. *Mater. Des.*, 204, 2021.
84. Moustafa, H., Mohamed, M., Mahmoud, A., Fatehy, M.A., Ultrafast response humidity sensors based on polyvinyl chloride/graphene oxide nanocomposites for intelligent food packaging. *Sens. Actuator Phys.*, 331, 2021.
85. Hazer, B., Karahaliloglu, Z., Ashby, R.D., Poly (vinyl chloride) derived food packaging applications with Antioxidative and Anticancer properties. *ACS Food Sci. Technol.*, 3, 4, 761–771, 2023.
86. Alhulw, H.F., Majed, A., Khulaif, A., Nageh, K.A., PVC/PVP/SrTiO3 polymer blend nanocomposites as potential materials for optoelectronic applications. *Results Phys.*, 44, 23, 106173, 2022.
87. Alshammari, A.H., Alshammari, M., Ibrahim, M., Alshammari, K., Taha, T., New hybrid PVC/PVP polymer blend modified with Er2O3 nanoparticles for optoelectronic applications. *Polymers*, 15, 3, 684, 2023.
88. Alshammari, A.H., Alshammari, M., Ibrahim, M., Alshammari, K., Processing polymer film nanocomposites of polyvinyl chloride – Polyvinylpyrrolidone and MoO3 for optoelectronic applications. *Opt. Laser Technol.*, 168, 3, 2024.
89. Alshammari, A.H. and Taha, T.A., Structure, thermal and dielectric insights of PVC/PVP/ZnFe2O4 polymer nanocomposites. *Eur. Phys. J. Plus*, 136, 12, 1201, 2021.
90. Chiellini, F., Perspectives on alternatives to phthalate plasticized poly(vinyl chloride) in medical devices applications. *Prog. Polym. Sci.*, 38, 7, 1067–1088, 2013.

7

PVC–Nanocellulose Biocomposites and Nanocomposites

Raghvendra Kumar Mishra[1,2,3*], **Kuruvilla Joseph**[4] **and Saurav Goel**[2]

[1]*School of Aerospace, Transport and Manufacturing, Cranfield University, Bedford, United Kingdom*
[2]*School of Engineering, London South Bank University, London, United Kingdom*
[3]*Department of Chemical Engineering, Indian Institute of Technology Delhi, Delhi, India*
[4]*Department of Chemistry, Indian Institute of Space Science and Technology, Thiruvananthapuram, Kerala, India*

Abstract

Polyvinyl chloride (PVC)–nanocellulose biocomposites and nanocomposites are advanced materials that combine the mechanical, thermal, and environmental benefits of nanocellulose with the versatility of PVC. This chapter explores the synthesis, characterization, and applications of these materials, highlighting their potential as sustainable, high-performance alternatives in various industries. The addition of nanocellulose enhances the mechanical strength, thermal stability, and biodegradability of PVC through improved interfaces and dispersion techniques. Surface modifications, such as esterification, silylation, and TEMPO-mediated oxidation, improve interfacial adhesion and dispersion stability by addressing hydrophilic–hydrophobic incompatibilities. Fabrication methods, including melt processing, solution casting, electrospinning, and 3D printing, offer precise control over the composite structure and properties. These materials demonstrate enhanced tensile properties, reduced oxygen permeability (20–50% reduction at low loadings), and maintained optical transparency (70–85% at 1 wt%), along with improved biodegradation rates compared to pure PVC. Challenges such as thermal processing issues, dispersion limitations, and environmental concerns are addressed through surface modification strategies, compatibilizing agents, and advanced manufacturing techniques. These composites have diverse applications,

**Corresponding author*: raghvendramishra4489@gmail.com

Shiv Kumari Panda and Visakh P. M (eds.) Polyvinyl Chloride-Based Biocomposites and Bionanocomposites, (185–232) © 2026 Scrivener Publishing LLC

including packaging, construction, automotive components, and biomedical devices, with a strong focus on areas requiring superior barrier properties, mechanical performance, and reduced environmental impact. By examining the structure–property–processing relationships, this chapter provides a framework for developing sustainable, high-performance PVC–nanocellulose composites, driving innovation in materials engineering.

Keywords: PVC–nanocellulose composites, interfacial engineering, mechanical reinforcement, barrier properties, sustainable materials

7.1 Introduction

Biocomposites and nanocomposites are pivotal in advancing sustainable materials development across a range of industrial sectors. Biocomposites consist of a polymeric matrix reinforced with organic or inorganic fillers, including natural fibers, polysaccharides, proteins, ceramics, metals, and nanostructured carbon materials [1]. These systems offer superior environmental performance, aligning with circular economy models by utilizing renewable resources, enabling recyclability, and lowering production energy demands [2, 3]. Nanocomposites, integrating nanoscale reinforcements such as metal oxides, ceramics, or nanopolymers, exhibit enhanced mechanical strength, thermal stability, and chemical resistance due to their high surface area-to-volume ratios [4]. Fiber type is a critical classification parameter in biocomposites, distinguishing between wood-based and non-wood fibers (e.g., flax, jute, hemp, sisal, and bamboo). These fibers, often derived from agricultural residues or recycled biomass, provide desirable mechanical performance and biodegradability [2, 3]. Compared to conventional glass fiber-reinforced plastics, biocomposites offer similar mechanical attributes with added advantages such as improved acoustic insulation, lower thermal expansion, and reduced carbon emissions [5]. Their production supports CO_2 sequestration and reduces reliance on petrochemical feedstocks [6]. Figure 7.1 presents a comparative schematic of biocomposites and nanocomposites, highlighting their classification, component structures, and multifunctional characteristics.

Nanocellulose, a bio-derived nanomaterial extracted from cellulose-rich sources like wood pulp, agricultural residues, and microbial cultures, has garnered considerable attention for its exceptional mechanical properties, high aspect ratio, large specific surface area, and modifiable surface chemistry [7, 8]. Its three primary forms, cellulose nanofibrils (CNFs), cellulose nanocrystals (CNCs), and bacterial nanocellulose (BNC), differ in

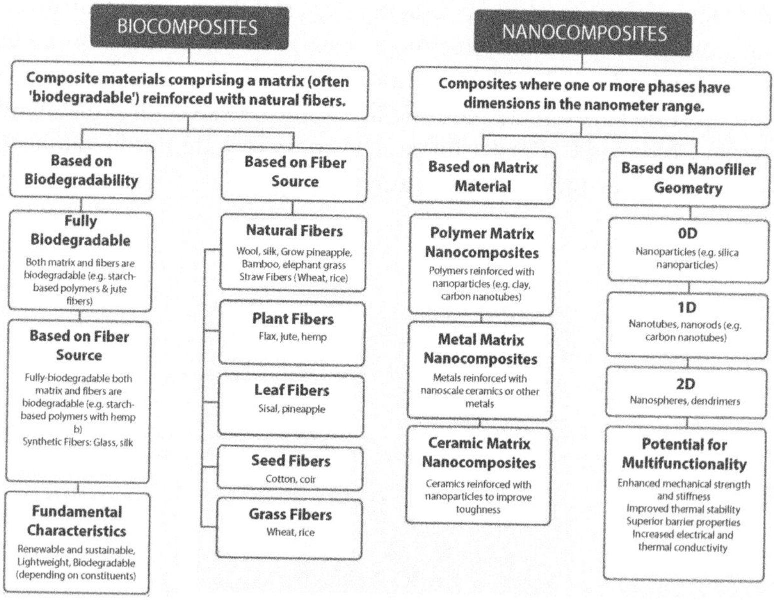

Figure 7.1 A comparative flowchart of biocomposites and nanocomposites, highlighting their definitions, categories, and key attributes: classified by biodegradability and fiber source, categorized by matrix type (polymer, metal, or ceramic) and nanofiller geometry (0D, 1D, 2D).

crystallinity, morphology, and processing routes [9, 10]. CNCs, characterized by high crystallinity, are typically produced *via* acid hydrolysis, while CNFs are obtained through high-shear mechanical techniques [11, 12]. BNC is synthesized through microbial fermentation, offering high purity and a three-dimensional nanofiber network [13]. Nanocellulose production can involve mechanical (e.g., homogenization, ultrasonication), chemical (e.g., acid hydrolysis), and enzymatic processes, each presenting trade-offs in terms of energy consumption, yield, and environmental impact [13, 14]. Enzymatic pretreatment, often in tandem with mechanical methods, represents a greener approach, minimizing chemical waste and facilitating energy-efficient nanofibrillation [15]. The physicochemical properties of nanocellulose including density, crystallinity, and surface functionality, are tunable, enabling tailored interactions with hydrophobic polymer matrices [15]. Life cycle assessments (LCAs) underscore the importance of regional energy mixes and processing strategies in determining nanocellulose's environmental footprint [16, 17]. Nanocellulose applications are diverse, spanning biomedicine (e.g., wound healing, tissue scaffolding, and drug delivery), food and packaging (e.g., stabilizers, encapsulants, and biodegradable films), water treatment (e.g., adsorbents, filtration media), and electronics (e.g., flexible substrates, supercapacitors) [18]. Figure 7.2a and b illustrates nanoscale cellulose fibers and the structural distinctions between CNFs and CNCs. Cellulose feedstocks include cotton, ramie, flax, sisal, wheat straw, and bacterial cultures.

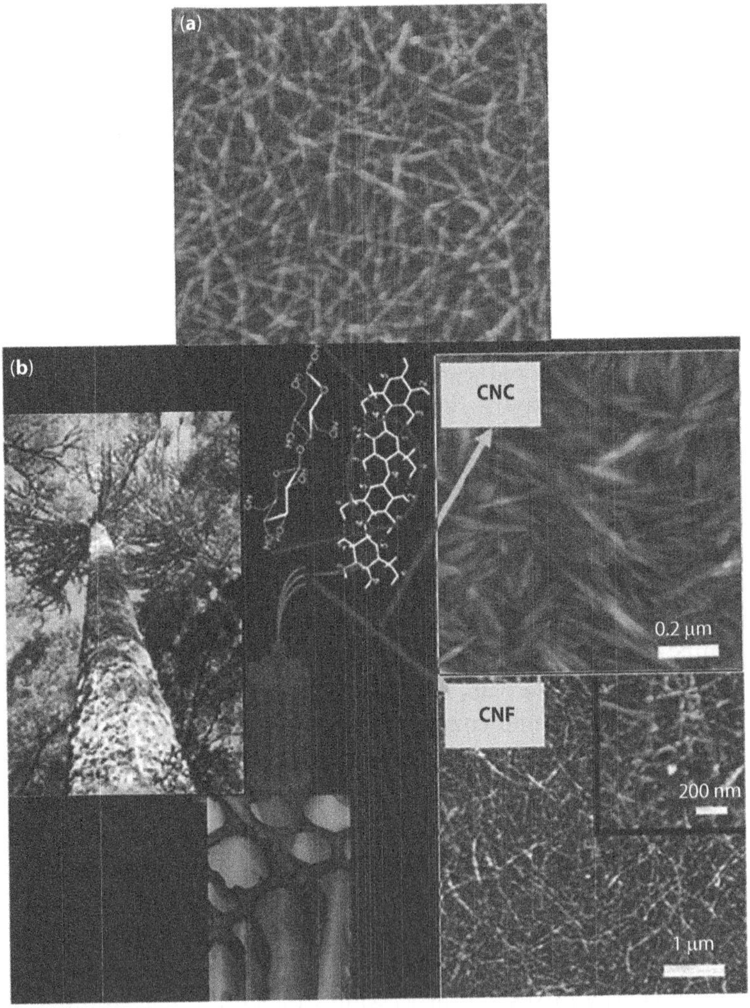

Figure 7.2 (a) Atomic force microscopy of cellulose fibers. (b) Schematic illustration of cellulose nanofibrils (CNF) and cellulose nanocrystals obtained from wood fibers after deconstruction of the highly hierarchical, multidimensional structure in wood [11].

PVC–nanocellulose composites leverage the synergistic integration of a widely used thermoplastic polymer with a sustainable, high-performance nanofiller [19]. Incorporating CNCs or CNFs into PVC matrices results in substantial improvements in mechanical behavior (e.g., tensile strength, Young's modulus), primarily due to the high aspect ratio and interfacial bonding capacity of nanocellulose [12]. The resultant stress transfer efficiency enhances the durability and load-bearing performance of the composite. Furthermore, nanocellulose incorporation contributes to improved thermal stability, expanding the potential application range in thermally demanding environments [19]. Scalable processing methods, such as extrusion, melt blending, and injection molding, facilitate the fabrication of PVC–nanocellulose composites. These techniques enable uniform filler dispersion and maintain composite homogeneity, critical for consistent performance [20]. Applications extend across automotive (e.g., interior components, dashboards), construction materials, and packaging sectors, where mechanical integrity, weight reduction, and environmental compliance are increasingly prioritized [21]. Despite PVC's inherent non-biodegradability, nanocellulose additions reduce the overall environmental impact of the composite, offering partial biodegradability and enhanced recyclability. Nevertheless, end-of-life disposal concerns persist due to potential toxic emissions associated with PVC degradation [22]. This chapter explores the synthesis, characterization, and application of PVC–nanocellulose composites, emphasizing their potential as sustainable, high-performance materials. By bridging conventional materials with sustainable solutions, these composites emerge as key contributors to next-generation material development.

7.2 Nanocellulose: Structure, Properties, and Production

Nanocellulose is a sustainable and versatile nanomaterial derived from cellulose, the most abundant biopolymer on Earth. It exhibits exceptional properties, including high tensile strength, low density, and a large specific surface area, making it valuable for various applications [10]. Nanocellulose is sourced from wood pulp, agricultural residues, and bacterial fermentation, with each source influencing its properties and effectiveness. Wood pulp, a traditional source, is processed mechanically or chemically to break down fibers into nanoscale dimensions, yielding high crystallinity and mechanical strength suitable for paper and packaging [7]. Agricultural residues, such as wheat straw and rice husks, provide an eco-friendly alternative, reducing dependence on forest resources while aiding in waste management. These residues offer unique properties due to their lignin and hemicellulose content, which affect their mechanical and thermal behavior [23]. Bacterial nanocellulose (BNC), synthesized extracellularly by bacteria such as *Komagataeibacter*, is a high-purity material free from lignin and hemicellulose. It exhibits superior properties, including high crystallinity, excellent water-holding capacity, and biocompatibility, making it ideal for biomedical applications [24]. The tensile strength of nanocellulose depends on its form and source. Cellulose nanocrystals (CNCs) demonstrate impressive tensile strengths of up to 7.5 GPa, attributed to their high crystallinity and robust hydrogen bonding. In contrast, cellulose nanofibrils (CNFs) generally exhibit tensile strengths ranging from 1 to 3 GPa, primarily due to the presence of amorphous regions [19]. The low density of this material is beneficial for lightweight applications within the aerospace and automotive sectors. Furthermore, its high surface area and adjustable surface chemistry facilitate chemical modifications, thereby enhancing compatibility with a variety of matrices [15]. Nanocellulose is categorized into three primary forms: cellulose nanocrystals (CNCs), cellulose nanofibers (CNFs), and bacterial nanocellulose (BNC). CNCs are derived from the acid hydrolysis of cellulose fibers, resulting in rod-like particles characterized by high crystallinity, with dimensions ranging from 100 to 500 nanometers in length and 5 to 30 nanometers in width. These structures function as reinforcing agents, rheology modifiers, and templates for electronic components. The process of sulfuric acid hydrolysis introduces negatively charged sulfate ester groups, which significantly enhance colloidal stability [25, 26]. CNCs exhibit high tensile strength, stiffness, and optical transparency [27]. Cellulose nanofibers (CNFs), recognized for their unique properties, are composed of elongated, flexible fibers that encompass both crystalline and amorphous regions. These fibers are produced *via* high-pressure homogenization or grinding techniques, resulting in a high aspect ratio and exceptional flexibility, which render them ideal for reinforcing various materials. Additionally, CNFs demonstrate biocompatibility, biodegradability, and an extensive surface area, thereby enhancing their interactions with other substances. Pretreatment methods, including TEMPO oxidation and enzymatic hydrolysis, are employed to minimize energy consumption during the fibrillation process [28, 29]. BNC, produced by bacteria like *Komagataeibacter xylinus*, forms a three-dimensional nanofibrous network with high purity, crystallinity, and excellent water-holding capacity, making it highly suitable for wound dressings, tissue engineering, and drug delivery [30]. Figure 7.3 illustrates

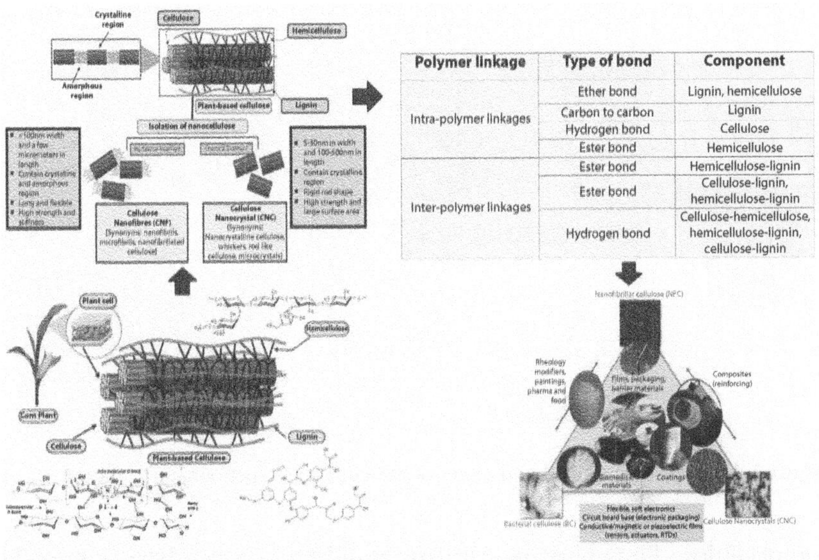

Figure 7.3 The main component of plant-based cellulose, isolation of nanocellulose through two distinct treatments leading to varying properties in nanocellulose, intra- and inter-polymer linkages between biopolymers [11, 31].

cellulose's fibrillar forms, including CNCs, CNFs, and BNC, highlighting production methods and structural characteristics applicable in fluidics, electronics, and sensing technologies [11].

Nanocellulose production methods include mechanical, chemical, and enzymatic processes. Mechanical treatments such as high-pressure homogenization and ultrasonication are effective but energy-intensive. Chemical processes, like acid hydrolysis, efficiently produce CNCs but require proper waste management. Enzymatic hydrolysis using cellulases offers an environmentally friendly alternative for breaking down cellulose's crystalline and amorphous regions [13, 15]. Surface modifications, including esterification, polymer grafting, and silane functionalization, improve dispersion and matrix compatibility, enhancing mechanical properties. Modification efficiency ranges from 40 to 80 percent, depending on reaction conditions and nanocellulose type [32, 33]. Pretreatment methods like TEMPO oxidation or ionic liquid processes further reduce energy consumption and environmental impact. Nanocellulose's biocompatibility and three-dimensional structure make it suitable for applications in wound dressings, tissue engineering, and drug delivery [34, 35]. It is also used as a bioadsorbent for heavy metal and contaminant removal from water.

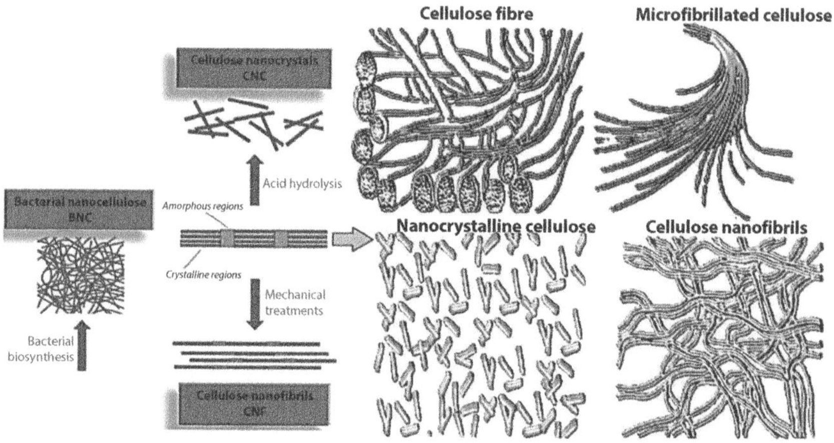

Figure 7.4 Preparation scheme and dimensions of different types of nanocelluloses.

Nanocellulose films provide excellent liquid and oxygen barrier properties for food and packaging applications [36]. In composites, it enhances mechanical properties and thermal stability, supporting applications in packaging, automotive, and construction [37, 38]. Advanced analytical tools, such as atomic force microscopy (AFM), X-ray diffraction (XRD), and dynamic light scattering (DLS), are used to study nanocellulose morphology and crystallinity. Its shear-thinning rheological behavior is beneficial for food and pharmaceutical applications [39, 40]. Figure 7.4 depicts the preparation scheme and dimensions of different types of nanocellulose, showcasing their unique characteristics and application suitability. In summary, nanocellulose stands out as a promising material due to its renewable nature, exceptional mechanical properties, and versatility across various applications. Its ability to be sourced from sustainable materials and modified for specific uses positions it as a key player in advancing eco-friendly technologies and materials.

7.3 Polyvinyl Chloride (PVC): Composition and Characteristics

Polyvinyl chloride (PVC) is a high-strength thermoplastic polymer known for its durability and versatility. It is synthesized through the polymerization of vinyl chloride monomers derived from petroleum or natural gas [41, 42]. The molecular structure of polyvinyl chloride (PVC) is defined by a two-carbon backbone, in which one hydrogen atom is substituted by a chlorine atom, resulting in the chemical formula $(C_2H_3Cl)_n$. This unique molecular configuration imparts PVC with distinctive properties, as demonstrated in Figure 7.5. In its pure form, PVC appears as a white, brittle solid, with a density ranging from 1.38 to 1.40 grams per cubic centimeter for rigid formulations. It exhibits a glass transition temperature between 80 and 85°C, with thermal degradation initiating at temperatures above 180°C. Rigid PVC generally shows tensile strength values between 40 and 45 MPa, depending on specific formulation and processing conditions [43, 44]. Due to its amorphous structure, polyvinyl chloride (PVC) lacks a specific melting point. Rather, its processing temperature range of 160°C to 220°C indicates the softening and flow characteristics observed during extrusion or molding. Additionally, its low thermal conductivity contributes to its exceptional insulation properties [45]. PVC exhibits excellent chemical resistance to a variety of substances, including acids, bases, salts, and aliphatic hydrocarbons. However, it is vulnerable to degradation when exposed to ketones, chlorinated hydrocarbons, and aromatic solvents. Furthermore, PVC is soluble in specific solvents such as dimethylformamide, tetrahydrofuran, and dimethylacetamide, which facilitates its processing across diverse applications [46].

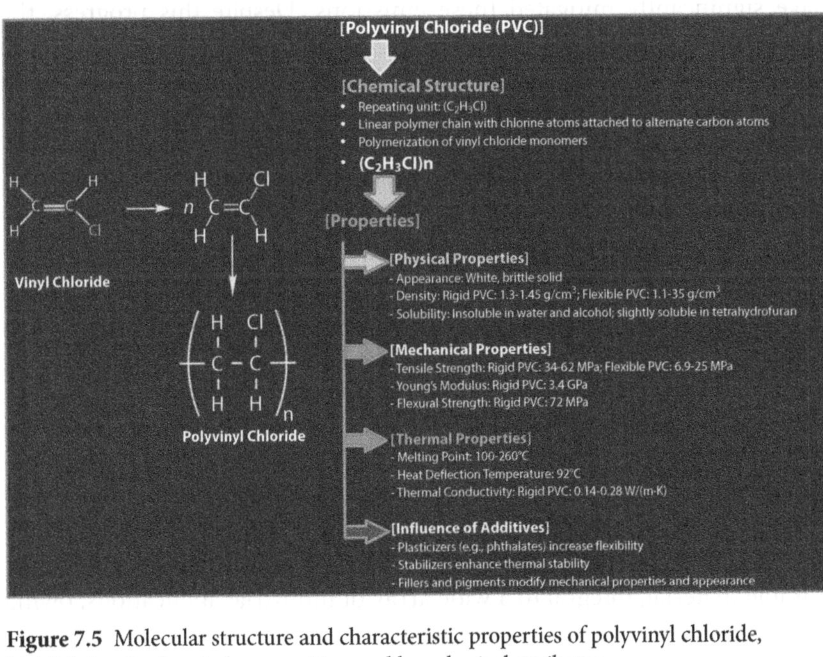

Figure 7.5 Molecular structure and characteristic properties of polyvinyl chloride, highlighting its chemical composition and key physical attributes.

The diverse properties of polyvinyl chloride (PVC) establish it as a fundamental material utilized across various industries. In the construction sector, PVC is extensively employed for applications such as piping, window framing, and flooring, attributed to its exceptional durability and resistance to environmental factors. In the medical field, PVC is critical for applications including blood bags, tubing, and catheters, due to its biocompatibility and effective sterilization capabilities. Additionally, the electrical insulation properties of PVC render it particularly suitable for wire and cable insulation [47]. PVC serves as a matrix material in composite applications, where it is integrated with nanofillers to enhance mechanical, thermal, and flame-retardant properties. These composites find applications in coatings, automotive components, and flame-retardant systems. The incorporation of fillers such as clay, wood flour, and plant-based materials significantly improves its mechanical and thermal performance, thereby expanding its applicability in advanced engineering domains [47, 48]. Despite its widespread utility, PVC poses environmental challenges. Its non-biodegradable nature and long lifespan contribute to significant waste management concerns [22]. The incineration of polyvinyl chloride (PVC) waste generates pollutants, including hydrogen chloride gas and, under suboptimal conditions, dioxins. However, contemporary waste-to-energy incinerators, which are equipped with advanced emission control systems, have significantly mitigated these emissions. Despite this progress, the widespread adoption of such technologies continues to pose a challenge [49–51]. Another issue arises from the migration of plasticizers in highly plasticized PVC products, which can affect performance and raise environmental safety concerns. Addressing these challenges requires the development of effective recycling methods and the use of eco-friendly additives to minimize environmental impact [52]. The current research is concentrated on the identification of appropriate fillers and additives aimed at improving the mechanical and thermal properties of polyvinyl chloride (PVC) while ensuring its processability. A significant challenge encountered in this domain is the limited compatibility between PVC and reinforcing agents, such as nanocellulose, which arises from inherent differences in hydrophilicity. These disparities frequently lead to issues with dispersion and result in weak interfacial bonding [31, 53]. Addressing the challenges associated with PVC composites necessitates the implementation of surface modification techniques and advanced processing methods, which are vital for enhancing their performance. PVC is recognized as a versatile and durable material, integral to a wide array of industrial applications, owing to its exceptional attributes, including high strength, chemical resistance, and thermal stability. The incorporation of advanced materials alongside innovative processing techniques presents significant potential for improving the performance of PVC while also tackling environmental concerns and promoting sustainability in contemporary applications [54].

7.4 Fabrication of Polyvinyl Chloride Nanocellulose Biocomposites and Nanocomposites

Polyvinyl chloride (PVC) is a widely utilized synthetic polymer, renowned for its durability and versatility. The mechanical properties and environmental impact of PVC can be significantly enhanced through the incorporation of nanocellulose, a renewable and biodegradable material derived from natural sources. This integration positions nanocellulose as an ideal reinforcing agent for nanocomposites, thereby improving both performance and sustainability [55]. Figure 7.6 presents a systematic approach to the development of sustainable PVC-based composites through the

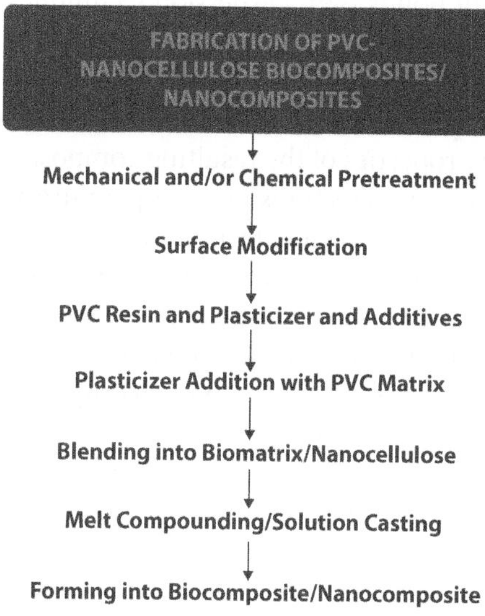

Figure 7.6 A schematic representation of the fabrication process for polyvinyl chloride nanocellulose biocomposites and nanocomposites, sequential steps involved, beginning with raw material preparation to the final product, including critical stages such as surface modification, blending techniques, and processing methodologies.

integration of eco-friendly nanocellulose. This process yields biocomposites characterized by enhanced strength, improved biodegradability, and a reduced environmental footprint. The fabrication process initiates with the pretreatment of raw cellulose materials, which may include plant fibers or bacterial cellulose. This pretreatment employs various methods such as grinding, homogenization, acid hydrolysis, or enzymatic treatments to effectively extract nanocellulose. Furthermore, surface modification techniques such as functionalization, grafting, or coating are employed to enhance the compatibility between the hydrophilic nanocellulose and the hydrophobic PVC matrix [56]. Meanwhile, the preparation phase involves mixing PVC resin with plasticizers and functional additives, such as stabilizers, ultraviolet absorbers, and fillers, to achieve a homogeneous composite [57]. Composite fabrication utilizes two principal processing techniques: thermomechanical blending, also known as melt compounding, conducted at elevated temperatures, and solution casting. The latter method involves dissolving components, followed by mixing and forming films through the process of solvent evaporation [58]. To address compatibility challenges, strategies such as the utilization of block copolymers, surfactants, and reactive compatibilizers (e.g., maleic anhydride) are employed to bridge the differences in material properties and facilitate the formation of covalent bonds [59]. Dispersion techniques, including ultrasonication, magnetic and electric field discharge, and electrospinning, are further employed to ensure a uniform distribution of nanocellulose within the composite [60]. Mechanical and thermomechanical analyses are conducted to assess the properties of the resulting composites, with a focus on key characteristics such as glass transition temperature and heat distortion behavior [61]. Advanced processing techniques are essential for achieving uniform nanocellulose dispersion, which enhances tensile strength, modulus, and thermal stability by facilitating efficient stress transfer within the composite [62]. Research indicates that even low loadings of nanocellulose (1 weight percent) can significantly improve composite properties when well dispersed [56]. For instance, Badi *et al.* demonstrated that the incorporation of cellulose nanocrystals into PVC through casting resulted in an increase in the polymer's decomposition temperature and an enhancement of its dielectric strength [63]. Furthermore, nanocellulose reinforcement contributes to improved stiffness, thermal stability, and barrier properties, effectively reducing water and vapor permeability while enhancing moisture and gas barrier performance in thin films [63, 64].

7.4.1 Methods of Polyvinyl Chloride Nanocellulose Nanocomposite Preparation

Polyvinyl chloride (PVC) composites reinforced with nanocellulose have garnered considerable attention due to their enhanced properties and environmental advantages. The incorporation of nanocellulose, a renewable and biodegradable material, into PVC matrices significantly improves mechanical properties and thermal stability and reduces environmental impact. However, the hydrophilic nature of nanocellulose, which is attributed to its abundant hydroxyl groups, presents compatibility challenges when combined with hydrophobic polymers such as PVC. These challenges can result in poor dispersion, agglomeration, diminished mechanical performance, and increased moisture sensitivity [31, 53]. To effectively address these compatibility issues, various surface modification techniques have been employed. Esterification and etherification processes reduce the hydrophilicity of nanocellulose, thereby enhancing its compatibility with hydrophobic polymers. Graft polymerization facilitates the attachment of hydrophobic polymer chains to the surfaces of nanocellulose, improving dispersion within hydrophobic matrices [65]. Furthermore, compatibilizers such as maleic anhydride form covalent bonds between nanocellulose and the polymer matrix. Hydrophilic substances, including gelatin, help prevent agglomeration and enhance redispersibility, as illustrated in Figure 7.7 [66]. Crosslinking within the polymer matrix minimizes water uptake and enhances dimensional stability. It is crucial to maintain the nanocellulose content below 7 weight percent to prevent excessive viscosity and ensure optimal mechanical performance [67]. The dispersion in hydrophobic matrices can be further improved through electrostatic adsorption and surface charge modifications [68]. TEMPO-mediated oxidation converts primary hydroxyl groups at the C6 positions to carboxylate groups, resulting in a negatively charged surface that enhances colloidal stability [69]. Functionalized nanocellulose demonstrates high adsorption capacities, with certain aerogels achieving copper(II) adsorption capacities of up to 452.49 milligrams per gram [70]. Advanced surface modifications, including silylation, amidation, and carbamation, introduce functional groups that improve polymer compatibility [71–73]. Sulfonation and phosphorylation add anionic groups, which enhance dispersibility and impart flame-retardant properties [74, 75]. Contemporary modification methods encompass enzymatic processes, deep eutectic solvents, and energy-efficient treatments such as microwave, cold plasma, and pulsed

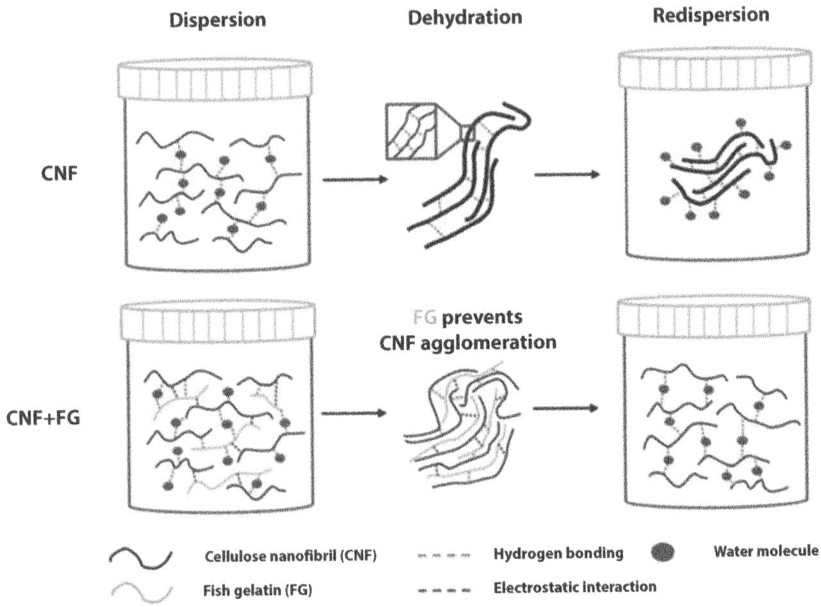

Figure 7.7 Prevention of cellulose nanofibril agglomeration during dehydration and enhancement of redispersibility by hydrophilic gelatin [66].

electric fields [76–78]. Cellulose nanocrystals (CNCs) exhibit an axial elastic modulus ranging from 110 to 150 GPa, as confirmed by both experimental and computational studies [79, 80]. These properties contribute to improved composite stiffness and stress transfer [81]. Various fabrication techniques, including solution casting, melt blending, and electrospinning, are employed in the production of these composites [54, 82]. While solution casting allows for controlled dispersion, melt blending offers industrial scalability, albeit with risks of thermal degradation [83]. Electrospinning facilitates the production of aligned nanofibers, whereas sol–gel processes create nanocellulose networks through solvent exchange. Dispersion is further enhanced by ultrasonication, which utilizes cavitation forces, and anionic surfactants that modify surface charge characteristics [84–86]. Chemical modifications significantly enhance interfacial adhesion, with single treatments typically achieving a 40–60% improvement. However, optimal results often necessitate combined treatments and specific processing conditions [87, 88].

7.4.1.1 Melt-Mixing Process of Polyvinyl Chloride Nanocellulose Nanocomposites

The melt-mixing process significantly enhances the mechanical properties and biodegradability of polyvinyl chloride (PVC) by integrating nanocellulose as a reinforcing agent. Achieving uniform dispersion of nanocellulose within the PVC matrix, facilitated by high shear forces during melt mixing, is essential for the effective fabrication of nanocomposites [89]. Precise thermal control is crucial in PVC processing, typically maintained between 160°C and 220°C, with an upper limit of 200°C to prevent thermal degradation. The primary degradation mechanism, dehydrochlorination, releases hydrogen chloride, leading to the formation of conjugated polyene sequences that result in discoloration and diminished mechanical performance [49]. The glass transition temperature (Tg) of rigid PVC is approximately 80–85°C, while plasticized PVC exhibits a lower Tg, contingent upon the plasticizer content. Heat stabilizers, including calcium stearate, zinc stearate, and epoxidized soybean oil, mitigate hydrogen chloride release, thereby addressing the poor thermal stability of PVC [90]. The thermal degradation of nanocellulose involves the breakdown of cellulose chains through glycosidic bond cleavage. Initial dehydration destabilizes the cellulose structure, resulting in the formation of char and volatile products, followed by depolymerization that releases carbon dioxide and carbon monoxide [91]. The onset of degradation varies with the type of nanocellulose: cellulose nanofibers (CNFs) and untreated nanocellulose degrade at temperatures between 220°C and 250°C, whereas sulfuric acid-treated cellulose nanocrystals (CNCs) degrade at lower temperatures, specifically between 150°C and 200°C, with significant weight loss occurring above 300°C [31, 92]. Dehydrochlorination of PVC initiates at 150–160°C and accelerates between 180°C and 250°C, involving the elimination of hydrogen chloride and the formation of conjugated double bonds along the polymer backbone. Stabilizers such as barium zinc and calcium zinc salts inhibit dehydrochlorination by reacting with labile chloride atoms, while organic-based stabilizers and eco-friendly alternatives, including phosphites and epoxidized oils, enhance thermal stability and recyclability [90, 93]. Melt-processing techniques, such as extrusion and injection molding, necessitate precise temperature control, which is achieved using modern extruders equipped with multiple heating zones. Dynamic mechanical analysis (DMA) is instrumental in optimizing processing conditions by providing real-time data on material response [61]. Twin-screw extruders improve the dispersion of nanocellulose through high shear forces; however, improper optimization of temperature, residence time, and screw configuration can lead to thermal degradation of both nanocellulose and PVC [94, 95]. Incremental addition of nanocellulose enhances dispersion and minimizes agglomeration [57]. Nanocellulose suspensions exhibit a shear-thinning behavior, characterized by a decrease in viscosity with increasing shear rate, facilitating flow and dispersion during extrusion. However, beyond a critical aspect ratio (typically between 50 and 100), further increases yield diminishing returns on viscosity and dispersion, with the percolation threshold and surface chemistry becoming more influential [96, 97]. Figure 7.8 illustrates the melt processing sequence for PVC nanocellulose composites, detailing the process from the feeding of nanocellulose fibers and PVC pellets through twin-screw extrusion to the formation of the final product. The materials are melted and thoroughly mixed to create a homogeneous composite, which is subsequently shaped using an extrusion die and cooled to produce the final product. This process is industrially viable due to its simplicity and compatibility with standard polymer processing equipment [89]. High shear mixing effectively disaggregates agglomerates; however, excessive shear or heat can lead to material degradation [28, 98]. Chemical modifications, such as acetylation and silane grafting, enhance compatibility between nanocellulose and PVC by reducing hydrophilicity and improving interfacial adhesion [87, 88].

7.4.1.2 Solution Casting Technique for Polyvinyl Chloride Nanocellulose Nanocomposites

The solution casting technique is a widely utilized method for fabricating composite films, particularly in the realm of polymer nanocomposites. This process entails the dispersion of polymers and nanomaterials in organic or water-based solvents to create homogeneous film-forming solutions through stirring or ultrasonication [89]. The uniform solution is subsequently cast onto substrates such as Petri dishes or glass sheets, followed by solvent evaporation in hot air ovens. Drying times are contingent upon film thickness, solvent type, and drying conditions, ranging from 2 to 72 h. Thin films (< 100 μm) utilizing volatile solvents may dry within 2–6 h, whereas thicker films necessitate extended drying periods [99]. Solution casting is characterized by its simplicity and minimal equipment requirements, allowing for precise control over nanocellulose dispersion and the formation of uniform films with enhanced mechanical and barrier properties. The selection of solvents and evaporation rates significantly impacts the internal structure and final properties of the composites [100]. Hydrophilic polymers typically employ water as a solvent,

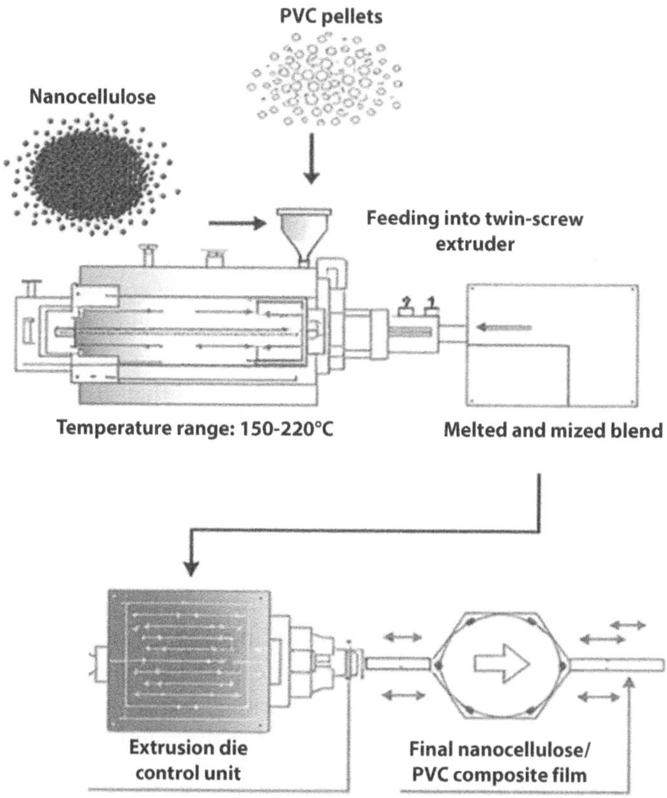

Figure 7.8 Schematic representation of the melt-processing sequence for PVC nanocellulose composites, showing progression from raw material feeding through twin-screw extrusion to final product formation, highlighting key processing stages and parameter controls.

while hydrophobic polymers necessitate organic solvents such as N,N-dimethylacetamide or tetrahydrofuran (THF) [101]. High boiling-point solvents facilitate slower evaporation, thereby providing improved control over film formation and minimizing nanocellulose aggregation. Controlled evaporation techniques, including temperature adjustments, ambient pressure modifications, and regulated solvent loss through covers, foster the development of ordered nanocellulose networks, which enhance mechanical strength and transparency [102]. This method accommodates various polymers and types of nanocellulose while allowing for precise control over film thickness [103]. Despite its advantages, solution casting presents certain limitations. Organic solvents necessitate efficient recovery systems to mitigate environmental impact. Extended drying times for thin films can hinder scalability, and improper solvent selection or rapid evaporation may lead to nanocellulose aggregation. Water-based systems are environmentally friendly and cost-effective for hydrophilic polymers, particularly in bio-based composites. Simple stirring suffices for low nanocellulose loadings; however, concentrations exceeding 5 weight percent can result in poor dispersion and increased surface roughness, as evidenced by scanning electron microscopy [104]. For polyvinyl chloride (PVC) nanocellulose composites, THF is the solvent of choice due to its efficacy in dissolving PVC and creating homogeneous casting solutions [105]. This moderately polar aprotic solvent, with a dielectric constant of 7.6, facilitates the incorporation of hydrophilic nanocellulose owing to its water miscibility [106, 107]. Sultana *et al.* demonstrated that dissolving PVC in THF at a concentration of 10 g per 30 mL of solvent ensures adequate dissolution for the formation of uniform films upon casting [108]. Research into bio-based nanocomposites has indicated that the incorporation of cellulose nanocrystals (CNCs) as fillers in PVC enhances dielectric strength and performance, even at low loadings of 1 weight percent. These composites are particularly well-suited for food packaging applications due to their superior barrier properties, including improved gas and moisture resistance [109]. Figure 7.9 illustrates the preparation of nanocellulose films utilizing solvent casting and super-critical impregnation technology, providing an overview of the steps involved in the solution casting method for preparing nanocomposites. The solution casting method remains an effective approach for producing PVC/nanocellulose nanocomposites with enhanced mechanical, thermal, and barrier properties. Its success is contingent upon the meticulous control of preparation parameters, solvent selection, and the type of nanocellulose employed to fulfill specific application requirements.

7.4.1.3 Advanced Manufacturing Methods for Nanocellulose/Polyvinyl Chloride Nanocomposites

Advanced manufacturing techniques are transforming the production of nanocellulose/polyvinyl chloride (PVC) nanocomposites by facilitating precise control over material properties and optimizing them for specific applications. Three primary methods, such as three-dimensional printing, electrospinning, and Pickering emulsion templating, are particularly effective in tailoring composite structures and enhancing performance. These techniques incorporate nanocellulose as a reinforcing agent, thereby improving strength-to-weight ratios and delivering superior

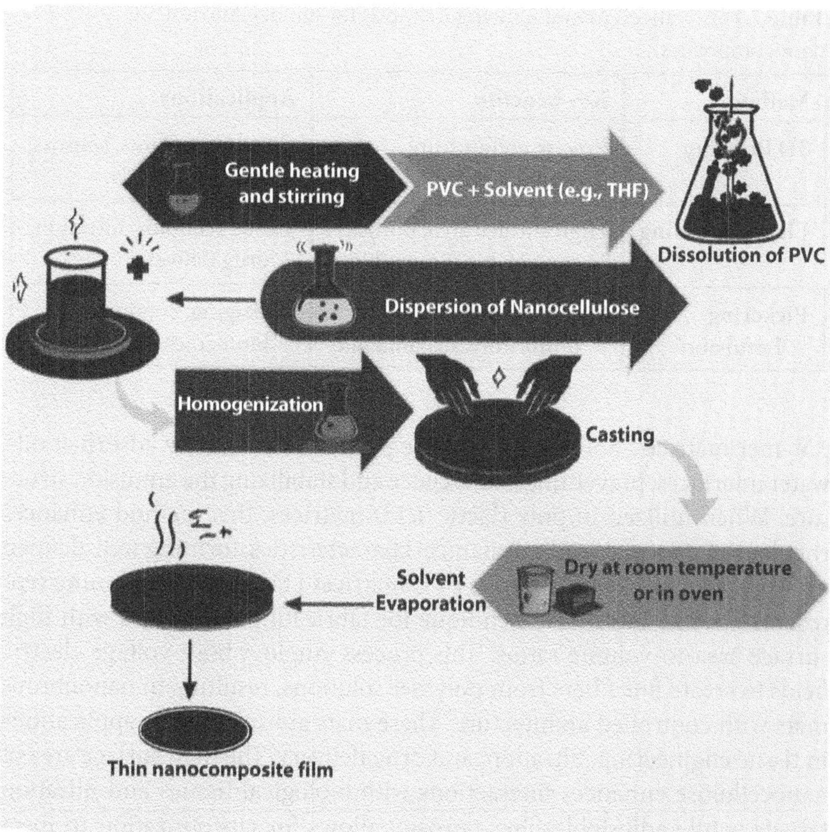

Figure 7.9 Schematic illustration of nanocellulose/PVC film preparation *via* solvent casting, showing sequential steps from solution preparation to film formation and the critical process parameters affecting final composite properties.

mechanical performance [110]. Additive manufacturing, particularly three-dimensional printing, offers unparalleled precision through layer-by-layer material placement. This capability enables the customization of porosity, density, and surface finish, which directly influences the mechanical, thermal, and barrier properties of the final products. Post-processing treatments, such as heat treatment, further enhance mechanical properties by modifying the material's microstructure. These treatments improve strength, dimensional stability, and reduce brittleness. The selection of post-processing techniques is contingent upon the material characteristics and desired properties, allowing for application-specific composite optimization [111]. Pickering emulsions, stabilized by nanocellulose, provide an effective method for enhancing cellulose nanocrystal dispersion within

Table 7.1 Advanced manufacturing methods for nanocellulose/PVC nanocomposites.

Method	Key benefits	Applications
3D Printing	Precise architecture, material efficiency	Custom implants, sensors, smart materials
Electrospinning	High surface area, porosity control, biocompatibility	Tissue scaffolds, filtration, drug delivery
Pickering Emulsion	Structural stability, uniformity, sustainability	Packaging, coatings, biomedical devices

polymer matrices. Solid nanocellulose particles irreversibly adsorb at oil-water interfaces, preventing coalescence and stabilizing the emulsion structure. When utilized in poly (lactic acid) matrices, this method enhances rheological, thermal, and mechanical characteristics, forming well-defined network structures that improve PVC matrices [112]. Electrospinning represents another advanced technique for fabricating nanofibers with high surface area-to-volume ratios. This process employs high-voltage electric fields to create fine fibers from polymer solutions, resulting in nanofibrous mats with controlled architecture. These mats are suitable for applications in tissue engineering, filtration, and drug delivery. The high surface area of nanocellulose enhances interactions with biological tissues and filtration targets, while adjustable fiber porosity allows for customization to meet specific requirements [113]. Collectively, these advanced manufacturing techniques enable the engineering of nanocellulose/PVC nanocomposites with improved mechanical properties, thermal stability, and biodegradability. The method of choice is determined by the desired characteristics and end-use application, as summarized in Table 7.1. Each technique presents unique advantages: three-dimensional printing offers precision and complexity, electrospinning achieves high surface area and porosity, and Pickering emulsions ensure stable nanocellulose dispersion within polymer matrices.

7.4.1.3.1 Electrospinning Methods for Nanocellulose/PVC Nanocomposites

Electrospinning is a highly versatile technique utilized for the production of nanofibers from polymer solutions or melts, leveraging electrohydrodynamic principles. The application of a high-voltage electric field to a polymer solution facilitates the generation of fine fibers, resulting in nanofibrous mats characterized by high surface area and porosity [114]. These intrinsic properties render electrospinning advantageous for a variety of applications, including tissue engineering, filtration, and drug delivery. Composites of nanocellulose and polyvinyl chloride (PVC) produced through electrospinning demonstrate enhanced mechanical and thermal properties, thereby making them suitable for advanced engineering and biomedical applications [114, 115]. Figure 7.10 illustrates a typical electrospinning setup, which comprises a syringe pump, a high-voltage power supply, a needle, and a grounded collector. The syringe pump regulates the flow rate of the polymer solution through the needle. Upon the application of high voltage at the needle tip, an electric field is established that forms a Taylor cone. Electrostatic forces then overcome the surface tension of the solution, ejecting a fine polymer jet that elongates and thins as the solvent evaporates, ultimately forming solid nanofibers. These fibers are deposited as a nonwoven mat on the grounded collector, ensuring electrical stability and precise deposition.

Electrospun PVC nanocellulose fibers typically exhibit diameters ranging from 100 to 800 nm. Processing conditions, including rotation speed,

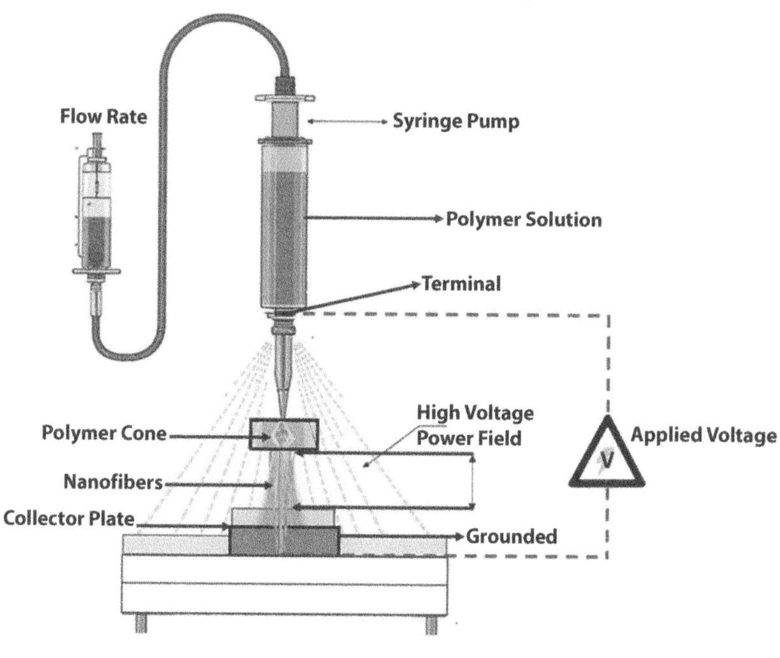

Figure 7.10 A typical electrospinning setup employed for the production of nanofibers from a polymer solution.

electric field strength, and solution viscosity, play a critical role in controlling fiber morphology and alignment. Lower rotation speeds yield random mats, whereas higher speeds facilitate the creation of aligned fibers [116]. Parameters such as flow rate, needle-to-collector distance, and solution viscosity significantly influence fiber quality and morphology [117]. Key factors, including viscosity, surface tension, and conductivity, are essential in determining jet stability and fiber formation. Solvents with high dielectric constants, such as N,N-dimethylformamide (DMF), enhance the responsiveness of the solution to the electric field. The use of binary solvent systems, such as DMF/tetrahydrofuran (THF) mixtures, effectively balances volatility and dielectric properties, ensuring uniform fiber production [118–120]. Electrospun nanocellulose/PVC nanocomposites exhibit significant potential in healthcare applications, including biosensors, tissue engineering, wound dressings, and drug delivery. Their high surface area promotes cellular adhesion and biocompatibility, while their porosity allows for efficient filtration with precise pore size control [121, 122].

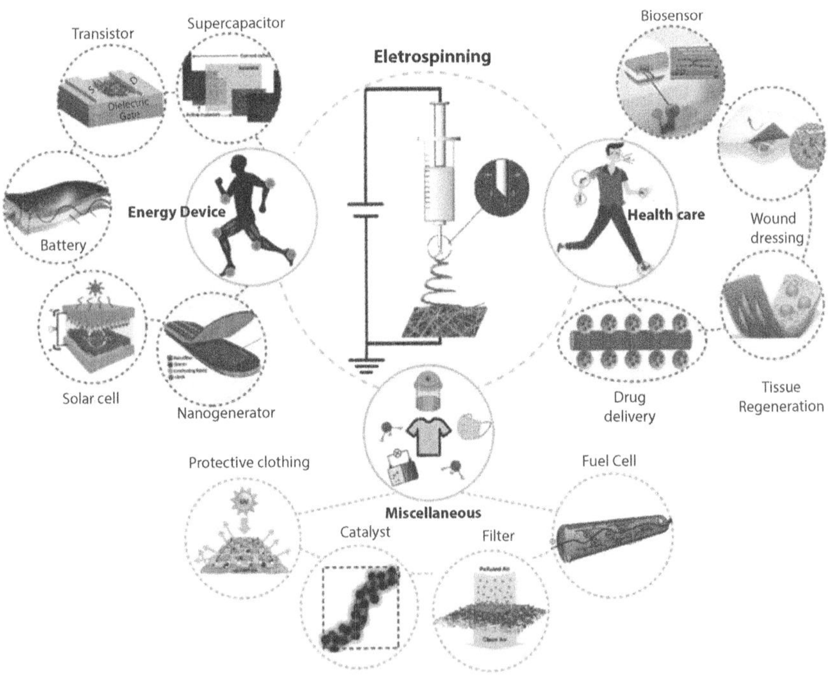

Figure 7.11 Applications of electrospun nanofibrous materials in healthcare (biosensors, tissue engineering, wound dressing, drug delivery), energy devices (batteries, supercapacitors, solar cells), and filtration systems [121].

Figure 7.11 depicts the diverse applications of these materials in healthcare, energy devices, and filtration systems [121]. The reinforcement of nanocellulose within PVC matrices during electrospinning enhances mechanical and thermal properties. The high aspect ratio and crystallinity of nanocellulose contribute to a reduction in fiber diameter and an increase in tensile strength and stiffness, rendering these nanocomposites ideal for applications in water treatment and air filtration [81, 121]. Despite their numerous advantages, electrospinning encounters challenges related to scalability, nanocellulose dispersion, and environmental impact [123]. Nevertheless, the combination of the biodegradability of nanocellulose and the mechanical strength of PVC positions these nanocomposites as high-performance materials suitable for a wide range of applications in filtration, biomedical devices, and advanced engineering [122].

7.4.1.3.2 3D Printing Methods for Nanocellulose/PVC Nanocomposites

3D printing, also known as additive manufacturing, represents a sophisticated technique for fabricating nanocellulose/PVC nanocomposites. In contrast to traditional subtractive manufacturing, this innovative method facilitates the construction of intricate geometries through a layer-by-layer deposition process based on digital models. The resolution and complexity of the printed structures are contingent upon material viscosity and the content of nanocellulose, with feature sizes typically ranging from 100 to 200 μm for PVC nanocellulose composites [124]. The process commences with a digital model, which can be designed using computer-aided design (CAD) software or acquired through 3D scanning [125]. This model is subsequently sliced into thin layers, resulting in a G-code file that directs the printer during material deposition. The selection of materials is determined by the desired properties, including strength, flexibility, durability, and cost [126, 127]. Fused deposition modeling (FDM) is the predominant 3D printing method employed for polymers, involving the extrusion of thermoplastic filaments through a heated nozzle to construct objects layer by layer [128]. Figure 7.12 illustrates the essential components of an FDM printer, which include the filament feed system, extruder head, hot-end nozzle, heated build plate, and control panel. Collectively, these components enable precise thermoplastic deposition. While FDM is recognized for its cost-effectiveness and durability, its material compatibility remains limited [129].

Advanced 3D printing techniques, such as stereolithography (SLA) and selective laser sintering (SLS), provide enhanced precision and superior

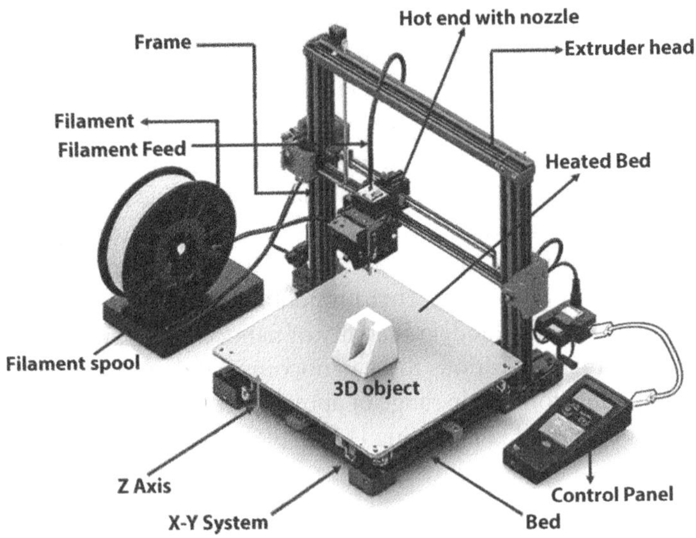

Figure 7.12 Key components of an FDM 3D printer.

surface finishes [131]. SLA utilizes laser technology to cure liquid resin into hardened plastic, rendering it suitable for intricate designs and medical applications [130]. Conversely, SLS employs laser sintering of powdered materials, facilitating the formation of support-free structures, which are ideal for aerospace and automotive applications [129, 130]. Digital light processing (DLP), akin to SLA, employs digital light projection for resin curing, offering high resolution and speed for complex designs [131]. Although vat polymerization methods like SLA and DLP excel in precision, their mechanical performance may be constrained [128, 130]. The integration of nanomaterials, such as nanocellulose, into polymer matrices significantly enhances the properties of printed components. The incorporation of nanocellulose markedly improves material characteristics. For example, FDM has been utilized to fabricate biodegradable bionanocomposite filaments, such as polylactic acid (PLA) reinforced with cellulose nanocrystals (CNCs). The presence of CNCs enhances PLA's crystallinity, thermal stability, and mechanical properties, with a 1 wt% CNC loading resulting in increases in tensile strength and modulus by 35.5% and 21.89%, respectively [132]. Similarly, cellulose nanofibrils (CNFs) enhance the mechanical properties and mitigate warping in high-density polyethylene (HDPE) composites [133]. In PVC matrices, silane-modified cellulose nanocrystals (SCNCs) improve tensile strength compared to unmodified CNCs, although higher filler content may reduce elongation [134]. Furthermore, nanocellulose-based materials excel in packaging applications, providing superior oxygen and water vapor barrier properties. When integrated into PVC, these nanocomposites enhance durability and sustainability, thereby reducing dependence on conventional petroleum-based plastics [128]. The advancement of 3D printing technologies for nanocellulose/PVC nanocomposites presents opportunities for the development of innovative materials with customizable properties. Tailoring mechanical, thermal, and barrier properties through 3D printing broadens potential applications across packaging, biomedical devices, and structural engineering sectors.

7.4.1.3.3 Pickering Emulsion Methods for Nanocellulose/PVC Nanocomposites

Pickering emulsions, first introduced by Percival Spencer Umfreville Pickering in 1907 and initially noted by Ramsden in 1903, use solid particles to stabilize emulsions by forming a protective layer around droplets that prevents coalescence [135]. Although research remained limited for much of the 20th century, interest surged in the late 1990s due to their diverse applications [136]. These emulsions stabilize two immiscible liquids, such as oil and water, by adsorbing solid particles at the interface, creating a barrier that enhances stability. Stability depends on particle concentration, size, and surface chemistry [135, 136]. Several methods are used to prepare Pickering emulsions, including rotor-stator homogenization, high-pressure homogenization, ultrasonic emulsification, and membrane emulsification [137]. Rotor-stator homogenization is simple and cost-effective but produces less stable emulsions compared to other methods [138]. High-pressure homogenization provides better control over droplet size and is scalable for industrial applications [137, 138]. Figure 7.13 illustrates pharmaceutical Pickering emulsions stabilized by nanoparticles, showing drug localization based on solubility and the parameters influencing emulsion stability. Pharmaceutical Pickering emulsions consist of two immiscible phases: an oil phase containing hydrophobic active pharmaceutical ingredients (APIs) like thymol, curcumin, and paclitaxel, and an aqueous phase with hydrophilic APIs such as dopamine, rutin, and silibinin. Nanoparticles at the interface form a protective barrier using materials such as natural and synthetic polymers, protein-based particles,

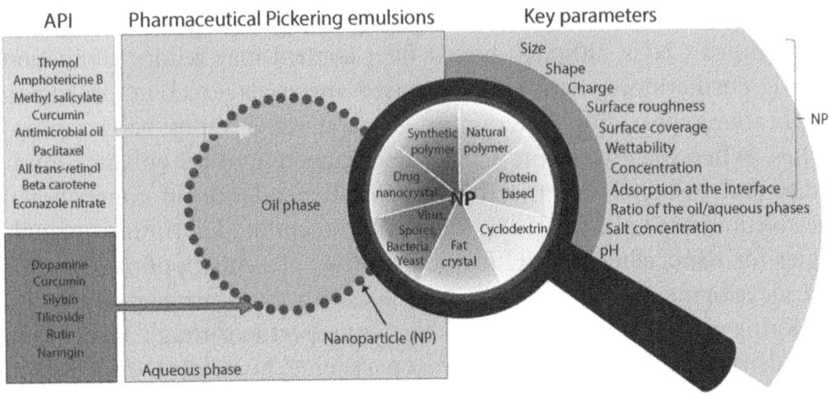

Figure 7.13 Illustration of pharmaceutical Pickering emulsions stabilized by nanoparticles (NPs), highlighting drug localization based on solubility and key parameters affecting emulsion stability [137].

cyclodextrins, fat crystals, and biological entities [136]. Ultrasonic emulsification generates nanoscaled droplets through cavitation, offering cost efficiency, though it may result in non-uniform distributions. Membrane emulsification provides precise droplet control and uniformity but operates at slower processing speeds [137, 139].

Natural solid particles like nanocellulose, starch, and proteins, along with synthetic particles such as silica and titanium dioxide (TiO_2), effectively stabilize Pickering emulsions [140]. Iron oxide nanoparticles (Fe_3O_4 NPs) also exhibit functionality in emulsion stabilization [141]. Particle wettability is a key factor, with intermediate wettability promoting strong adsorption at the oil–water interface [142]. Reduced interfacial tension caused by particle adsorption supports the formation of smaller droplets and enhances stability [143]. Nanocellulose particles used for Pickering emulsions typically range from 20 to 200 nm in size, with a contact angle range of 40–75° for stable oil-in-water emulsions [144]. Their irreversible adsorption at the interface prevents coalescence and Ostwald ripening. Lower surface charge density aids oil droplet dispersion, and hydrophobic surface modifications increase affinity for the oil phase. Particle concentration and oil-to-water ratios also influence emulsion type and stability [145]. These emulsions are particularly useful in pharmaceuticals, enabling controlled drug release and targeted delivery. The solid particle barrier improves biocompatibility and formulation stability, effectively accommodating both hydrophobic and hydrophilic APIs in separate phases [136, 146].

7.4.2 Strategies for Optimizing Performance of PVC–Nanocellulose Nanocomposites

The optimization of polyvinyl chloride (PVC)-nanocellulose composites requires careful evaluation of mechanical properties, thermal stability, and interfacial adhesion. Improving compatibility between the hydrophobic PVC matrix and hydrophilic nanocellulose is essential and is achieved through surface treatments and chemical modifications. These approaches enhance interfacial bonding, resulting in improved tensile strength and stiffness [147]. Uniform nanocellulose dispersion within the PVC matrix prevents agglomeration, which could otherwise weaken the composite's mechanical properties [148]. The size and concentration of nanocellulose particles play a critical role in determining composite performance. Smaller particles provide greater surface area interaction, enhancing mechanical strength, but excessive loading may lead to brittleness [147, 149]. Additional reinforcements, such as graphene or glass fibers, synergistically improve rigidity and strength through enhanced interfacial interactions. Surface modifications, including TEMPO-mediated oxidation, acetylation, and hydrophobic moiety grafting, improve compatibility between nanocellulose and PVC. The solvent casting technique allows precise control over nanocellulose dispersion, producing films with superior mechanical and barrier properties [16, 100]. Ultrasonication, which uses high-frequency sound waves, effectively disperses nanocellulose by generating localized energy through cavitation. Temperature control during ultrasonication, such as using ice or water baths, prevents material degradation [150, 151]. Emerging environmentally friendly techniques, such as magnetic and electric field discharge, align nanocellulose particles along field lines, improving dispersion [152–154]. High-pressure homogenization facilitates the defibrillation of cellulose fibers, separating microfibrils using high shear forces rather than merely reducing particle size [155]. Chemical grafting, such as the use of silanes like 3-aminopropyl triethoxysilane (APTES), covalently attaches hydrophobic functional groups to nanocellulose, enhancing dispersion and mechanical properties [156]. Coupling agents typically form covalent bonds with nanocellulose hydroxyl groups while physically entangling with PVC chains, rather than forming actual chemical bonds with the PVC matrix [157]. The layer-by-layer (LbL) assembly technique provides precise control over interfacial architecture by sequentially depositing oppositely charged species, improving composite structure and functionality [19, 158]. Processing conditions, such as shear rate and temperature control, are essential to achieving optimal composite properties. *In situ* polymerization of PVC monomers in the presence of dispersed nanocellulose enhances interfacial bonding through molecular entanglement and grafting, further improving the mechanical and thermal properties of the composite [15, 19].

7.5 Properties of PVC–Nanocellulose Nanocomposites

Polyvinyl chloride (PVC)–nanocellulose composites have garnered considerable attention due to their enhanced mechanical properties, thermal stability, and sustainability. The interfacial bonding between PVC and nanocellulose is pivotal for optimizing composite performance. Strategies such as the use of coupling agents, compatibilizers, and advanced assembly methods significantly improve interfacial adhesion. Compatibilizers effectively reduce surface energy mismatches between the hydrophobic PVC matrix and the hydrophilic nanocellulose, thereby enhancing mechanical performance [19]. The Halpin-Kardos equation serves as a model for modulus behavior above the glass transition temperature, accounting for immobilized phases surrounding nanocellulose whiskers [159]. Achieving a uniform dispersion of nanocellulose within the PVC matrix is essential to prevent agglomeration, which can compromise mechanical strength. However, excessive mixing may lead to degradation of both nanocellulose and PVC [157]. Processing temperatures for PVC–nanocellulose composites typically range from 160°C to 190°C, considering that nanocellulose begins to degrade at 220–250°C and PVC at 180–200°C. The incorporation of stabilizers extends the processing window by permitting brief exposure to elevated temperatures [15]. Surface modifications, including silane coupling agents and layer-by-layer assembly, address the hydrophilic nature of nanocellulose, thereby improving interfacial bonding and mechanical stability [158]. For instance, the addition of 2 wt% nanocellulose has been shown to increase tensile strength while simultaneously reducing elongation at break due to the stiffening effect and stress concentration at the filler–matrix interface. Improvements in tensile strength range from 15 to 40%, contingent upon processing conditions and surface treatments [109]. Dynamic mechanical analysis indicates enhanced viscoelastic properties with filler fractions up to 12.4 volume percent, although higher filler content may result in agglomeration and diminished performance. The high aspect ratio and surface area of nanocellulose contribute to improved energy absorption and impact resistance [52]. Furthermore, nanocellulose influences the thermal degradation behavior of PVC composites [56]. Neat PVC undergoes degradation in two stages: dehydrochlorination (200–370°C) and polymer chain cracking (370–500°C). The presence of nanocellulose reduces the onset degradation temperature to approximately 250°C due to its lower thermal stability, with sulfate groups from sulfuric acid hydrolysis accelerating initial degradation. However, convergence occurs at 280–320°C, revealing differences in residual char content between neat PVC and its composites [56]. The incorporation of nanocellulose has a minimal impact on the crystallinity of PVC, attributed to its amorphous nature, with thermal peak shifts indicating restricted chain mobility near filler surfaces [109]. Additionally, nanocellulose enhances the barrier properties of PVC by creating tortuous paths for gas diffusion, resulting in a reduction of oxygen permeability by 20–50% at 0.3–1 wt%, depending on the quality of dispersion [81]. However, its hydrophilic nature increases water uptake, which can compromise moisture barrier performance. Surface treatments or hydrophobic coatings can mitigate this limitation [160]. Nanocellulose also improves dielectric properties; at 1 wt%, it enhances dielectric strength while reducing dielectric loss, making PVC/nanocellulose composites suitable for high-voltage insulation. Fourier transform infrared spectroscopy (FTIR) analyses reveal strong interfacial interactions, with increased O–H peak intensity (3500–3200 cm^{-1}) correlating with improved bonding [109]. Furthermore, nanocellulose contributes to the sustainability of composites by enhancing biodegradability. Composites containing 3–8 wt% nanocellulose exhibit 10–20% mass loss over several months due to microbial degradation of exposed nanocellulose, leading to the formation of micropores that increase water penetration and physical erosion. Nevertheless, the bulk PVC matrix remains largely unaffected, showing negligible degradation [56, 160, 161]. These composites are well-suited for applications in packaging, biomedical devices, and wastewater treatment. For example, PVC–nanocellulose membranes modified with titanium aluminate achieve removal efficiencies of 85–95%, contingent upon operating conditions and contaminant concentration [162]. Figure 7.14 illustrates the improved thermal stability of PVC–nanocellulose composites. The thermogravimetric analysis (TGA) curve indicates that incorporating approximately 2 wt% nanocellulose raises the decomposition onset temperature to 210°C, while neat PVC begins degrading at 180–200°C. Peak decomposition temperatures also shift higher, signifying enhanced thermal stability [56, 108].

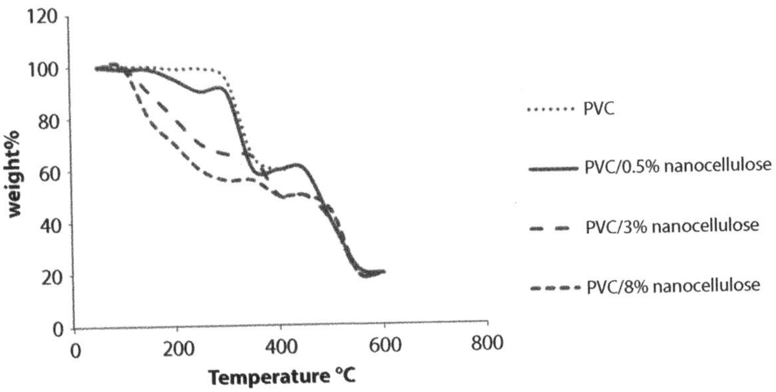

Figure 7.14 TGA curve of PVC and PVC/nanocellulose composites [56, 108].

Figure 7.15(a) presents nanocellulose particles with diameters ranging from 30 to 45 nm, exhibiting some aggregation due to high surface energy and ionic charges. Figure 7.15(b) demonstrates the dispersion of nanocellulose in PVC, revealing uniform dispersion at low concentrations (~0.5 wt%) but increased aggregation at higher concentrations (8 wt%) [56].

Figure 7.16 highlights the morphological changes induced by nanocellulose. Figure 7.16(a) depicts rod-like crystalline nanocellulose following acid hydrolysis, while Figure 7.16(b) illustrates the porous structure of neat PVC formed through solvent evaporation. Figures 7.16(c)–(e) showcase fracture surfaces of PVC/nanocellulose composites at 0.5, 1, and 3 wt%, respectively. At 0.5 wt%, nanocellulose is uniformly dispersed, whereas 1 wt% results in reduced pore areas and improved reinforcement. At 3 wt%, increased roughness indicates enhanced interfacial bonding. The quality of

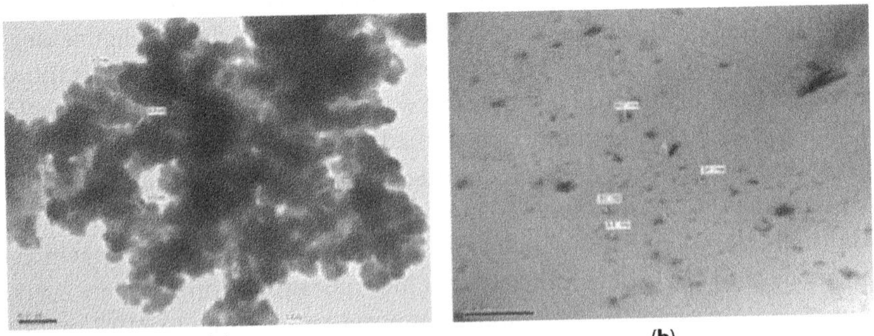

Figure 7.15 TEM images of (a) nanocellulose and (b) PVC/nanocellulose with 8 wt% nanocellulose [56].

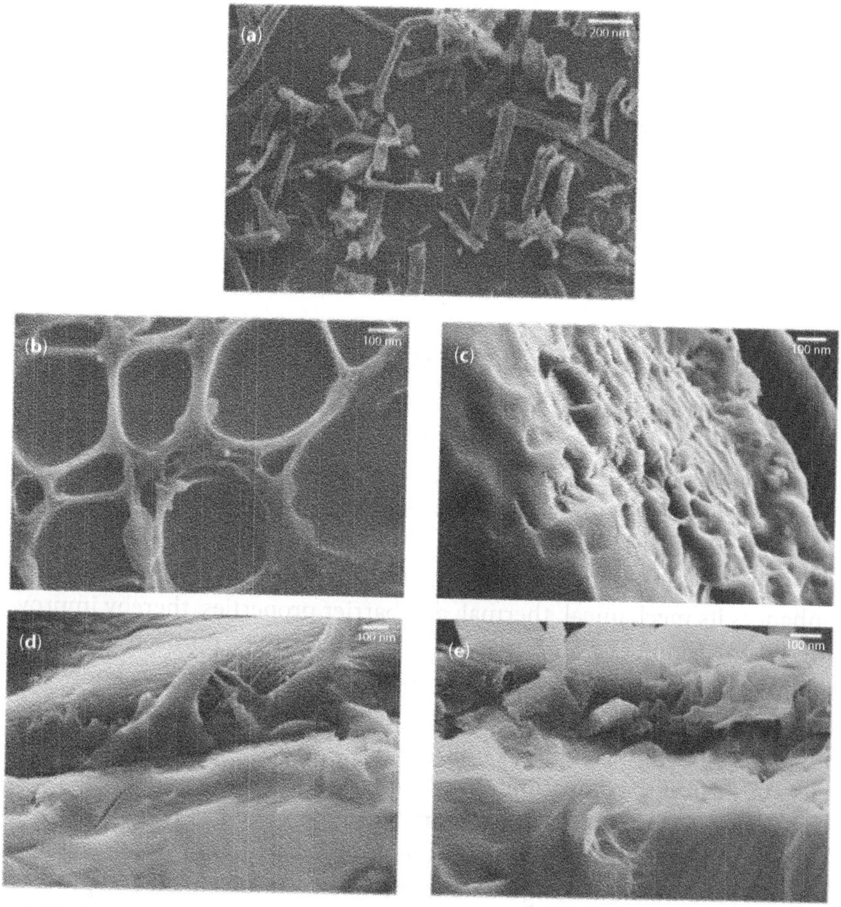

Figure 7.16 SEM images of (a) CNC, (b) neat PVC, (c) PVC/CNC 0.5, (d) PVC/CNC 1, and (e) PVC/CNC 3 [109].

dispersion is influenced by processing methods and surface modifications, with poor dispersion observed even at low loadings without optimization [109]. Nanocellulose enhances oxygen barrier properties by forming a rigid network, reducing gas permeability by 40–60% at 0.3 wt%, contingent upon humidity and processing methods [81]. PVC–nanocellulose composites maintain transparency at low nanocellulose loadings (75–85% at 1 wt%), provided that dispersion is uniform. Aggregation diminishes clarity, while proper dispersion minimizes haze. Unmodified nanocellulose offers limited UV protection; however, functionalized nanocellulose with UV-absorbing groups achieves 60–70% UV absorption at 0.1 wt% [161]. The barrier properties, transparency, and UV resistance of these composites render them ideal for protective films, although surface modifications are necessary to address moisture sensitivity in humid conditions. The biodegradability of nanocellulose presents a sustainable alternative to traditional PVC materials. Overall, nanocellulose enhances the dielectric strength and sustainability of PVC composites, with its renewable nature facilitating biodegradation and addressing the environmental challenges associated with PVC. These composites exhibit significant potential for applications in packaging, UV-stable coatings, and high-voltage insulation [109].

7.6 Applications of PVC–Nanocellulose Nanocomposites

Polyvinyl chloride (PVC) is a versatile material extensively utilized across various sectors, including construction, packaging, automotive, electronics, and medical devices [12]. The incorporation of nanocellulose (NC), a renewable and biodegradable nanomaterial, into PVC significantly enhances its mechanical, thermal, and barrier properties, thereby improving its environmental sustainability. This innovative approach effectively addresses the limitations associated with pure PVC, which is known for its persistence as a plastic [25, 26]. Recent research has focused on optimizing dispersion techniques utilizing functionalized cellulose nanocrystals (CNCs) and compatibilizers, facilitating the development of PVC/CNC composites derived from biomass sources such as rice straw, sawdust, and jute. Optimal properties are achieved with a low nanocellulose content (1–5 wt%), striking a balance between reinforcement and processability [56, 160, 161]. Figure 7.17 illustrates the key applications of PVC–nanocellulose nanocomposites, which encompass four primary sectors: (1) packaging materials, which enhance biodegradability and mechanical strength; (2) construction materials, which improve durability and reduce emissions; (3) biomedical devices and implants, which offer biocompatible solutions; and (4) automotive parts, which contribute to lightweight, high-strength components. PVC/CNC composites exhibit significant potential in packaging due to their enhanced mechanical strength, barrier properties, and biodegradability. The incorporation of nanocellulose creates tortuous diffusion paths, resulting in a reduction of oxygen permeability by 25–40% at a CNC loading of 0.3 wt%. Furthermore, reductions exceeding 70% can be achieved with higher CNC loadings (2–5 wt%) and optimal dispersion [81]. However, the hydrophilic nature of nanocellulose can lead to increased water vapor permeability. Surface treatments and hydrophobic coatings have proven effective in mitigating this limitation, with

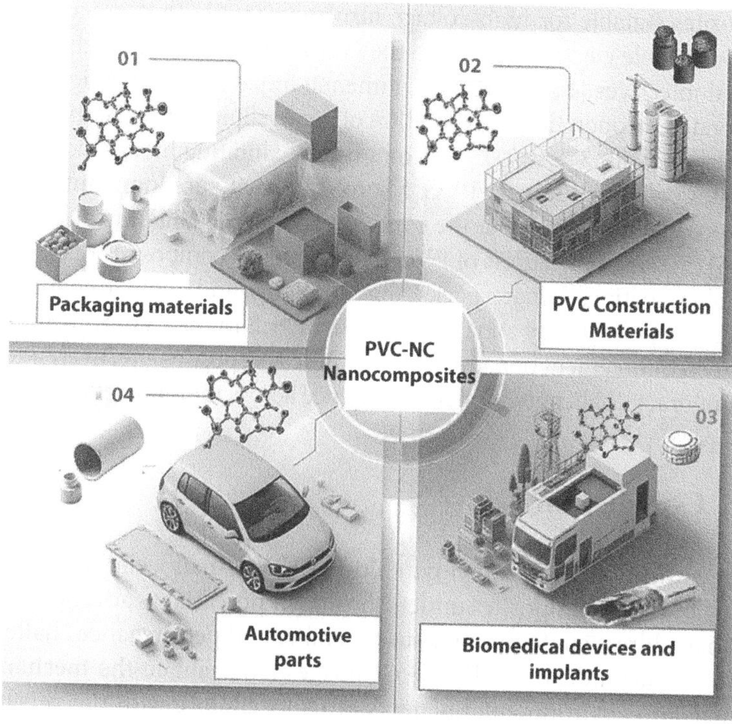

Figure 7.17 Applications of PVC–nanocellulose nanocomposites.

increases in water vapor permeability ranging from 20 to 100% depending on humidity and surface modification. These composites also facilitate active packaging functionalities, as CNC can carry antibacterial agents or UV absorbers. Additionally, PVC/CNC composites maintain high transparency while effectively blocking UVA/UVB radiation [161].

In the construction sector, these composites are ideally suited for applications such as flooring, wall panels, window frames, and pipes, owing to their enhanced mechanical properties and resistance to environmental stress. Wood-PVC composites reinforced with nanocellulose demonstrate increased rigidity and reduced emissions, further aligning with sustainable construction practices [163]. In the automotive industry, cellulose nanofiber composites have been commercialized in polypropylene-based components. For instance, Toyota Gosei has developed cellulose-reinforced plastics for seats and panels. PVC/CNC systems show particular promise for non-structural automotive parts that require flexibility and toughness [16]. In electrical applications, PVC's established role as an insulator is further enhanced by the addition of nanocellulose. Incorporating 1 wt% CNC improves dielectric strength while reducing dielectric loss, rendering these composites suitable for high-voltage insulation [109]. They are also applicable in flexible electronics, where they function as dielectric substrates or thermal interfaces [164]. The environmental impact of PVC is significantly ameliorated through the incorporation of nanocellulose. CNC enhances soil degradation, with 3–8 wt% CNC composites exhibiting higher biodegradation rates compared to neat PVC. Microbial activity predominantly targets the exposed nanocellulose at the composite surface, facilitating degradation [16, 56]. Future applications of PVC–nanocellulose composites may include UV-stable coatings, smart packaging solutions, and high-strength automotive components. By synergizing the practicality of PVC with the eco-friendly reinforcement of nanocellulose, these composites effectively address contemporary challenges in material science across multiple industries.

7.7 Conclusion

The investigation into polyvinyl chloride (PVC)-nanocellulose biocomposites and nanocomposites underscores their potential as advanced materials that address contemporary sustainability and performance challenges across various industries. These composites amalgamate the mechanical, thermal, and environmental benefits of nanocellulose with the versatility and broad applicability of PVC. Through systematic research encompassing synthesis, characterization, and optimization, these materials demonstrate promise in advancing materials science while providing sustainable solutions. The integration of nanocellulose into PVC matrices significantly enhances mechanical properties, including tensile strength, modulus, and toughness. Enhanced interfacial adhesion, achieved through chemical grafting and surface modifications, facilitates efficient stress transfer under mechanical loads. Such advancements enable applications in sustainable packaging, durable construction materials, and lightweight automotive components. The biodegradability of nanocellulose, coupled with its reliance on renewable resources, markedly reduces the environmental impact of PVC composites, aligning with global sustainability objectives. In biomedical contexts, these materials exhibit exceptional biocompatibility and functional versatility, rendering them suitable for medical devices and implants. Despite challenges related to dispersion, processing, and scalability, advanced manufacturing techniques such as 3D printing, electrospinning, and Pickering emulsion methods offer innovative solutions for precise control over material properties. PVC–nanocellulose biocomposites represent a promising class of materials that harmonize high performance with sustainability. By leveraging the unique properties of nanocellulose while addressing existing limitations, these composites possess the potential to revolutionize multiple industries and contribute to environmental sustainability. Ongoing research and technological advancements are essential for achieving large-scale industrial adoption and unlocking their full potential. These composites provide a combination of exceptional mechanical properties and environmental benefits, offering innovative solutions across packaging, construction, automotive, and biomedical sectors. As research continues to evolve, PVC–nanocellulose biocomposites are strategically positioned to play a transformative role in addressing contemporary materials science challenges while supporting global sustainability initiatives.

AI disclosure statement: The author acknowledges the use of AI-based tools for minor language editing or grammatical correction. However, all intellectual contributions, analyses, and conclusions presented in this chapter are entirely the author's own.

References

1. Ita-Nagy, D., Vázquez-Rowe, I., Kahhat, R., Quispe, I., Chinga-Carrasco, G., Clauser, N.M., Area, M.C., Life cycle assessment of bagasse fiber reinforced biocomposites. *Sci. Total Environ.*, 720, 2020, https://doi.org/10.1016/j.scitotenv.2020.137586.
2. Haraguchi, K., Biocomposites, in: *Encyclopedia of Polymeric Nanomaterials*, pp. 1–8, Springer Berlin Heidelberg, Berlin, Heidelberg, 2014, https://doi.org/10.1007/978-3-642-36199-9_316-1.
3. Mishra, R.K., Shaji, A., Thomas, S., Processing, rheology, barrier properties, and theoretical study of microfibrillar and nanofibrillar *in situ* composites, 2017, https://doi.org/10.1016/B978-0-08-101991-7.00010-8.
4. Yaragalla, S., Mishra, R., Thomas, S., Kalarikkal, N., Maria, H.J., Carbon-based nanofillers and their rubber nanocomposites, 2018, https://doi.org/10.1016/C2016-0-03648-3.
5. Riedel, U., Biocomposites: Long Natural Fiber-Reinforced Biopolymers, in: *Polymer Science: A Comprehensive Reference*, K. Matyjaszewski and M. Möller (eds.), pp. 295–315, Elsevier, The Netherlands, 2012, ISBN 9780080878621, https://doi.org/10.1016/B978-0-444-53349-4.00268-5.
6. Kumar Mishra, R., *Nano-food Engineering*, Springer International Publishing, Cham, 2020, https://doi.org/10.1007/978-3-030-44552-2.
7. Mishra, R.K., Sabu, A., Tiwari, S.K., Materials chemistry and the futurist eco-friendly applications of nanocellulose: Status and prospect. *J. Saudi Chem. Soc.*, 22, 2018, https://doi.org/10.1016/j.jscs.2018.02.005.
8. Kargarzadeh, H., Mariano, M., Gopakumar, D., Ahmad, I., Thomas, S., Dufresne, A., Huang, J., Lin, N., Advances in cellulose nanomaterials. *Cellulose*, 25, 2151–2189, 2018, https://doi.org/10.1007/s10570-018-1723-5.

9. Mishra, R.K., Remya, V.R., Mufeeda, K., Thomas, S., Kalarikkal, N., Microscopy and Spectroscopy Characterization of Eco-Friendly Composites and Nanocomposites, in: *Biopolymers and Biomaterials*, vol. 2018, pp. 297–344, Apple Academic Press, Toronto New Jersey, 2018, https://doi.org/10.1201/9781315161983-24.
10. Kargarzadeh, H., Huang, J., Lin, N., Ahmad, I., Mariano, M., Dufresne, A., Thomas, S., Gałęski, A., Recent developments in nanocellulose-based biodegradable polymers, thermoplastic polymers, and porous nanocomposites. *Prog. Polym. Sci.*, 87, 197–227, 2018, https://doi.org/10.1016/j.progpolymsci.2018.07.008.
11. Zhang, Y., Nypelö, T., Salas, C., Arboleda, J., Hoeger, I.C., Rojas, O.J., Cellulose Nanofibrils. *J. Renew Mater.*, 1, 195–211, 2013, https://doi.org/10.7569/JRM.2013.634115.
12. Sultana, S., Reja, L.N., Mohammad Mahbub Hilary, Saiful Islam, M., Zamsed Uddin, M., Ajmotgir, WM., Preparation and characterization of nanocellulose from Albizia lebbeck sawdust and their application in nanocomposites using poly (vinyl chloride) (PVC). *Res. Sq*, 2021.
13. Das, R., Lindström, T., Khan, M., Rezaei, M., Hsiao, B.S., Nanocellulose preparation from diverse plant feedstocks, processes, and chemical treatments: A review emphasizing non-woods. *Bioresources*, 19, 2023, https://doi.org/10.15376/biores.19.1.Das.
14. Elfaleh, I., Abbassi, F., Habibi, M., Ahmad, F., Guedri, M., Nasri, M., Garnier, C., A comprehensive review of natural fibers and their composites: An eco-friendly alternative to conventional materials. *Results Eng.*, 19, 101271, 2023, https://doi.org/10.1016/j.rineng.2023.101271.
15. Trache, D., Tarchoun, A.F., Derradji, M., Hamidon, T.S., Masruchin, N., Brosse, N., Hussin, M.H., Nanocellulose: From Fundamentals to Advanced Applications. *Front. Chem.*, 8, 2020, https://doi.org/10.3389/fchem.2020.00392.
16. Norizan, M.N., Shazleen, S.S., Alias, A.H., Sabaruddin, F.A., Asyraf, M.R.M., Zainudin, E.S., Abdullah, N., Samsudin, M.S., Kamarudin, S.H., Norrrahim, M.N.F., Nanocellulose-Based Nanocomposites for Sustainable Applications: A Review. *Nanomaterials*, 12, 3483, 2022, https://doi.org/10.3390/nano12193483.
17. Gu, H., Reiner, R., Bergman, R., Rudie, A., LCA Study for Pilot Scale Production of Cellulose Nano Crystals (CNC) from Wood Pulp. *LCA XV Paper Proceedings – A "Bright" Green" Future*, 2015.
18. Arockiasamy, F.S., Manoharan, B., Santhi, V.M., Prakalathan, K., Periasamy, D., Dhandapani, A., Natarajan, V., Krishnasamy, S., Thiagamani, S.M.K., Ilyas, R.A., Navigating the nano-world future: Harnessing cellulose nanocrystals from green sources for sustainable innovation. *Heliyon*, 11, e41188, 2025, https://doi.org/10.1016/j.heliyon.2024.e41188.
19. Khalid, M.Y., Al Rashid, A., Arif, Z.U., Ahmed, W., Arshad, H., Recent advances in nanocellulose-based different biomaterials: types, properties,

and emerging applications. *J. Mater. Res. Technol.*, 14, 2601–2623, 2021, https://doi.org/10.1016/j.jmrt.2021.07.128.
20. Boran, S., Kiziltas, A., Kiziltas, E.E., Gardner, D.J., Characterization of Ultrafine Cellulose-filled High-Density Polyethylene Composites Prepared using Different Compounding Methods. *Bioresources*, 11, 2016, https://doi.org/10.15376/BIORES.11.4.8178-8199.
21. Abdollahi Boraei, S.B., Bakhshandeh, B., Mohammadzadeh, F., Haghighi, D.M., Mohammadpour, Z., Clay-reinforced PVC composites and nanocomposites. *Heliyon*, 10, e29196, 2024, https://doi.org/10.1016/j.heliyon.2024.e29196.
22. Brebu, M., Environmental Degradation of Plastic Composites with Natural Fillers—A Review. *Polym. (Basel)*, 12, 166, 2020, https://doi.org/10.3390/polym12010166.
23. Kaur, P., Sharma, N., Munagala, M., Rajkhowa, R., Aallardyce, B., Shastri, Y., Agrawal, R., Nanocellulose: Resources, Physio-Chemical Properties, Current Uses and Future Applications. *Front. Nanotechnol.*, 3, 2021, https://doi.org/10.3389/fnano.2021.747329.
24. Gupta, G.K. and Shukla, P., Lignocellulosic Biomass for the Synthesis of Nanocellulose and Its Eco-Friendly Advanced Applications. *Front. Chem.*, 8, 2020, https://doi.org/10.3389/fchem.2020.601256.
25. Grishkewich, N., Mohammed, N., Tang, J., Tam, K.C., Recent advances in the application of cellulose nanocrystals. *Curr. Opin. Colloid Interface Sci.*, 29, 32–45, 2017, https://doi.org/10.1016/j.cocis.2017.01.005.
26. Lagerwall, J.P.F., Schütz, C., Salajkova, M., Noh, J., Hyun Park, J., Scalia, G., Bergström, L., Cellulose nanocrystal-based materials: from liquid crystal self-assembly and glass formation to multifunctional thin films. *NPG Asia Mater.*, 6, e80–e80, 2014, https://doi.org/10.1038/am.2013.69.
27. Bin Rashid, A., Hoque, M.E., Kabir, N., Rifat, F.F., Ishrak, H., Alqahtani, A., Chowdhury, M.E.H., Synthesis, Properties, Applications, and Future Prospective of Cellulose Nanocrystals. *Polym. (Basel)*, 15, 4070, 2023, https://doi.org/10.3390/polym15204070.
28. Misra, M., Vivekanandhan, S., Mohanty, A.K., Denault, J., Nanotechnologies for Agricultural Bioproducts, in: *Comprehensive Biotechnology (Second Edition)*, M. Moo-Young (ed.), pp. 111–119, Academic Press, Amsterdam, The Netherlands, 2011, ISBN 9780080885049, https://doi.org/10.1016/B978-0-08-088504-9.00260-9.
29. Kim, H.-J., Roy, S., Rhim, J.-W., Effects of various types of cellulose nanofibers on the physical properties of the CNF-based films. *J. Environ. Chem. Eng.*, 9, 106043, 2021, https://doi.org/10.1016/j.jece.2021.106043.
30. Barja, F., Bacterial nanocellulose production and biomedical applications. *J. Biomed. Res*, 35, 2021, https://doi.org/10.7555/JBR.35.20210036.
31. Manimaran, M., Norizan, M.N., Kassim, M.H.M., Adam, M.R., Norrrahim, M.N.F., Knight, V.F., Critical assessment of the thermal stability and degradation of chemically functionalized nanocellulose-based

polymer nanocomposites. *Nanotechnol. Rev.*, 13, 2024, https://doi.org/10.1515/ntrev-2024-0005.

32. Hu, W., Chen, S., Yang, J., Li, Z., Wang, H., Functionalized bacterial cellulose derivatives and nanocomposites. *Carbohydr. Polym.*, 101, 1043–1060, 2014, https://doi.org/10.1016/j.carbpol.2013.09.102.
33. Fumagalli, M., Sanchez, F., Molina-Boisseau, S., Heux, L., Surface-restricted modification of nanocellulose aerogels in gas-phase esterification by di-functional fatty acid reagents. *Cellulose*, 22, 1451–1457, 2015, https://doi.org/10.1007/s10570-015-0585-3.
34. Sun, F., Nordli, H.R., Pukstad, B., Kristofer Gamstedt, E., Chinga-Carrasco, G., Mechanical characteristics of nanocellulose-PEG bionanocomposite wound dressings in wet conditions. *J. Mech. Behav. Biomed. Mater.*, 69, 377–384, 2017, https://doi.org/10.1016/j.jmbbm.2017.01.049.
35. Jorfi, M. and Foster, E.J., Recent advances in nanocellulose for biomedical applications. *J. Appl. Polym. Sci.*, 132, 2015, https://doi.org/10.1002/app.41719.
36. Berglund, L., Cellulose-based nanocomposites. *Nat. Fibers Biopolymers Biocomposites*, 807–832, 2005, https://doi.org/10.1201/9780203508206.ch1.
37. Mathew, A.P., Thielemans, W., Dufresne, A., Mechanical properties of nanocomposites from sorbitol plasticized starch and tunicin whiskers. *J. Appl. Polym. Sci.*, 109, 4065–4074, 2008, https://doi.org/10.1002/app.28623.
38. Ladhar, A., Bendahou, A., Arous, M., Dufresne, A., Kaddami, H., Dielectric Spectroscopy: An Efficient Tool to Study the Interfacial Adhesion and Properties of Natural Rubber/Nanocellulose-Based Green Nanocomposites, in: *Handbook of Nanocellulose and Cellulose Nanocomposites*, pp. 627–648, Wiley-VCH Verlag GmbH & Co. KGaA, Weinheim, Germany, 2017, https://doi.org/10.1002/9783527689972.ch19.
39. Jia, X., Chen, Y., Shi, C., Ye, Y., Abid, M., Jabbar, S., Wang, P., Zeng, X., Wu, T., Rheological properties of an amorphous cellulose suspension. *Food Hydrocoll.*, 39, 27–33, 2014, https://doi.org/10.1016/j.foodhyd.2013.12.026.
40. Li, M.C., Wu, Q., Song, K., Lee, S., Qing, Y., Wu, Y., Cellulose Nanoparticles: Structure-Morphology-Rheology Relationships. *ACS Sustain Chem. Eng.*, 3, 821–832, 2015, https://doi.org/10.1021/acssuschemeng.5b00144.
41. Ait-Touchente, Z., Khellaf, M., Raffin, G., Lebaz, N., Elaissari, A., Recent advances in polyvinyl chloride (PVC) recycling. *Polym. Adv. Technol.*, 35, 2024, https://doi.org/10.1002/pat.6228.
42. Asadinezhad, A., Lehocký, M., Sáha, P., Mozetič, M., Recent progress in surface modification of polyvinyl chloride. *Materials*, 5, 2012, https://doi.org/10.3390/ma5122937.
43. Bakar, N.A., Chee, C.Y., Abdullah, L.C., Ratnam, C.T., Ibrahim, N.A., Thermal and dynamic mechanical properties of grafted kenaf filled poly (vinyl chloride)/ethylene vinyl acetate composites. *Mater. Des.*, 65, 204–211, 2015, https://doi.org/10.1016/j.matdes.2014.09.027.
44. Miller, R.W., Vinyl Chloride. *Pediatrics*, 54, 138–138, 1974, https://doi.org/10.1542/peds.54.2.138.

45. Mobarak, Y., Bassyouni, M., Almutawa, M., Materials Selection, Synthesis, and Dielectrical Properties of PVC Nanocomposites. *Adv. Mater. Sci. Eng.*, 2013, 1–6, 2013, https://doi.org/10.1155/2013/149672.
46. Khan, S.M., Gull, N., Khan, R.U., Butt, M.T.Z., Polyvinylchloride (PVC): Structure and Properties Relationship, in: 2022: pp. 19–47, https://doi.org/10.1007/978-3-030-78455-3_2.
47. Elgharbawy, A.S., Poly Vinyl Chloride Additives and Applications - A Review. *J. Risk Anal. Crisis Response*, 12, 2022, doi: 10.54560/jracr.v12i3.335.
48. Hashmi, S.U.M., Iqbal, M.A., Malik, M., Qamar, M.T., Khan, M., Zahid, A., Md.R. Islam, M., Morsy, K., Lai, W.-C., Synthesis and Characterization of Polyvinyl Chloride Matrix Composites with Modified Scrap Iron for Advanced Electronic, Photonic, and Optical Systems. *Nanomater.*, 12, 3147, 2022, https://doi.org/10.3390/nano12183147.
49. Tomaszewska, J., Sterzyński, T., Walczak, D., Thermal Stability of Nanosilica-Modified Poly(vinyl chloride). *Polym. (Basel)*, 13, 2057, 2021, https://doi.org/10.3390/polym13132057.
50. Andrew, L.J., Lizundia, E., MacLachlan, M.J., Designing for Degradation: Transient Devices Enabled by (Nano)Cellulose. *Adv. Mater.*, 2024, https://doi.org/10.1002/adma.202401560.
51. Baird, D.G., Polymer Processing, in: *Encyclopedia of Physical Science and Technology (Third Edition)*, R.A. Meyers (ed.), pp. 611–643, Academic Press, San Diego, California, United States, 2003, ISBN 9780122274107, https://doi.org/10.1016/B0-12-227410-5/00593-7.
52. Chazeau, L., Cavaillé, J.Y., Perez, J., Plasticized PVC reinforced with cellulose whiskers. II. Plastic behavior. *J. Polym. Sci. B Polym. Phys.*, 38, 2000, https://doi.org/10.1002/(SICI)1099-0488(20000201)38:3<383::AID-POLB5>3.0.CO;2-Q.
53. Chu, Y., Sun, Y., Wu, W., Xiao, H., Dispersion Properties of Nanocellulose: A Review. *Carbohydr. Polym.*, 250, 116892, 2020, https://doi.org/10.1016/j.carbpol.2020.116892.
54. Shen, R., Xue, S., Xu, Y., Liu, Q., Feng, Z., Ren, H., Zhai, H., Kong, F., Research Progress and Development Demand of Nanocellulose Reinforced Polymer Composites. *Polym. (Basel)*, 12, 2113, 2020, https://doi.org/10.3390/polym12092113.
55. Parthasarathy, V., Kumar, P.S., Aureen Albert, A., Krishnasamy, S., Chandrasekar, M., Recent progress in nanocellulose-based biocomposites for bone tissue engineering and wound healing applications. *Carbohydr. Polym.*, 357, 123455, 2025, https://doi.org/10.1016/j.carbpol.2025.123455.
56. Kadry, G. and Abd El-Hakim, A.E.F., Effect of nanocellulose on the biodegradation, morphology and mechanical properties of polyvinylchloride/nanocellulose nanocomposites. *Res. J. Pharm. Biol. Chem. Sci.*, 6, 2015.
57. Chanda, S. and Bajwa, D.S., A review of current physical techniques for dispersion of cellulose nanomaterials in polymer matrices. *Rev. Adv. Mater. Sci.*, 60, 325–341, 2021, https://doi.org/10.1515/rams-2021-0023.

58. Zhang, J., Wu, Q., Li, W., Surface and Reducing End Modification of Nanocellulose to Tailor Miscibility and Mechanical Performance of Reinforced Elastomer Sustainable Composites, in: 2024, https://doi.org/10.5772/intechopen.114105.
59. Cichosz, S., Masek, A., Rylski, A., Cellulose Modification for Improved Compatibility with the Polymer Matrix: Mechanical Characterization of the Composite Material. *Materials*, 13, 5519, 2020, https://doi.org/10.3390/ma13235519.
60. Shimpi, N.G., Verma, J., Mishra, S., Dispersion of Nano CaCO3 on PVC and its Influence on Mechanical and Thermal Properties. *J. Compos. Mater.*, 44, 211–219, 2010, https://doi.org/10.1177/0021998309344637.
61. Gan, P.G., Sam, S.T., bin Abdullah, M.F., Omar, M.F., Thermal properties of nanocellulose-reinforced composites: A review. *J. Appl. Polym. Sci.*, 137, 2020, https://doi.org/10.1002/app.48544.
62. Geng, S., Wei, J., Aitomäki, Y., Noël, M., Oksman, K., Well-dispersed cellulose nanocrystals in hydrophobic polymers by *in situ* polymerization for synthesizing highly reinforced bio-nanocomposites. *Nanoscale*, 10, 11797–11807, 2018, https://doi.org/10.1039/C7NR09080C.
63. Ullah, M.W., Ul-Islam, M., Wahid, F., Yang, G., Editorial: Nanocellulose: A Multipurpose Advanced Functional Material, Volume II. *Front. Bioeng. Biotechnol.*, 10, 2022, https://doi.org/10.3389/fbioe.2022.931256.
64. He, Y., Zhu, J., Wang, W., Ni, H., Surface modification of cellulose nanocrystals with different acid anhydrides for improved dispersion in poly(butylene succinate). *RSC Adv.*, 8, 38305–38314, 2018, https://doi.org/10.1039/C8RA07597B.
65. Solhi, L., Guccini, V., Heise, K., Solala, I., Niinivaara, E., Xu, W., Mihhels, K., Kröger, M., Meng, Z., Wohlert, J., Tao, H., Cranston, E.D., Kontturi, E., Understanding Nanocellulose–Water Interactions: Turning a Detriment into an Asset. *Chem. Rev.*, 123, 1925–2015, 2023, https://doi.org/10.1021/acs.chemrev.2c00611.
66. Kwak, H.W., You, J., Lee, M.E., Jin, H.-J., Prevention of cellulose nanofibril agglomeration during dehydration and enhancement of redispersibility by hydrophilic gelatin. *Cellulose*, 26, 4357–4369, 2019, https://doi.org/10.1007/s10570-019-02387-z.
67. Cho, M.-J. and Park, B.-D., Tensile and thermal properties of nanocellulose-reinforced poly(vinyl alcohol) nanocomposites. *J. Ind. Eng. Chem.*, 17, 36–40, 2011, https://doi.org/10.1016/j.jiec.2010.10.006.
68. Qiao, A., Cui, M., Huang, R., Ding, G., Qi, W., He, Z., Klemeš, J.J., Su, R., Advances in nanocellulose-based materials as adsorbents of heavy metals and dyes. *Carbohydr. Polym.*, 272, 118471, 2021, https://doi.org/10.1016/j.carbpol.2021.118471.
69. Liu, P., Oksman, K., Mathew, A.P., Surface adsorption and self-assembly of Cu(II) ions on TEMPO-oxidized cellulose nanofibers in aqueous media.

J. Colloid Interface Sci., 464, 175–182, 2016, https://doi.org/10.1016/j.jcis.2015.11.033.

70. Mo, L., Tan, Y., Shen, Y., Zhang, S., Highly compressible nanocellulose aerogels with a cellular structure for high-performance adsorption of Cu(II). *Chemosphere*, 291, 132887, 2022, https://doi.org/10.1016/j.chemosphere.2021.132887.
71. Robles, E., Csóka, L., Labidi, J., Effect of Reaction Conditions on the Surface Modification of Cellulose Nanofibrils with Aminopropyl Triethoxysilane. *Coatings*, 8, 139, 2018, https://doi.org/10.3390/coatings8040139.
72. Ghasemlou, M., Daver, F., Ivanova, E.P., Habibi, Y., Adhikari, B., Surface modifications of nanocellulose: From synthesis to high-performance nanocomposites. *Prog. Polym. Sci.*, 119, 101418, 2021, https://doi.org/10.1016/j.progpolymsci.2021.101418.
73. Liu, S., Tian, Z., Ji, X.-X., Ma, M.-G., Preparation and modification of nanocellulose using deep eutectic solvents and their applications. *Cellulose*, 31, 2175–2205, 2024, https://doi.org/10.1007/s10570-024-05738-7.
74. Colquhoun, H.M., Lewis, D.F., Herbertson, P.L., Wade, K., Polyetherketones based on para-carborane: Synthesis, sulfonation, and membrane-forming characteristics. *Polym. (Guildf)*, 1997, https://doi.org/10.1016/S0032-3861(96)01016-6.
75. Agostoni, V., Anand, R., Monti, S., Hall, S., Maurin, G., Horcajada, P., Serre, C., Bouchemal, K., Gref, R., Impact of phosphorylation on the encapsulation of nucleoside analogues within porous iron(iii) metal–organic framework MIL-100(Fe) nanoparticles. *J. Mater. Chem. B*, 1, 4231, 2013, https://doi.org/10.1039/c3tb20653j.
76. Kaczmarek, M.B., Struszczyk-Swita, K., Li, X., Szczęsna-Antczak, M., Daroch, M., Enzymatic modifications of chitin, chitosan, and chitooligosaccharides. *Front. Bioeng. Biotechnol.*, 7, 2019, https://doi.org/10.3389/fbioe.2019.00243.
77. Dinodia, M., Microwave-promoted Synthesis of Novel Bioactive N-based Heterocycles. *Mini-Rev. Org. Chem.*, 20, 136–155, 2023, https://doi.org/10.2174/1570193X19666220420133723.
78. Chu, P., Plasma-surface modification of biomaterials. *Mater. Sci. Eng.: R: Rep.*, 36, 143–206, 2002, https://doi.org/10.1016/S0927-796X(02)00004-9.
79. Meng, X., Bocharova, V., Tekinalp, H., Cheng, S., Kisliuk, A., Sokolov, A.P., Kunc, V., Peter, W.H., Ozcan, S., Toughening of nanocelluose/PLA composites *via* bio-epoxy interaction: Mechanistic study. *Mater. Des.*, 139, 188–197, 2018, https://doi.org/10.1016/j.matdes.2017.11.012.
80. Bárta, J., Hájková, K., Sikora, A., Jurczyková, T., Popelková, D., Kalous, P., Effect of a Nanocellulose Addition on the Mechanical Properties of Paper. *Polym. (Basel)*, 16, 73, 2023, https://doi.org/10.3390/polym16010073.
81. Chakrabarty, A. and Teramoto, Y., Recent Advances in Nanocellulose Composites with Polymers: A Guide for Choosing Partners and How to Incorporate Them. *Polym. (Basel)*, 10, 517, 2018, https://doi.org/10.3390/polym10050517.

82. Yang, X., Biswas, S.K., Han, J., Tanpichai, S., Li, M., Chen, C., Zhu, S., Das, A.K., Yano, H., Surface and Interface Engineering for Nanocellulosic Advanced Materials. *Adv. Mater.*, 33, 2021, https://doi.org/10.1002/adma.202002264.
83. Dufresne, A., Cellulose nanomaterial reinforced polymer nanocomposites. *Curr. Opin. Colloid Interface Sci.*, 29, 2017, https://doi.org/10.1016/j.cocis.2017.01.004.
84. Mohd Ishak, N.A., Abdullah, F.Z., Muhd Julkapli, N., Production and characteristics of nanocellulose obtained with using of ionic liquid and ultrasonication. *J. Nanopart. Res.*, 24, 171, 2022, https://doi.org/10.1007/s11051-022-05549-6.
85. Shojaeiarani, J., Bajwa, D., Holt, G., Sonication amplitude and processing time influence the cellulose nanocrystals morphology and dispersion. *Nanocomposites*, 6, 2020, https://doi.org/10.1080/20550324.2019.1710974.
86. de Souza, A.G., de Lima, G.F., Colombo, R., Rosa, D.S., A new approach for the use of anionic surfactants: nanocellulose modification and development of biodegradable nanocomposites. *Cellulose*, 27, 5707–5728, 2020, https://doi.org/10.1007/s10570-020-03160-3.
87. Patil, S.P., Baikar, P., Tayade, R.R., Surface modification of Nano cellulose: The path to advanced uses of smart and sustainable bio-material. *GSC Adv. Res. Rev.*, 17, 127–137, 2023, https://doi.org/10.30574/gscarr.2023.17.1.0400.
88. Dhali, K., Daver, F., Cass, P., Adhikari, B., Surface modification of the cellulose nanocrystals through vinyl silane grafting. *Int. J. Biol. Macromol.*, 200, 397–408, 2022, https://doi.org/10.1016/j.ijbiomac.2022.01.079.
89. Darie-Nita, R.N., Râpă, M., Visakh, P.M., Bio-Based Polyvinylchloride (PVC)-Related Blends, in: 2022: pp. 211–234, https://doi.org/10.1007/978-3-030-78455-3_10.
90. Jubsilp, C., Asawakosinchai, A., Mora, P., Saramas, D., Rimdusit, S., Effects of Organic Based Heat Stabilizer on Properties of Polyvinyl Chloride for Pipe Applications: A Comparative Study with Pb and CaZn Systems. *Polym. (Basel)*, 14, 133, 2021, https://doi.org/10.3390/polym14010133.
91. Zheng, C., Li, D., Ek, M., Mechanism and kinetics of thermal degradation of insulating materials developed from cellulose fiber and fire retardants. *J. Therm. Anal. Calorim.*, 135, 3015–3027, 2019, https://doi.org/10.1007/s10973-018-7564-5.
92. López-Beceiro, J., Álvarez-García, A., Sebio-Puñal, T., Zaragoza-Fernández, S., Álvarez-García, B., Díaz-Díaz, A., Janeiro, J., Artiaga, R., Kinetics of thermal degradation of cellulose: Analysis based on isothermal and linear heating data. *Bioresources*, 11, 2016, https://doi.org/10.15376/biores.11.3.5870-5888.
93. Pasek, R.J., Chang, D.P.Y., Jones, A.D., Investigation of Thermal Decomposition of Chlorinated Polymers. *Hazard. Waste Hazard. Mater.*, 13, 23–38, 1996, https://doi.org/10.1089/hwm.1996.13.23.
94. Venkatraman, P., Gohn, A.M., Rhoades, A.M., Foster, E.J., Developing high performance PA 11/cellulose nanocomposites for industrial-scale

melt processing. *Compos. B Eng*, 174, 2019, https://doi.org/10.1016/j.compositesb.2019.106988.
95. Zheng, X.G., Tang, L.H., Zhang, N., Gao, Q.H., Zhang, C.F., Bin Zhu, Z., Dehydrochlorination of PVC materials at high temperature. *Energy Fuels*, 17, 2003, https://doi.org/10.1021/ef020131g.
96. Benchikh, L., Aouissi, T., Aitferhat, Y., Chorfi, H., Abacha, I., Kebaili, M., Guessoum, M., Merzouki, A., Grohens, Y., Carraro, M., Djellali, S., Effect of different compatibilization routes on the mechanical, thermal and rheological properties of polypropylene/cellulose nanocrystals nanocomposites. *Polym. Bull.*, 81, 11007–11026, 2024, https://doi.org/10.1007/s00289-024-05225-w.
97. Moberg, T., Sahlin, K., Yao, K., Geng, S., Westman, G., Zhou, Q., Oksman, K., Rigdahl, M., Rheological properties of nanocellulose suspensions: effects of fibril/particle dimensions and surface characteristics. *Cellulose*, 24, 2499–2510, 2017, https://doi.org/10.1007/s10570-017-1283-0.
98. Loranger, E., Piché, A.-O., Daneault, C., Influence of High Shear Dispersion on the Production of Cellulose Nanofibers by Ultrasound-Assisted TEMPO-Oxidation of Kraft Pulp. *Nanomaterials*, 2, 286–297, 2012, https://doi.org/10.3390/nano2030286.
99. Kramar, A., Rodríguez Ortega, I., González-Gaitano, G., González-Benito, J., Solution casting of cellulose acetate films: influence of surface substrate and humidity on wettability, morphology and optical properties. *Cellulose*, 30, 2023, https://doi.org/10.1007/s10570-022-05026-2.
100. Aigaje, E., Riofrio, A., Baykara, H., Processing, Properties, Modifications, and Environmental Impact of Nanocellulose/Biopolymer Composites: A Review. *Polym. (Basel)*, 15, 1219, 2023, doi: 10.3390/polym15051219.
101. Soeta, H., Fujisawa, S., Saito, T., Isogai, A., Controlling Miscibility of the Interphase in Polymer-Grafted Nanocellulose/Cellulose Triacetate Nanocomposites. *ACS Omega*, 5, 23755–23761, 2020, https://doi.org/10.1021/acsomega.0c02772.
102. Rebouillat, S. and Pla, F., State of the Art Manufacturing and Engineering of Nanocellulose: A Review of Available Data and Industrial Applications. *J. Biomater. Nanobiotechnol.*, 04, 165–188, 2013, https://doi.org/10.4236/jbnb.2013.42022.
103. Hubbe, M.A., Ferrer, A., Tyagi, P., Yin, Y., Salas, C., Pal, L., Rojas, O.J., Nanocellulose in Thin Films, Coatings, and Plies for Packaging Applications: A Review. . *Bioresour.*, 12, 2143–2233, 2017, https://doi.org/10.15376/biores.12.1.2143-2233.
104. Balea, A., Blanco, A., Delgado-Aguilar, M., Concepcion Monte, M., Tarrés, Q., Fuente, E., Mutjé, P., Negro, C., Nanocellulose Characterization Challenges. *Bioresources*, 16, 2021, https://doi.org/10.15376/biores.16.2.Balea.
105. Wu, Q., Xie, W., Wu, H., Wang, L., Liang, S., Chang, H., Liu, B., Effect of volatile solvent and evaporation time on formation and performance of PVC/PVC-*g*-PEGMA blended membranes. *RSC Adv.*, 9, 34486–34495, 2019, https://doi.org/10.1039/C9RA05454E.

106. Gu, X., Song, X., Shao, C., Zeng, P., Lu, X., Shen, X., Yang, Q., Electrospinning of poly(butylene-carbonate):Effect of Solvents on the Properties of the Nanofibers Film. *Int. J. Electrochem. Sci.*, 9, 8045–8056, 2014, https://doi.org/10.1016/S1452-3981(23)11025-X.
107. Wannatong, L., Sirivat, A., Supaphol, P., Effects of solvents on electrospun polymeric fibers: preliminary study on polystyrene. *Polym. Int.*, 53, 1851–1859, 2004, https://doi.org/10.1002/pi.1599.
108. Sultana, S., Reja, M.M., Hilary, L.N., Islam, M.S., Uddin, M.Z., Ajmo, WM., Preparation and characterization of nanocellulose from Albizia lebbeck sawdust and their application in nanocomposites using poly(vinyl chloride) (PVC), (2021), https://doi.org/10.21203/rs.3.rs-386714/v1.
109. Badi, N.S.A., Makmud, M.Z.H., Jamain, Z., Mohd Amin, K.N., Illias, H.A., Enhancement of structural and dielectric properties of PVC/CNC nanocomposites as electrical insulation materials. *Polym. Polym. Compos.*, 32, 2024, https://doi.org/10.1177/09673911241299196.
110. Abdulhameed, O., Al-Ahmari, A., Ameen, W., Mian, S.H., Additive manufacturing: Challenges, trends, and applications. *Adv. Mech. Eng.*, 11, 2019, https://doi.org/10.1177/1687814018822880.
111. Diniță, A., Neacșa, A., Portoacă, A.I., Tănase, M., Ilinca, C.N., Ramadan, I.N., Additive Manufacturing Post-Processing Treatments, a Review with Emphasis on Mechanical Characteristics. *Materials*, 16, 4610, 2023, https://doi.org/10.3390/ma16134610.
112. Zhang, Y., Cui, L., Xu, H., Feng, X., Wang, B., Pukánszky, B., Mao, Z., Sui, X., Poly(lactic acid)/cellulose nanocrystal composites *via* the Pickering emulsion approach: Rheological, thermal and mechanical properties. *Int. J. Biol. Macromol.*, 137, 197–204, 2019, https://doi.org/10.1016/j.ijbiomac.2019.06.204.
113. Xue, J., Wu, T., Dai, Y., Xia, Y., Electrospinning and Electrospun Nanofibers: Methods, Materials, and Applications. *Chem. Rev.*, 119, 5298–5415, 2019, https://doi.org/10.1021/acs.chemrev.8b00593.
114. Pasaoglu, M.E., Vatanpour, V., Koyuncu, I., Electrospinning of nanocellulose, in: *Woodhead Publishing Series in Composites Science and Engineering, Elastomeric Nanocellulose Composites,* S. Thomas, P.K. Mohamed, J. Kim, M. Tom (eds.), pp. 67–92, Woodhead Publishing, United Kingdom, 2024, ISBN 9780443186080, https://doi.org/10.1016/B978-0-443-18608-0.00005-X.
115. Saebnazar, A., Ahdeno, N., Nazemiyeh, A., Dadashi, H., Jaymand, M., Eskandani, M., Vandghanooni, S., Electrospinning of Nanocellulose–Nanoparticle Composites: Multifunctional Biomaterials for Advanced Tissue Engineering, in: *Nanocellulose-Based Hybrid Systems for Tissue Engineering, Royal Society of Chemistry*, pp. 170–197, 2024, https://doi.org/10.1039/9781837673094-00170.
116. Islam, M.S., Ang, B.C., Andriyana, A., Afifi, A.M., A review on fabrication of nanofibers *via* electrospinning and their applications. *SN Appl. Sci.*, 1, 2019, https://doi.org/10.1007/s42452-019-1288-4.

117. Ahmadi Bonakdar, M. and Rodrigue, D., Electrospinning: Processes. *Structures Materials Macromol.*, 4, 58–103, 2024, https://doi.org/10.3390/macromol4010004.
118. Leach, M.K., Feng, Z.-Q., Tuck, S.J., Corey, J.M., Electrospinning Fundamentals: Optimizing Solution and Apparatus Parameters. *J. Visualized Exp.*, 2011, https://doi.org/10.3791/2494.
119. Grause, G., Hirahashi, S., Toyoda, H., Kameda, T., Yoshioka, T., Solubility parameters for determining optimal solvents for separating PVC from PVC-coated PET fibers. *J. Mater Cycles Waste Manag.*, 19, 612–622, 2017, https://doi.org/10.1007/s10163-015-0457-9.
120. Lee, K.H., Kim, H.Y., La, Y.M., Lee, D.R., Sung, N.H., Influence of a mixing solvent with tetrahydrofuran and *N,N* -dimethylformamide on electrospun poly(vinyl chloride) nonwoven mats. *J. Polym. Sci. B Polym. Phys.*, 40, 2259–2268, 2002, https://doi.org/10.1002/polb.10293.
121. Reddy, V.S., Tian, Y., Zhang, C., Ye, Z., Roy, K., Chinnappan, A., Ramakrishna, S., Liu, W., Ghosh, R., A Review on Electrospun Nanofibers Based Advanced Applications: From Health Care to Energy Devices. *Polym. (Basel)*, 13, 3746, 2021, https://doi.org/10.3390/polym13213746.
122. Hosseini Ravandi, S.A., Sadrjahani, M., Valipouri, A., Dabirian, F., Ko, F.K., Recently developed electrospinning methods: a review. *Text. Res. J.*, 92, 5130–5145, 2022, https://doi.org/10.1177/00405175211069880.
123. Angel, N., Li, S., Yan, F., Kong, L., Recent advances in electrospinning of nanofibers from bio-based carbohydrate polymers and their applications. *Trends Food Sci. Technol.*, 120, 308–324, 2022, https://doi.org/10.1016/j.tifs.2022.01.003.
124. Chen, Z., Li, Z., Li, J., Liu, C., Lao, C., Fu, Y., Liu, C., Li, Y., Wang, P., He, Y., 3D printing of ceramics: A review. *J. Eur. Ceram. Soc.*, 39, 2019, https://doi.org/10.1016/j.jeurceramsoc.2018.11.013.
125. Horvath, J., *Slicing a 3D Model, in: Mastering 3D Printing*, pp. 47–63, Apress, Berkeley, CA, 2014, https://doi.org/10.1007/978-1-4842-0025-4_5.
126. Molenaar, J. and Ingrassia, D., 3D Printers, in: 2024: pp. 153–198, https://doi.org/10.1007/978-1-4842-9849-7_5.
127. 3D printing, *in: Up and Running with AutoCAD® 2020*, E.J. Gindis and R.C. Kaebisch (eds.), pp. 549–551, Academic Press, Elsevier, United States, 2020, ISBN 9780128198629, https://doi.org/10.1016/B978-0-12-819862-9.15010-9.
128. Finny, A.S., Popoola, O., Andreescu, S., 3D-Printable Nanocellulose-Based Functional Materials: Fundamentals and Applications. *Nanomaterials*, 11, 2358, 2021, https://doi.org/10.3390/nano11092358.
129. Thomas, D.J. and Claypole, T.C., 3-D Printing, in: *Printing on Polymers*, J. Izdebska and S. Thomas (eds.), pp. 293–306, William Andrew Publishing, Elsevier, Amsterdam, Netherlands, 2016, ISBN 9780323374682, https://doi.org/10.1016/B978-0-323-37468-2.00018-X.

130. Klar, V., Pere, J., Turpeinen, T., Kärki, P., Orelma, H., Kuosmanen, P., Shape fidelity and structure of 3D printed high consistency nanocellulose. *Sci. Rep.*, 9, 3822, 2019, https://doi.org/10.1038/s41598-019-40469-x.
131. Arey, B., Barrett, C., Arslan, I., Kennedy, Z., Warner, M., Schroeder, H., Investigating 3D Printing with Microscopy and Spectroscopy Techniques. *Microsc. Microanal.*, 23, 312–313, 2017, https://doi.org/10.1017/S1431927617002240.
132. Trivedi, A.K. and Gupta, M.K., PLA based biodegradable bionanocomposite filaments reinforced with nanocellulose: development and analysis of properties. *Sci. Rep.*, 14, 23819, 2024, https://doi.org/10.1038/s41598-024-71619-5.
133. Dalloul, F., Mietner, J.B., Navarro, J.R.G., Production and 3D Printing of a Nanocellulose-Based Composite Filament Composed of Polymer-Modified Cellulose Nanofibrils and High-Density Polyethylene (HDPE) for the Fabrication of 3D Complex Shapes. *Fibers*, 10, 91, 2022, https://doi.org/10.3390/fib10100091.
134. Sheltami, R.M., Kargarzadeh, H., Abdullah, I., Effects of silane surface treatment of cellulose nanocrystals on the tensile properties of cellulose-polyvinyl chloride nanocomposite. *Sains Malays*, 44, 2015, https://doi.org/10.17576/jsm-2015-4406-05.
135. Sun, Z., Yan, X., Xiao, Y., Hu, L., Eggersdorfer, M., Chen, D., Yang, Z., Weitz, D.A., Pickering emulsions stabilized by colloidal surfactants: Role of solid particles. *Particuology*, 64, 2022, https://doi.org/10.1016/j.partic.2021.06.004.
136. de Carvalho-Guimarães, F.B., Correa, K.L., de Souza, T.P., Rodríguez Amado, J.R., Ribeiro-Costa, R.M., Silva-Júnior, J.O.C., A Review of Pickering Emulsions: Perspectives and Applications. *Pharmaceuticals*, 15, 1413, 2022, https://doi.org/10.3390/ph15111413.
137. Albert, C., Beladjine, M., Tsapis, N., Fattal, E., Agnely, F., Huang, N., Pickering emulsions: Preparation processes, key parameters governing their properties and potential for pharmaceutical applications. *J. Controlled Release*, 309, 302–332, 2019, https://doi.org/10.1016/j.jconrel.2019.07.003.
138. Chen, L., Ao, F., Ge, X., Shen, W., Food-Grade Pickering Emulsions: Preparation. *Stabilization Applications Mol.*, 25, 3202, 2020, https://doi.org/10.3390/molecules25143202.
139. Pan, J., Chen, J., Wang, X., Wang, Y., Fan, J., Pickering emulsion: From controllable fabrication to biomedical application. *Interdiscip. Med.*, 1, 2023, https://doi.org/10.1002/INMD.20230014.
140. Marto, J., Nunes, A., Martins, A.M., Carvalheira, J., Prazeres, P., Gonçalves, L., Marques, A., Lucas, A., Ribeiro, H.M., Pickering Emulsions Stabilized by Calcium Carbonate Particles: A New Topical Formulation. *Cosmetics*, 7, 62, 2020, https://doi.org/10.3390/cosmetics7030062.
141. Udoetok, I.A., Wilson, L.D., Headley, J.V., Stabilization of Pickering Emulsions by Iron oxide Nano-particles (Fe3O4 NPs). *Adv. Mater. Sci.*, 1, 2016, https://doi.org/10.15761/AMS.1000107.

142. Costa, C., Medronho, B., Filipe, A., Mira, I., Lindman, B., Edlund, H., Norgren, M., Emulsion Formation and Stabilization by Biomolecules: The Leading Role of Cellulose. *Polym. (Basel)*, 11, 1570, 2019, https://doi.org/10.3390/polym11101570.
143. Tian, Y., Zhou, J., He, C., He, L., Li, X., Sui, H., Formation, T., Stabilization and Separation of Oil–Water Emulsions: A Review. *Processes*, 10, 738, 2022, https://doi.org/10.3390/pr10040738.
144. Ji, C. and Wang, Y., Nanocellulose-stabilized Pickering emulsions: Fabrication, stabilization, and food applications. *Adv. Colloid Interface Sci.*, 318, 2023, https://doi.org/10.1016/j.cis.2023.102970.
145. Morais, J.P.S., de F, M., de Brito, E.S., de Azeredo, H.M.C., de Figueirêdo, M.C.B., Sustainable Pickering Emulsions with Nanocellulose: Innovations and Challenges. *Foods*, 12, 3599, 2023, https://doi.org/10.3390/foods12193599.
146. Fujisawa, S., Togawa, E., Kuroda, K., Nanocellulose-stabilized Pickering emulsions and their applications. *Sci. Technol. Adv. Mater.*, 18, 959–971, 2017, https://doi.org/10.1080/14686996.2017.1401423.
147. Hammiche, D., Benslimane, A., Boukerrou, A., Mechanical Properties of PVC Composites and Nanocomposites, in: 2024: pp. 261–273, https://doi.org/10.1007/978-3-031-45375-5_12.
148. Abdel-Hakim, A. and Mourad, R., Nanocellulose and its polymer composites: preparation, characterization, and applications. *Russ. Chem. Rev.*, 92, 2023, https://doi.org/10.57634/rcr5076.
149. Lee, K.-Y., Aitomäki, Y., Berglund, L.A., Oksman, K., Bismarck, A., On the use of nanocellulose as reinforcement in polymer matrix composites. *Compos. Sci. Technol.*, 105, 15–27, 2014, https://doi.org/10.1016/j.compscitech.2014.08.032.
150. Li, W., Yue, J., Liu, S., Preparation of nanocrystalline cellulose *via* ultrasound and its reinforcement capability for poly(vinyl alcohol) composites. *Ultrason. Sonochem.*, 19, 479–485, 2012, https://doi.org/10.1016/j.ultsonch.2011.11.007.
151. Girard, M., Vidal, D., Bertrand, F., Tavares, J.R., Heuzey, M.-C., Evidence-based guidelines for the ultrasonic dispersion of cellulose nanocrystals. *Ultrason. Sonochem.*, 71, 105378, 2021, https://doi.org/10.1016/j.ultsonch.2020.105378.
152. Zhu, K., Zhou, X., Wang, D., Li, D., Lei, W., Fang, C., Hu, J., Luo, R., The Correlations between Microstructures and Color Properties of Nanocrystalline Cellulose: A Concise Review. *Polym. (Basel)*, 16, 2774, 2024, https://doi.org/10.3390/polym16192774.
153. Babaei-Ghazvini, A., Cudmore, B., Dunlop, M.J., Acharya, B., Bissessur, R., Ahmed, M., Whelan, W.M., Effect of magnetic field alignment of cellulose nanocrystals in starch nanocomposites: Physicochemical and mechanical properties. *Carbohydr. Polym.*, 247, 116688, 2020, https://doi.org/10.1016/j.carbpol.2020.116688.

154. Dhar, P., Kumar, A., Katiyar, V., Magnetic Cellulose Nanocrystal Based Anisotropic Polylactic Acid Nanocomposite Films: Influence on Electrical, Magnetic, Thermal, and Mechanical Properties. *ACS Appl. Mater. Interfaces*, 8, 18393–18409, 2016, https://doi.org/10.1021/acsami.6b02828.
155. Mesa, J., Hinestroza-Córdoba, L.I., Barrera, C., Seguí, L., Betoret, E., Betoret, N., High Homogenization Pressures to Improve Food Quality. *Functionality Sustainability Mol.*, 25, 3305, 2020, https://doi.org/10.3390/molecules25143305.
156. Bangar, S.P., Harussani, M.M., Ilyas, R.A., Ashogbon, A.O., Singh, A., Trif, M., Jafari, S.M., Surface modifications of cellulose nanocrystals: Processes, properties, and applications. *Food Hydrocoll.*, 130, 107689, 2022, https://doi.org/10.1016/j.foodhyd.2022.107689.
157. Wang, L., Gardner, D.J., Wang, J., Yang, Y., Tekinalp, H.L., Tajvidi, M., Li, K., Zhao, X., Neivandt, D.J., Han, Y., Ozcan, S., Anderson, J., Towards industrial-scale production of cellulose nanocomposites using melt processing: A critical review on structure-processing-property relationships. *Compos. B Eng.*, 201, 108297, 2020, https://doi.org/10.1016/j.compositesb.2020.108297.
158. Wågberg, L. and Erlandsson, J., The Use of Layer-by-Layer Self-Assembly and Nanocellulose to Prepare Advanced Functional Materials. *Adv. Mater.*, 33, 2021, https://doi.org/10.1002/adma.202001474.
159. Chazeau, L., Cavaillé, J.Y., Canova, G., Dendievel, R., Boutherin, B., Viscoelastic Properties of Plasticized PVC Reinforced with Cellulose Whiskers. *J. Appl. Polym. Sci.*, 71, 1999, https://doi.org/10.1002/(sici)1097-4628(19990314)71:11<1797::aid-app9>3.0.co;2-e.
160. Wu, Y., Liang, Y., Mei, C., Cai, L., Nadda, A., Van Le, Q., Peng, Y., Lam, S.S., Sonne, C., Xia, C., Advanced nanocellulose-based gas barrier materials: Present status and prospects. *Chemosphere*, 286, 131891, 2022, https://doi.org/10.1016/j.chemosphere.2021.131891.
161. Sirviö, J.A., Visanko, M., Heiskanen, J.P., Liimatainen, H., UV-absorbing cellulose nanocrystals as functional reinforcing fillers in polymer nanocomposite films. *J. Mater Chem. A Mater.*, 4, 6368–6375, 2016, https://doi.org/10.1039/C6TA00900J.
162. Elfiky, A.A.E.A., Mubarak, M.F., Keshawy, M., El Sayed, I.E.T., Moghny, T.A., Removing of cationic dyes using self-cleaning membranes-based PVC/nano-cellulose combined with titanium aluminate. *Environ. Sci. Pollut. Res.*, 30, 79091–79105, 2023, https://doi.org/10.1007/s11356-023-27691-x.
163. Virtanen, S., Vartianen, J., Setälä, H., Tammelin, T., Vuoti, S., Modified nanofibrillated cellulose–polyvinyl alcohol films with improved mechanical performance. *RSC Adv.*, 4, 11343, 2014, https://doi.org/10.1039/c3ra46287k.
164. Sharma, A., Thakur, M., Bhattacharya, M., Mandal, T., Goswami, S., Commercial application of cellulose nano-composites – A review. *Biotechnol. Rep*, 21, e00316, 2019, https://doi.org/10.1016/j.btre.2019.e00316.

8

Biodegradation of Polyvinyl Chloride (PVC)-Based Different Fiber Bionanocomposites

Udeshna Changmai[1,2], Sahana S.K.[1,2], Rituparna Duarah[1,3] and Tridip Phukan[1,2]*

[1]Agro-Technology and Rural Development Division, CSIR-North East Institute of Science and Technology (CSIR-NEIST), Jorhat, Assam, India
[2]Academy of Scientific and Innovative Research (AcSIR), Ghaziabad, Uttar Pradesh, India
[3]Polymer and Petroleum Group, Materials Sciences and Technology Division, CSIR North-East Institute of Science and Technology (CSIR-NEIST), Jorhat, Assam, India

Abstract
Polyvinyl chloride (PVC) is one of the most widely used synthetic polymers due to its versatility, durability, and cost-effectiveness. However, its environmental impact has raised significant concerns, primarily due to its resistance to natural degradation processes, contributing to long-term pollution in both terrestrial and marine ecosystems, making it essential to explore alternative solutions. In recent years, PVC-based natural fiber–reinforced bionanocomposites have emerged as a promising alternative. By incorporating natural fibers like jute, rice straw, sisal, silk fiber, wool fiber, and cashmere and nanoparticles such as nanoclays and titanium dioxide (TiO_2) to these PVC bionanocomposites, it enhances the material's strength, thermal stability, and environmental benefits. This chapter explores the potential of these composites as sustainable alternatives to conventional PVC, focusing on their biodegradation processes and associated mechanisms. The biodegradation mechanism of PVC-based bionanocomposites is a complex process influenced by the interaction between environmental and biological factors with the biocomposites. Key stages including colonization, biodeterioration, biofragmentation, assimilation, and mineralization are discussed along with types

**Corresponding author*: tridip@neist.res.in

Shiv Kumari Panda and Visakh P. M (eds.) Polyvinyl Chloride-Based Biocomposites and Bionanocomposites, (233–270) © 2026 Scrivener Publishing LLC

of microbes and enzymes associated with the biodegradation process. The role of microbial activity, enzymatic reactions, and external conditions (like physical, chemical, and synergistic effects) in breaking down the polymer matrix and natural fibers is highlighted. Since PVC-based bionanocomposites have various applications in automobile, construction, packaging, and medical fields, a better understanding of the biodegradation mechanism will facilitate the development of highly biodegradable and superior quality PVC-based bionanocomposites which will aid in the management of such plastic wastes for a sustainable environment.

***Keywords*:** Polyvinyl chloride (PVC), fiber-reinforced bionanocomposites, nanoparticles, biodegradation, assimilation, mineralization

8.1 Introduction

Poly(vinyl chloride), commonly known as PVC, is the third-most widely produced plastic globally, following polyethylene and polypropylene [1]. Widely utilized in the construction industry, PVC outperforms traditional materials like copper, iron, and wood, particularly in pipe and profile applications. Its versatility extends further when plasticizers, primarily phthalates, are added, rendering it softer and more flexible. This modified form of PVC finds extensive use in clothing, upholstery, electrical cable insulation, inflatable products, and various other applications where it often serves as a substitute for rubber [2]. Its utility is largely due to its resistance to both fire and water, properties that make it suitable for manufacturing raincoats, shower curtains, and water pipes [3]. The inherent flame resistance of PVC is attributed to its chlorine content, which inhibits combustion by releasing chlorine atoms when burned. Structurally, PVC is a vinyl polymer, similar to polyethylene, but with a chlorine atom replacing one hydrogen atom on every other carbon in the backbone chain. This polymer is typically produced *via* the free radical polymerization of vinyl chloride, though it can also be synthesized by reacting acetylene with hydrochloric acid (HCl) [4]. Furthermore, plasticized PVC is used to manufacture coated fabrics, including materials for tarpaulins, large tents, and halls, as well as artificial leather for the clothing, automotive, and furniture industries. In the automotive sector, PVC is predominantly used for cable insulation, fuel hoses, soundproofing mats, and anti-corrosion coatings. The extensive and varied use of PVC has led to the generation of significant waste streams that require safe management.

PVC can undergo weathering degradation which causes embrittlement and microcracking and leads to the generation of nanoplastics and microplastics which persists for a longer duration of time in the environment.

It was discovered that farmland soil samples in China were consistently polluted with various forms of microplastics, especially in open-field agriculture soils (4308 items/kg) with a 40.9% PVC concentration [5, 6]. The environmental concerns associated with PVC are further exacerbated by the additives used in its production, such as plasticizers, stabilizers, and flame retardants. PVC contains the highest concentration of plasticizer (up to 50%) and stabilizers in comparison to all major kinds of synthetic plastics, such as polypropylene (PP), polystyrene (PS), and polyethylene (PE) [7]. These additives can leach out of the material over time, causing soil and water contamination and posing risks to both wildlife and human health. The traditional techniques for getting rid of PVC include recycling, landfilling, and incineration; however, these approaches have a number of drawbacks, including poor efficiency, high costs, and the possibility of secondary contamination [8].

The need for sustainable materials that can degrade more readily in the environment has led to the development of PVC-based bionanocomposites, which incorporate natural fibers and nanoparticles into the polymer matrix. Bionanocomposites (BNCs) formed by PVC polymer matrix reinforcement with natural fibers are more environmentally friendly than traditional PVCs, which are known to contain high chlorine components and considered as potential hazardous materials to the ecosystem [9]. Since, natural fiber is "ecologically friendly," combining PVC with natural fibers provides an intriguing substitute. Utilizing the synergistic assembly of biopolymers with inorganic nanometer-sized materials, the production of innovative bionanocomposites with multifunctionality is a promising area of research [10]. These bionanocomposites not only retain the desirable properties of PVC but also offer the potential for enhanced biodegradability. Therefore, the use of PVC-based natural fiber bionanocomposites offers eco-friendliness and sustainability due to its biodegradable properties.

Microbial biodegradation of polymers is nowadays gaining attention worldwide due to its economic and environmental benefits that reduce the negative impact of polymers in an environment-friendly manner [11]. Despite a common misconception that PVC is difficult to degrade due to its complex composition, microbe-based solutions have been gaining attention for PVC waste management. This chapter aims to highlight the PVC-based natural fiber–reinforced bionanocomposites and potential biodegradation of such materials, particularly for readers who may not be familiar with the intricacies of PVC biodegradation mechanisms by microbes. Understanding this biodegradation mechanism is crucial in advancing towards developing superior PVC-based bionanocomposites with high biodegradability potential for better waste management and a cleaner environment.

8.2 PVC-Based Natural Fiber–Reinforced Bionanocomposites

Bionanocomposites are advanced composite materials synthesized by combining natural fibers and biopolymers or traditional polymers with nanoparticles within the range of 1–100 nm [12]. In a PVC-based fiber bionanocomposites, the polymer matrix, i.e., synthetic polymer PVC acts as the primary component, which is majorly chosen for its durability, chemical resistance, and widespread use in industries. The reinforcing agents in bionanocomposites are often natural fibers derived from plants, such as jute, hemp, flax, and sisal. A diagrammatic representation of PVC-based natural fiber–reinforced bionanocomposite has been given in Figure 8.1. Natural fibers contribute significantly to the mechanical properties of bionanocomposites by providing high tensile strength, stiffness, impact

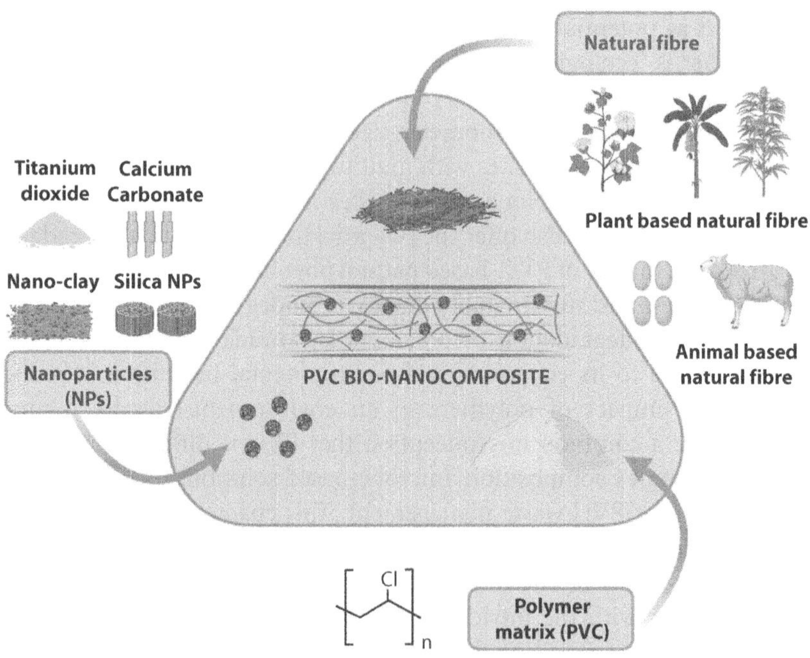

Figure 8.1 Diagrammatic representation of the composition in constructing PVC-based natural fiber–reinforced bionanocomposites.

resistance, and improved mechanical properties and gas barrier properties, which are essential for different applications. These fibers are more renewable, biodegradable, and cost-effective than synthetic fibers like glass or carbon, making them ideal for both automotive and construction materials and also in applications requiring lightweight materials [13, 14].

In recent years, several biopolymers have been used in combination with PVC and nanoparticles to create varieties of bionanocomposites. A growing number of synthetic and naturally occurring polymers, such as poly(lactic-co-glycolic acids) (PLGA), poly(ε-caprolactone) (PCL), poly(butylene succinate), and naturally occurring starch, cellulose, poly(hydroxyalkanoates), pectin, and chitosan, are being utilized in bionanocomposite technology. Over the past few decades, other synthetic biodegradable polymers have been produced, especially those made of harmless building components such as fatty diols, carbonic acid, or dicarboxylic acids [10]. Additionally, animal and silk fibers are also applied to produce PVC bionanocomposites with distinct mechanical, chemical, and biocompatible qualities. The most common fibers used are silk fiber, wool fiber, cashmere, mohair fiber, and alpaca fiber, etc., which provide a vast range of functionality and applications. These are frequently utilized in the production of biomedical devices, in construction materials, automotive industry, and also in the textile sector. After plant-based fiber, they are the second most often utilized fiber. However, animal fibers are not commonly used in the commercial manufacturing of naturally reinforced fiber composites (NFRCs) due to the higher cost of protein fibers compared to plant fibers [15, 16].

Nanoparticles are added to the composites to further improve its properties. These can include nanoclays, carbon nanotubes, or metal oxide nanoparticles. The inclusion of nanoparticles at the nanoscale level leads to significant improvements in thermal stability, barrier properties, resistance to moisture and chemicals, and mechanical strength, allowing better stress transfer between the polymer matrix and the reinforcing agents often without adding much weight to the composite [17]. Certain nanoparticles can impart additional functionalities to the bionanocomposite, such as antimicrobial activity, UV protection, or electrical conductivity [13]. Following the addition of inorganic nanofiller, a polymer's biodegradation rate likewise varies, either increasing or decreasing, but its biodegradation capacity remains intact.

The production of different types of PVC-based fiber bionanocomposites possesses multiple advantages over traditional PVC. The addition of hemp fibers integrated to PVC enhances the tensile strength and biodegradability, while nanoclays or titanium dioxide (TiO_2) helps in improving thermal stability and barrier properties [18, 19]. This combination makes the composite suitable for automotive and construction applications where durability and environmental sustainability are highly essential. PVC is reinforced with jute fibers and nanoparticles like calcium carbonate ($CaCO_3$), where jute fibers contribute to the tensile and flexural strength of the composite, while $CaCO_3$ nanoparticles enhance its stiffness and thermal resistance [20, 21]. These composites also show better biodegradability compared to pure PVC. Sisal fibers and silica nanoparticles are also used to make PVC-based bionanocomposites, where sisal fibers provide high impact resistance, while silica nanoparticles improve the composite's resistance to UV degradation and enhance its barrier properties. This makes the composite suitable for outdoor applications such as fencing, decking, and outdoor furniture [22]. Different types of natural fibers used in the production of PVC-based fiber-reinforced bionanocomposites are discussed below and a list has been given in Table 8.1.

Table 8.1 Various natural fibers and nanomaterials/coupling agents associated with the PVC polymer matrix.

Natural fiber	Nanomaterials and coupling agents	References
Wood composites	Maleated polypropylene (MAPP)	[46]
Wood-fiber composites	Chlorinated polyethylene (CPE), ethylene vinyl acetate (EVA), methacrylate–butadiene styrene (MBS), and all acrylic elastomer	[47]
Wood flour	Organic acids, anhydrides, isocyanates, and silanes	[48]
Wood flour	Calcium carbonate	[49]
Pine flour	Isocyanate copper EA, thermos plastic starch (TPS)	[50]
Bagasse fiber	Benzoic acid	[51]
Bagasse fibers	Sodium hydroxide, benzoic	[52]
Palm fiber and sawdust	Calcium carbonate, lubricants, stabilizers and impact modifiers	[53]

(*Continued*)

Table 8.1 Various natural fibers and nanomaterials/coupling agents associated with the PVC polymer matrix. (*Continued*)

Natural fiber	Nanomaterials and coupling agents	References
Palm fiber	NaOH, 3-aminopropyltriethoxysilane, combined alkali-silane, bleaching with $NaClO_2$, and $H_2O_2+HNO_3$	[54, 55]
Sawdust	Silane	[56]
Oil palm empty fruit bunch (OPEFB) composites	Silane, Zirconate; benzoyl chloride and NaOH	[57, 58]
Rice husk	Microcrystalline cellulose (MCC) and montmorillonite (MMT)	[59]
Rice husk ash	γ-aminopropyltrimethoxysilane	[60]
Rice straw	SDBS-modified graphene oxide	[61]
Bamboo flour	Block copolymers based on NMP P[(SMA)-b-(SAN)]	[62]
Bamboo	Calcium/zinc, acrylic acid, dicarboxylic acid ester, fatty acid, glycerol ester, polyethylene wax, *B. vulgaris* particles (fillers)	[63]
Bamboo	Sodium bisulfate (oxidant agent), sodium hydroxide (alkali agent), sodium silicate (silicate agent)	[64]
Kenaf fiber	Poly [methylene poly (phenyl isocyanate)]	[65]
Kenaf fiber	(With thermoplastic polyurethane)	[66]
Olive husk	Calcium carbonate, titanium dioxide	[67]
Jute fabric		[68]
Cotton stalk fiber	Copper ethanolamine	[69]

(*Continued*)

Table 8.1 Various natural fibers and nanomaterials/coupling agents associated with the PVC polymer matrix. (*Continued*)

Natural fiber	Nanomaterials and coupling agents	References
Areca sheath	Dioctyl phthalate, stearic acid, sodium hydroxide, benzoyl chloride	[70]
Coconut shell	Calcium stearate, acrylic polymer, titanium oxide, tin, stearic acid	[71]
Sisal fiber	Benzoyl peroxide, acetic anhydride, stearic acid, permanganate, silane, maleic anhydride	[22]
Citrus maxima	Tetrahydrofuran	[72]

8.2.1 Clay-Reinforced Polyvinyl Chloride (PVC) Fiber Composites

Clay-reinforced polyvinyl chloride (PVC) nanocomposites represent a significant advancement in material science, finding utility across a broad range of applications, including the biomedical sector, automotive industry, water purification, packaging, fire protection, and construction [23]. Key clays utilized in the fabrication of these composites include bentonite, montmorillonite, kaolinite, and illite. Incorporating these nanoclays into the PVC matrix enhances its mechanical strength, thermal stability, and yellowness index [24]. Clay minerals, composed of layered aluminosilicates, have shown significant potential as sorbents in environmental cleanup, food packaging, biomedicine, construction, and various other applications. This broad utility is attributed to their availability, cost-effectiveness, extensive surface area, cation-exchange capacity, and layered structure. Even a small amount of nanoclay as reinforcing agent can significantly enhance the polymer's physicochemical properties. The production of clay-based polymer nanocomposites can lead to the formation of intercalated or exfoliated structures, that improve attributes such as thermal stability, mechanical strength, and barrier performance [25–27].

Rani *et al.* synthesized PVC/MMT/CuO nanocomposite films using the Solution Blending method [28]. As the polymer is mixed with the swollen clay, aided by agitation techniques (sheer mixing, stirring, ultrasonication, etc.), it replaces the solvent molecules within the clay galleries, followed by removal of the solvent, which leads to the reassembly of the polymer-intercalated clay to form the nanocomposite. The films, with thicknesses ranging from 60 to 80 μm, were prepared by dispersing PVC in dimethylformamide, heating the mixture to 60°C while stirring, and then separately dispersing CuO nanoparticles and MMT in the same solvent through ultrasonication, followed by stirring for 12 h, and finally cast onto a Petri dish and dried for removal of solvent [28]. Melt blending is another technique for preparing clay/PVC nanocomposites, where the polymer and nanoclay are processed above the polymer's melting temperature (T_m) [29] followed by secondary processing, such as extrusion, injection molding, or compression molding. In comparison to Solution Blending, Melt Blending is more suitable for industrial applications, as it achieves better dispersion of the polymer and clay. Bouchareb *et al.* reported the synthesis of PVC/montmorillonite nanocomposites by a melt-mixing process using three types of montmorillonite (MMT): sodium montmorillonite (Na+-MMT), raw montmorillonite (R-MMT), and organically modified montmorillonite (O-MMT), in presence of dioctyl phthalate (DOP) [30]. Again, *in situ* polymerization has been reported as an effective approach for preparing clay/PVC nanocomposites which offers significant advantages in terms of material properties and processing control [31]. *In situ* polymerization entails the insertion of monomers into the interlayer spaces of clay, with subsequent polymerization triggered by heat, radiation, or chemical initiators [32]. This results in the production of highly dispersed polymer composites, as the polymer chains grow on the surface of the nanoclays, facilitating clay exfoliation. Yang *et al.* reported *in situ* intercalated polymerization of vinyl chloride (VC) for the synthesis of exfoliated PVC/montmorillonite (MMT) nanocomposites, in presence of organic-intercalated montmorillonite (OMMT) [33]. Again, Haiyan *et al.* reported the synthesis of PVC–organoclay nanocomposites *via* melt blending intercalation and *in situ* polymerization intercalation [34].

8.2.2 Lignin-Polyvinyl Chloride (PVC) Blends

Lignin and its derivatives have found great utilization as functional, environmentally friendly fillers in polyvinyl chloride (PVC). Lignin, a complex natural polymer found in plant cell walls, provides essential structural support, rigidity, and protection against microbial attacks [29]. As the second most abundant organic polymer on Earth, after cellulose, lignin is crucial in forming wood and bark, thereby enhancing the plant's strength and durability. Research has explored the partial substitution of PVC with various types of lignin to develop cost-effective materials. These studies have examined both plasticized (soft) and rigid PVC blends [35]. Despite this, there is limited literature on the use of hybrid materials for PVC composites where the organic component includes either unmodified or modified lignin. Lignin's complex, cross-linked 3D network structure offers strong compatibility with PVC. It has a naturally low glass transition temperature and forms part of a block copolymer with amorphous polysaccharides. Ping *et al.* have reported that while lignin and PVC show a certain degree of compatibility, reducing the lignin content tends to improve this compatibility [36]. However, there is a decrease in impact strength, tensile strength, and bending strength of PVC/lignin composites with increase in the lignin content. Maintaining adequate thermal stability during the processing of molten PVC remains a common challenge. Zhang *et al.* have reported enhanced thermal stability and flame retardancy of PVC-based composites with the incorporation of magnesium borate hydrate with mechanically activated lignin while still preserving desirable tensile and flexural properties [37]. Further investigations have examined the effects of different lignin types—such as Alcell lignin, Indulin lignin, hardwood lignin, softwood lignin, and softwood soda lignin—on the properties of PVC blends. Feldman *et al.* studied the incorporation of kraft and Alcell® lignins into PVC composites to evaluate their impact on mechanical and thermal properties [38]. Additionally, Klapiszewski *et al.* developed a novel SiO_2-lignin hybrid material as a new filler for PVC [39]. They processed PVC compounds as a dry blend with this silica-lignin hybrid material and separately with the individual components used to create the hybrid, utilizing a Brabender mixing chamber for the blending process.

8.2.3 Polysaccharide-Based Polyvinyl Chloride (PVC) Fiber Bionanocomposites

Polysaccharides such as cellulose, chitin, glycogen, and starch have been incorporated into PVC to create various blends, offering opportunities to enhance both the properties and sustainability of PVC materials. Cellulose, for example, when blended with PVC, can significantly improve mechanical strength and thermal stability, making it ideal for applications such as biodegradable films and coatings [40]. Cellulose nanocrystals (CNCs), which are derived from renewable sources like wood pulp, are renowned for their high aspect ratio, mechanical strength, and biodegradability. When CNCs are melt-blended with PVC, the resulting composite benefits from enhanced tensile strength and stiffness due to the strong interfacial bonding between the nanocrystals and the PVC matrix.

Such bionanocomposites are particularly useful in applications requiring superior mechanical properties, such as construction materials and automotive components. Chitosan, a biopolymer sourced from chitin in crustacean exoskeletons, is valued for its antimicrobial properties and biocompatibility [41]. When chitosan nanoparticles are melt-blended with PVC, the resulting bionanocomposite exhibits improved antimicrobial activity alongside enhanced mechanical properties, including increased impact resistance and tensile strength. These characteristics make chitosan-PVC composites suitable for medical applications, such as antimicrobial surfaces and devices, as well as for packaging materials where resistance to microbial growth is essential. Starch, a naturally occurring polysaccharide, can also be converted into nanoparticles through various chemical and physical processes [42]. When these starch nanoparticles are incorporated into the PVC matrix through melt blending and extrusion, they enhance the biodegradability of the composite while maintaining its mechanical properties. This makes starch-PVC blends particularly valuable in the packaging industry, where there is growing demand for eco-friendly materials that decompose more readily. These blends improve flexibility and reduce density, benefiting applications like packaging materials and agricultural films [3]. In addition, innovative techniques have been employed to modify PVC with polysaccharides. For instance, Bigot *et al.* used click chemistry to graft antibacterial seaweed polysaccharides onto PVC surfaces [43]. Sobahi *et al.* prepared PVC–chitosan blends using dithizone nanoparticles and various solvents, achieving good compatibility and investigating their use for removing metal ions from aqueous solutions [44]. Badr *et al.* explored the incorporation of heparin-modified chitosan membranes to enhance sodium-selective membrane electrodes based on PVC. Heparin, a naturally occurring anticoagulant, was chemically modified to create a heparin–chitosan conjugate, which was then blended with PVC [45]. Overall, incorporating polysaccharides into PVC blends not only enhances performance characteristics such as mechanical strength and thermal stability but also improves the environmental sustainability of PVC. This aligns with the growing demand for more sustainable and eco-friendly materials.

8.3 Biodegradation Processes of PVC-Based Bionanocomposites

The biodegradation of PVC-based natural fiber–reinforced bionanocomposites is a complex process involving various mechanisms that break

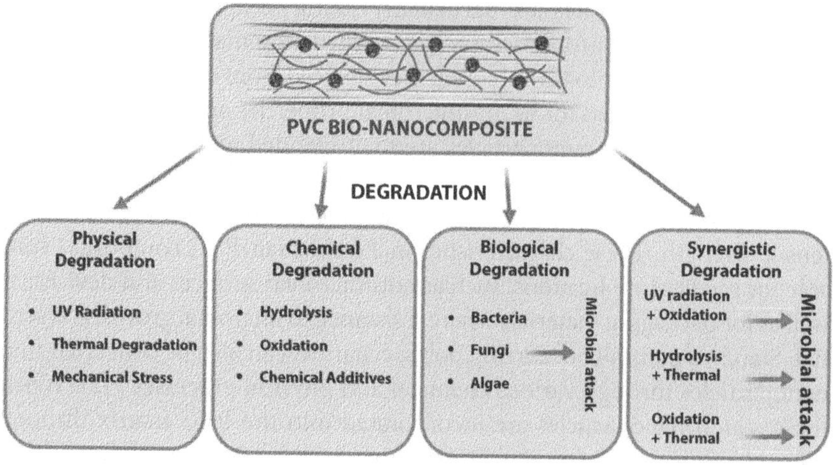

Figure 8.2 Various processes involved in the degradation of PVC-based bionanocomposites.

down the composite materials into simpler compounds. The degradation of biocomposites is influenced by the chemical, physical, biochemical, and synergistic interactions of the polymer matrix, natural fibers, and nanoparticles [73]. It includes the breakdown of covalent bonds in the polymer's backbone, which causes irreversible changes in the chemical and molecular structure of the material, thus breaking down the composite material into simpler byproducts [74]. The geographical variations and the presence of microbes are examples of extrinsic circumstances that impact a polymer's breakdown process, in addition to the polymer's intrinsic qualities. Graphical representation of biodegradation factors is shown in Figure 8.2.

8.3.1 Types of Biodegradations

8.3.1.1 Physical Degradation

Physical degradation refers to the mechanical and environmental factors that lead to the breakdown of the composite's structure. Abiotic factors including heat, light, radiation, humidity, the medium's pH, mechanical stress, and chemical attack can all function as catalysts to start the breakdown process of polymers. In the context of PVC-based bionanocomposites, this can include the decomposition of PVC due to elevated temperatures, leading to the release of hydrochloric acid (HCl) and other volatile compounds. This process can initiate further degradation by breaking the polymer chains, leading to embrittlement and loss of mechanical integrity [75]. Repeated cycles of heating and cooling can also induce thermal stress within the composite, causing cracks and fissures, which can act as pathways for further degradation by chemical or biological means. Mechanical forces, such as abrasion, impact, or flexural stress, can physically break down the composite material. This degradation is often localized, leading to microcracks that propagate over time and weaken the material. Natural fibers are hydrophilic, meaning they absorb moisture. This can lead to swelling and subsequent mechanical stress within the composite, further promoting physical degradation. Ultraviolet (UV) irradiation UV-A (319–400 nm) and UV-B (290–315 nm) can also play a major role in inducing photochemical reactions in the PVC matrix, leading to chain scission [76]. UV exposure causes PVC to dehydrochlorinate quickly, resulting in short fragments of conjugated unsaturation in the polymer. These unsaturated C=C double bonds can be further photodegraded and are less stable. PVC contains chromophores, which can absorb UV light and produce free radicals. These radicals then react with oxygen to make peroxides and other oxidative products through the abstraction of hydrogen. Aldehydes, ketones, carboxylic acids, esters, and alcohols are produced when the peroxide moiety dissociates into macro-alkoxy and hydroxyl radicals, which catalyzes the next chemical sequence. This accelerates the polymeric artifact's degrading process by causing chain scission and cross-linking of polymers [77–79]. Natural fibers, such as flax or jute, are also susceptible to UV degradation, which can weaken the fiber–matrix interface and contribute to the overall degradation of the composite.

8.3.1.2 Chemical Degradation

Chemical degradation involves the interaction of the composite material with various chemicals present in the atmosphere that directly or indirectly catalyze the formation of radicals by photochemical reactions, which lead to the breakdown of its structure [80]. The presence of water leads to hydrolytic degradation, where water molecules react with the polymer chains, breaking ester or amide bonds within the matrix [81]. This process is particularly relevant for biodegradable polymers, but in PVC, it can be accelerated by the presence of natural fibers, which can absorb and retain water. PVC degradation releases hydrochloric acid (HCl), further catalyzing hydrolysis reactions, particularly in the presence of moisture. This acid can also interact with the natural fibers, leading to their breakdown and further compromising the integrity of the composite. Free radicals formed during thermal or UV degradation reacts with oxygen present in the environment, leading to the formation of peroxides, ketones, and carboxylic acids. This oxidative process weakens the polymer chains, leading to embrittlement and loss of mechanical properties [82], and also makes the fibers prone to oxidative degradation.

8.3.1.3 *Biological Degradation*

Biotic degradation occurs due to microbial action, such as bacteria, fungi, and algae [83, 84], which modify and consume the polymer or polymeric monomers, producing molecules of low molar mass (terpenes, aldehydes, and acids) and gases (N_2, CH_4, and CO_2). However, surface topography, crystallinity, hydrophobicity, and molecular size of synthetic polymers play a major factor in the biodegradation of the building blocks of plastics [85]. Natural fibers within the composite provide an organic substrate that can be colonized by microorganisms. These microorganisms secrete specific enzymes, such as cellulases, cutinases, ligninases, proteases, carboxylesterases, and laccases, which can break down the cellulose and lignin components of the fibers, leading to fiber degradation. These enzymes are also studied to attack the PVC matrix to a lesser extent [86–88]. The presence of natural fibers can also promote the formation of biofilms on the composite surface. These biofilms can create localized environments with different pH levels or oxygen concentrations, further promoting the degradation process. During the biodegradation process, microorganisms adhere to the polymer surface, colonizes the exposed surface followed by enzymatic degradation of the polymers *via* hydrolytic cleavage. This produces molecules of lower molecular weight and finally mineralizes into CO_2 and H_2O. There are different types of plastic-degrading enzymes available in nature. A list of these enzymes is given in Table 8.2.

8.3.1.4 *Synergistic Degradation*

Two or more degradation mechanisms interact to cause synergistic degradation, leading to an accelerated or enhanced breakdown of the composite material. The combination of thermal and oxidative degradation can lead to a more rapid breakdown of the polymer matrix while also causing thermal and oxidative degradation of the natural fibers [95]. Heat accelerates the formation of free radicals, while oxygen promotes the formation of oxidative products, leading to a synergistic effect that weakens the fiber–matrix interface, hence triggering rapid breakdown of the composite [96]. The presence of microorganisms on the composite surface can create localized environments that enhance chemical degradation processes, such as

Table 8.2 Different types of plastic degrading enzymes and their mode of action.

Enzyme name	Type of polymer	Biodegradation process	Mode of action
Cellulases	Cellulose	Hydrolysis	Cellulases are enzymes that cleave -1,4-glycosidic linkages in cellulose to produce glucose and play an essential role in the breakdown of synthetic polymers [89].
Proteases	Proteins	Hydrolysis	Protease enzymes degrade proteins by breaking peptide links and converting them into smaller peptides and, eventually, amino acids. Proteases are the only enzymes that can hydrolyze amide or urethane bonds [90].
Lipases	Lipids	Hydrolysis	Lipases are responsible for breaking ester bonds in lipids, creating free fatty acids and glycerol and are known to degrade a variety of plastics [91].
Amylases	Starch	Hydrolysis	Amylase degrades 1, 4-glycosidic bonds in starch, and study reveals that in the presence of starch, alpha-amylase has a non-specific effect on polyethylene (PE) molecules, degrading them [92].

(*Continued*)

Table 8.2 Different types of plastic degrading enzymes and their mode of action. (*Continued*)

Enzyme name	Type of polymer	Biodegradation process	Mode of action
Laccases	Lignin	Oxidation	Break down the polymer into smaller fragments by oxidizing the phenolic and non-phenolic components in lignin [91]
Chitinases	Chitin	Hydrolysis	Capable of breaking down 1,4-glycosidic bonds and releasing N-acetylglucosamine from chitin, a complex polymer found in the exoskeletons of arthropods, fungus, and other organisms [93]. Even though reports suggest that chitinases do not directly degrade plastic, it is thought that chitinases may be involved in the biodegradation of plastic.
Peroxidases	Lignin	Oxidation	Catalyze the oxidation of lignin, converting it into smaller particles using hydrogen peroxide or oxygen. Some peroxidases, such as lignin peroxidases found in fungus, can break down PVC [94].

hydrolysis or oxidation. For example, microbial activity can produce acidic or basic byproducts that catalyze chemical reactions, leading to a synergistic effect [97]. The absorption of moisture by natural fibers can also create a favorable environment for microbial growth. The combined effects of water-induced swelling, microbial enzyme production, and chemical interactions can lead to accelerated degradation of the composite material.

8.3.2 Factors Affecting the Biodegradation of PVC-Based Bionanocomposites

The biodegradation of PVC-based bionanocomposites is a complex and multifaceted process influenced by various factors like material composition, environmental conditions, and microbial activity. These factors determine the rate, extent, and nature of the degradation, ultimately affecting the material's environmental impact and longevity.

The biodegradation processes are markedly affected by the inherent properties of the PVC matrix, the type and content of natural fibers, and the presence of nanoparticles in the composite. Factors such as fiber-matrix compatibility, the hydrophilicity of the natural fibers, and the biodegradability of any additives or nanoparticles contribute to how the composite material breaks down under various conditions [98]. Natural fibers like jute, hemp, sisal, and flax have high cellulose content, which is more readily biodegradable by microorganisms. Whereas fibers with higher lignin content, such as hardwood fibers, are more resistant to biodegradation, thus slowing down the overall degradation of the composite. The microstructure of the fibers also plays a role in the biodegradation process. For example, fibers with a porous structure allow easier access for microorganisms and enzymes, enhancing biodegradability. Chemical treatments like alkali treatment (mercerization) are often used to improve fiber–matrix adhesion in composites. These treatments can also alter the biodegradability of the fibers, making the fibers more susceptible to microbial attack [99]. Some fibers may be coated with hydrophobic materials to improve their compatibility with the PVC matrix. While this improves mechanical properties, it can reduce the fiber's biodegradability by limiting microbial access. The amount of natural fiber in the composite (fiber loading) also affects biodegradability. Higher fiber content generally increases the overall biodegradability of the composite since a larger proportion of the material is composed of biodegradable components. However, beyond a certain threshold, the amount of fiber can lead to poor dispersion in the matrix, resulting in weaker mechanical properties and potentially inconsistent biodegradation [100].

External factors such as temperature, moisture, UV exposure, and the presence of oxygen or other chemicals in the environment play a pivotal role in the degradation process. The type and activity of microorganisms present in the environment, along with the availability of nutrients, influence the rate and extent of biodegradation. Temperature is a critical factor influencing the rate of biodegradation. Higher temperatures generally enhance microbial activity and enzyme production, accelerating the biodegradation process. Extremely high temperatures can however denature enzymes or kill microorganisms; while very low temperatures can slow down microbial metabolism, significantly reducing the biodegradation rate. In addition to biological processes, temperature can also induce thermal degradation of the PVC matrix and fibers, leading to the formation of microcracks that facilitate microbial colonization. Adequate moisture levels are essential for microbial growth, and it is also necessary for hydrolytic degradation, thus breaking down the chemical bonds within the polymer matrix and fibers. In dry conditions, biodegradation is significantly slowed down as the activity of microorganisms is hindered. The pH of the environment can affect microbial growth and enzyme activity. Most microorganisms thrive in neutral to slightly acidic conditions (pH 5–7). In highly acidic or alkaline environments, the rate of biodegradation may decrease due to the inhibition of microbial activity or denaturation of enzymes involved in the degradation process. The pH can also affect the chemical stability of the PVC matrix and natural fibers. Acidic conditions can lead to hydrolysis of the polymer chains, weakening the material and making it more susceptible to biodegradation. The diversity and concentration of microorganisms in the environment directly influence the rate of biodegradation. Environments rich in cellulose-degrading bacteria and fungi, such as soil rich in organic matter, will see faster degradation of natural fibers in the composite. Microorganisms can form biofilms on the surface of the composite, where they secrete extracellular enzymes that break down the polymer matrix and fibers. The formation of biofilms can enhance the biodegradation process, especially in natural environments like soil or compost [16, 101].

Understanding these factors is crucial for designing bionanocomposites with desirable degradation properties tailored to specific applications while minimizing their environmental impact.

8.3.3 Stages of Biodegradation

The biodegradation of PVC-based natural fiber–reinforced bionanocomposites take place in a multi-stage process that involves the gradual breakdown of the composite materials into simpler compounds by environmental and biological factors. It can happen over time (which usually approximates 6 months) in a variety of ways. Ideally, it should occur spontaneously and without any human involvement. Five stages are involved in the biodegradation process: colonization, biodeterioration, biofragmentation, assimilation, and mineralization (Figure 8.3). Each stage plays a crucial role in the overall degradation of the composite, leading to its eventual

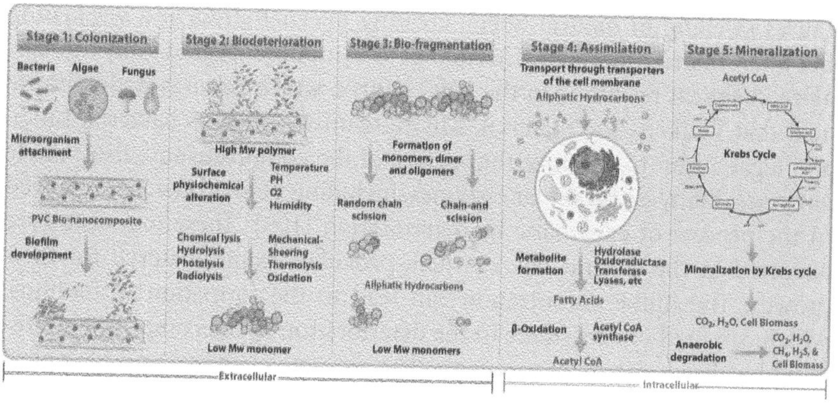

Figure 8.3 Stages of biodegradation of PVC-based bionanocomposites.

breakdown and assimilation into the environment; however, plastic biodegradation is only confirmed after verifying mineralization [102, 103].

8.3.3.1 *Colonization*

The colonization of microbial species on polymer surfaces and the formation of microbial consortia are the initial stages of microbial biodegradation. This leads to the development of biofilms and serious damage to the polymer surface [97]. The production of various proteins and polysaccharides, which pierce into the material pores and alter their size, enables microbial adhesion to the polymer surface [104]. As a result, microbial adhesion and activity on the plastic surface cause the biodeterioration stage. Surface degradation occurs at this stage, changing the physicochemical properties of the plastic polymer.

8.3.3.2 *Biodeterioration*

Biodeterioration, which happens when many microorganisms and/or abiotic stressors combine to break down large materials into minute pieces (micro, sub-micro), acts as the first indicator of biodegradation. Deterioration is a visible degradation that affects the material's mechanical, physical, and chemical qualities. The material's surface is deteriorated by the action of abiotic factors such as UV radiation, temperature fluctuations, moisture, and mechanical stress (e.g., abrasion). UV radiations induce photochemical reactions in the PVC matrix, leading to the formation of free radicals, chain scission, and surface cracks whereas moisture causes swelling in natural fibers, creating internal stress and weakening the fiber–matrix interface. Thermal or UV-induced degradation can catalyze chemical reactions in the PVC matrix, which releases hydrochloric acid that weakens the material. Microorganisms such as bacteria, fungi, and algae begin to colonize the surface of the composite and also secrete enzymes. Many extracellular enzymes, including peroxidases, are involved in the process of enzymatic biodeterioration [105]. These organisms are attracted to the natural fibers, which provide organic material for their growth. The initial colonization can lead to the formation of biofilms, which can further contribute to the deterioration of the composite by creating localized microenvironments that promote chemical and biological reactions [13, 86, 106]. Acidic byproducts from microbial activity can also contribute to the chemical deterioration of both the PVC matrix and the natural fibers.

8.3.3.3 *Biofragmentation*

In this process, microorganisms excrete catalytic agents (e.g., enzymes), gradually lowering the polymer's molecular weight. At this point, the polymers are broken into tiny fragments. The terms biofragmentation or, in certain circumstances, depolymerization are used to describe situations in which macromolecular size is reduced without affecting the chemical content or structure of the monomer unit. Enzymatic assays are performed to determine polymer biofragmentation propensity. Catalytic agents like enzymes, are excreted by the microorganisms, which targets the natural fibers in the composites and also progressively reduces the molecular weight of the polymers. At this moment, the polymers are cleaved until the production of small molecules (dimers and monomers). Enzymes such as cellulases, hemicellulases, and ligninases break down the cellulose, hemicellulose, and lignin components of the fibers, leading to the fragmentation of the fibers into smaller units like oligosaccharides, monosaccharides, and lignin monomers [107]. Microbial activity and the presence of oxidative agents can lead to the scission of polymer chains within the PVC matrix. This can result in the formation of smaller polymer fragments, which are more susceptible to further degradation [108]. The scission process can be initiated by oxidative reactions, UV-induced free radical formation, or hydrolytic reactions. The biofragmentation process of bionanocomposites can also lead to the release of embedded nanoparticles, which may then undergo further degradation or dispersion into the environment [109]. The behavior and fate of these nanoparticles depend on their chemical composition and the surrounding environmental conditions.

8.3.3.4 *Assimilation*

Assimilation of plastic happens in the cytoplasm when tiny fragments produced during depolymerization assimilate with microbial metabolism to create energy, biomass, and other metabolites. The small organic molecules, such as monosaccharides, fatty acids, and small polymer fragments, produced during biofragmentation are assimilated by microorganisms and transported into the cytoplasm of microbial cells, where they are catabolized to undergo further metabolic processes. During the assimilation process, the polyfragments enter the cells *via* three common catabolic routes: fermentation, anaerobic respiration, or aerobic respiration. Microorganisms assimilate when they employ polymers as sources of carbon and nitrogen, transforming into the building blocks of cells [110, 111]. Monosaccharides derived from cellulose degradation from natural fibers associated with PVC matrix may enter the glycolysis pathway, leading to the production of energy (ATP) and intermediate compounds that are further processed in the citric acid cycle (Kreb's cycle). The energy and carbon obtained from the catabolism of degradation products allow microbes to develop, grow, and consume new energy packages (substrates) from the environment. This results in the formation of microbial biomass, which can continue to colonize and degrade the composite material [102].

8.3.3.5 *Mineralization*

Mineralization, the final step in biodegradation, happens simultaneously with assimilation, turning organic matter into minerals by excreting metabolites and simple molecules that the environment and microbes may absorb [73, 102]. Since many microorganisms involved in biodegradation have intricate relationships and symbioses, it is difficult to replicate natural degradation in a lab setting. Some microorganisms, for example, break down polymers and produce CO_2 (mineralization), while others reduce the polymer into its constituent monomers, and some use these monomers to excrete simpler residual compounds that serve as a source of energy, while others use the excreted residues as substrates. Given the intricacy of the metabolic pathways and the formation of new products, it is worth noting that CO_2 and H_2O gases are formed during aerobic biodegradation and may be used to monitor the activity at this stage. In contrast to aerobic processes, which produce CO_2, the anaerobic process results in the generation of both CO_2 and CH_4 as primary end products [111, 112]. The mineralization process results in the release of inorganic substances, which are then absorbed by the surrounding environment. These inorganic compounds, including CO_2, H_2O, and mineral salts, are essential components of the global carbon and nutrient cycles. The final degradation products are fully integrated into the environment, contributing to soil and atmospheric components.

Each stage of the biodegradation process of PVC-based natural fiber–reinforced bionanocomposites is crucial in the complete breakdown of the composite material, leading to its eventual integration into the environment. Biotechnological advances in the field of biodegradation of these composite matters have also shown increased degradation rate, making it essential for designing more sustainable bionanocomposites. Scientists have studied the usage of stimulants and inducers (e.g., 0.2% sodium lactate) that could help in increasing the enzyme activities of microorganisms to enhance the MP degradation rate [85]. Another process involves the utilization of pure and/or consortia polymer-degrading cultures and the addition of genetically engineered microorganisms (i.e., bioaugmentation), which enhances the biodegradation activities of microorganisms due to the presence of a wider enzymatic catabolic activity [113].

8.4 Microbial and Enzymatic Degradation on PVC-Based Natural Fiber–Reinforced Bionanocomposites

PVC-based bionanocomposites including natural fibers usually have higher quantities of lignocellulosic components, which consist of cellulose, hemicellulose, and lignin. Different enzymes generated by bacteria, fungi, and other microbes aid in the breakdown of these fibers into smaller, more assimilable components. Cellulose is a main target for microbial breakdown and enzymes involved in cellulolysis, such as endoglucanases, exoglucanases (cellobiohydrolases), and β-glucosidases that depolymerizes cellulose into shorter cellulose fragments, cellobiose (disaccharides), or glucose monomer. Numerous microbes, such as *Clostridium thermocellum* and *Trichoderma reesei*, provide the enzymes necessary for the cellulolysis of the fibrous component of the composite [114, 115]. Hemicellulose is less crystalline than cellulose and is made up of several sugar monomers such as xylose, mannose, glucose, and arabinose. This makes it more vulnerable to enzymatic attacks and lead to the formation of different sugar monomers and oligosaccharides. Studies have shown that hemicellulose is broken down by xylanases, mannases, and arabinofurosidases, which are produced by microorganisms like *Bacillus subtilis* and *Aspergillus niger* [116].

Major ligninolytic enzymes that can break down both the phenolic and non-phenolic components of lignin include laccases, peroxidases (e.g., manganese peroxidase and lignin peroxidase), and various other peroxidases. These are mostly generated by certain bacterial strains, such as *Panibacillus* sp., *Aneurinibacillus aneurinilyticus*, and *Bacillus* sp., and white-rot fungi like *Phanerochaete chrysosporium* and *Trametes versicolour*, which produces smaller aromatic compounds due to lignin breakdown [117]. These compounds can be further metabolized by other microorganisms while facilitating the access of cellulolytic and hemicellulolytic enzymes to the cellulose and hemicellulose within the fibers. As natural fibers degrade, they weaken the composite's structural integrity, creating more surface area and pathways for environmental factors and microorganisms to penetrate the PVC matrix. The degradation of fibers and PVC can lead to the release of any embedded nanoparticles. These nanoparticles may also undergo degradation or contribute to the overall biofragmentation process by catalyzing certain reactions.

Although natural fibers are more susceptible to microbial degradation, the PVC matrix is inherently resistant due to its synthetic, non-biodegradable nature. PVC is composed of repeating vinyl chloride monomers, each containing a chlorine atom. There have been various reports on the biodegradability of PVC by microbes that rely solely on carbon for growth. *Micrococcus* species with PVC biodegradation capacity that use PVC as their only carbon source were identified by their ability to liberate chloride from the PVC polymer, produce carbon dioxide, grow on the surface of the PVC film, and increase cell density in test conditions. Over 70 days of experiment, the *Micrococcus* species demonstrated 0.36% chloride release and 8.87% mineralization as determined by carbon dioxide evolution [118]. According to Rozana *et al.* waste landfill isolates *Staphylococcus capitis, Bacillus subtilis,* and *Acinetobacter pittii* may degrade PVC [119]. The physicochemical and structural investigations of PVC by *Citrobacter koseri* isolated from the gut of the superworm, *Z. atratus* larvae, established the gut microbiota's capability to degrade PVC. Chemical and structural alterations demonstrated that *C. koseri* caused the oxidation of the PVC surface by forming carbonyl groups (C=O) and hydroxyl groups (-OH). Furthermore, the gut microbiota, which included a variety of microbial species, oxidized PVC at the same rate as *C. koseri* [120]. Polyvinyl chloride (PVC) breakdown in *T. molitor* larvae utilizing PVC microplastic powders (MPs) showed that 80% of *T. molitor* larvae could live on PVC alone after 5 weeks. About 2.9% of the chloride consumed was liberated, showing that the PVC MPs were only partially mineralized in the body. The chlorine element mass flow (%, w/w) contains 0.03% as HCl in air, 2.08% as chloride ions in larval biomass, 0.77% as dissolved chloride in frass, 34.60% as residual PVC, and 62.51% as putative chlorinated intermediates. Gel permeation chromatography (GPC) examination revealed a drop in Mw, Mn, and Mz of 33.4%, 32.8%, and 36.4%, respectively, indicating widespread depolymerization. Biodegradation and oxidation of PVC MPs were confirmed by the presence of hydroxyl and –C=C– functional groups as well and significant changes in thermal properties was noted [121].

Although various studies have revealed the involvement of microorganisms in PVC breakdown, a partly isolated lignin peroxidase from *Phanerocheate chrysosporium* validated the involvement of an enzyme in PVC degradation [94]. Numerous holes and alterations in the surface structure of the plastic film were found, indicating that PVC biodegradation occurs as a result of enzyme treatment. The enzyme has a molecular weight of 46 kDa and could lower the weight of PVC film by up to 31% in a shake flask experiment. The 3D structure of the lignin peroxidase has been given in Figure 8.4. The FTIR spectra of the enzyme-treated plastic sheet indicated structural alterations in chemical composition associated with alkenyl C–H and O–H stretching. Recently, PVC breakdown in the stomach of insect larvae was discovered where a variety of bacteria, including *Enterococcus*, *Klebsiella*, and others, were found to be responsible for the degradation. A multiomic approach (proteomics, metabolomics, and

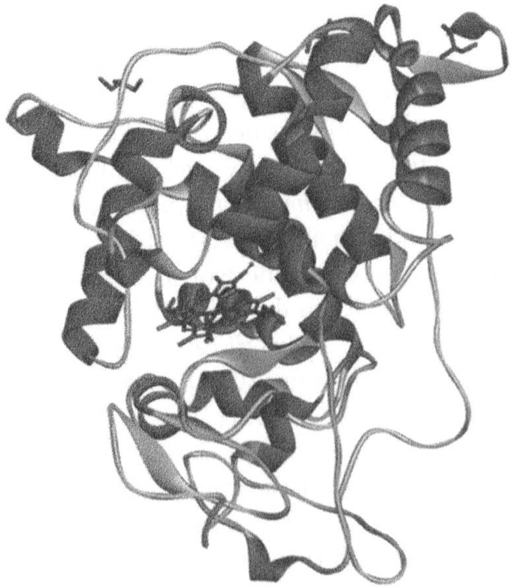

Figure 8.4 3D structure of lignin peroxidase from *Phanerocheate chrysosporium*.

transcriptomics) revealed that enzymes such as oxygenase, dehalogenases, enolases, aldehyde dehydrogenases, and catalase-peroxidase were involved in the PVC degradation pathway [86].

Therefore, the degradation of the PVC matrix by microorganisms is primarily facilitated through two main mechanisms: dechlorination and oxidative scission. Both processes reduce the polymer's recalcitrance, making it more susceptible to microbial assimilation. The removal of chlorine atoms can be facilitated by dechlorinating enzymes produced by bacterial species, (such as *Pseudomonas aeruginosa*, *Achromobacter* sp., *Pseudomonas citronellolis*, and *Bacillus flexus*), which makes the remaining carbon backbone more susceptible to other microbial attack [122, 123]. Certain anaerobic bacteria like *Clostridium bifermentans*, *Sporobacter* sp., and *Dethiosulfovibrio* sp. have the potential to remove chlorine atoms from the PVC chain through the process called reductive dehalogenation. This process involves the reduction of the carbon-chlorine (C–Cl) bonds, resulting in the formation of a less chlorinated or completely dechlorinated carbon backbone. Dechlorination reduces the toxicity and chemical stability of PVC, making the resulting polymer fragments more amenable to microbial attack [124, 125]. Some microorganisms (e.g., *Nocardia asteroids*, *Rhodococcus ruber*) have been reported to degrade polymers by initiating oxidative degradation by producing reactive oxygen species (ROS), such as superoxide radicals, hydrogen peroxide, and hydroxyl radicals. These have the potential to initiate the oxidative degradation of the PVC matrix, thus leading to the formation of small, oxidized polymer fragments that can be metabolized. These ROS can attack the carbon–carbon (C–C) bonds in the PVC backbone, leading to chain scission. ROS formation can be accelerated in the presence of UV light, leading to the photodegradation of PVC, which further weakens the polymer structure. The oxidative process generates shorter polymer fragments, often containing functional groups such as alcohols, aldehydes, or carboxylic acids, which are more reactive and can be further metabolized by microorganisms [126].

The oxidized and dechlorinated fragments of PVC can be small enough to be transported across microbial cell membranes through passive diffusion or active transport mechanisms. Specific membrane proteins may also recognize and transport these fragments into the cell, where they enter metabolic pathways [127]. Once inside the cell, the degraded PVC fragments, particularly those containing oxygenated functional groups, can be funneled into central metabolic pathways. For example, fragments resembling fatty acids or alcohols can enter β-oxidation or alcohol dehydrogenation pathways, respectively. Enzymes such as oxygenases, dehydrogenases, and reductases may further modify the PVC fragments, converting them into intermediates that can be fully assimilated into the microbial metabolic network. The oxidized fragments of PVC are metabolized *via* pathways such as the citric acid cycle (Kreb's cycle). In aerobic environments, the mineralization of PVC fragments involves complete oxidation, primarily resulting in the production of carbon dioxide (CO_2) and water (H_2O). Enzymes such as monooxygenases and dioxygenases catalyze the incorporation of oxygen into the PVC fragments, facilitating their breakdown into intermediates that can be fully mineralized. These enzymes remove hydrogen atoms from the PVC fragments, generating electron donors (e.g., NADH, $FADH_2$) for the electron transport chain [121, 128, 129].

In anaerobic environments, where oxygen is scarce, microorganisms like *Methanosarcina* species and *Sulfolobus solfataricus*, utilize alternative electron acceptors such as nitrate (NO_3^-), sulfate (SO_4^{2-}), or carbon dioxide (CO_2) itself. In methanogenic environments, archaea can convert CO_2 and hydrogen (H_2) into methane (CH_4) and various other mineral salts. This process is common in anaerobic environments such as landfills, where PVC-based composites might end up. In some anaerobic environments, PVC fragments might be fermented into organic acids, alcohols, or hydrogen gas, which are then utilized by other microorganisms for further mineralization [124]. The intermediate products of PVC degradation can be transferred among microbial species in a syntrophic relationship, leading to complete mineralization. In anaerobic conditions, the mineralization process may result in the production of methane (CH_4) and CO_2, as well as other reduced compounds such as ammonia (NH_3) and hydrogen sulfide (H_2S). These byproducts are released into the environment, contributing to the global carbon and nitrogen cycles [130]. However, compared to other plastic polymers, very limited reports on the mechanism of PVC degradation are available and therefore extensive studies needed for better understanding of such biodegradation processes. These studies will help in finding alternatives to recycle such polymers as well as the intermediate product may be utilized as substrate for industrial applications. A detailed pictorial representation of microbes and enzymes associated with the degradation of PVC-based natural fiber–reinforced nanocomposites has been given in Figure 8.5.

Furthermore, the future prospects for these materials are promising. With ongoing advancements in material science and biotechnology, there is potential to further optimize the compatibility between natural fibers and the polymer matrix, accelerate biodegradation rates, and scale up production for wider commercial use. In particular, research focused on improving fiber–matrix adhesion, harnessing advanced nanomaterials, and refining microbial degradation pathways will be key to unlocking the

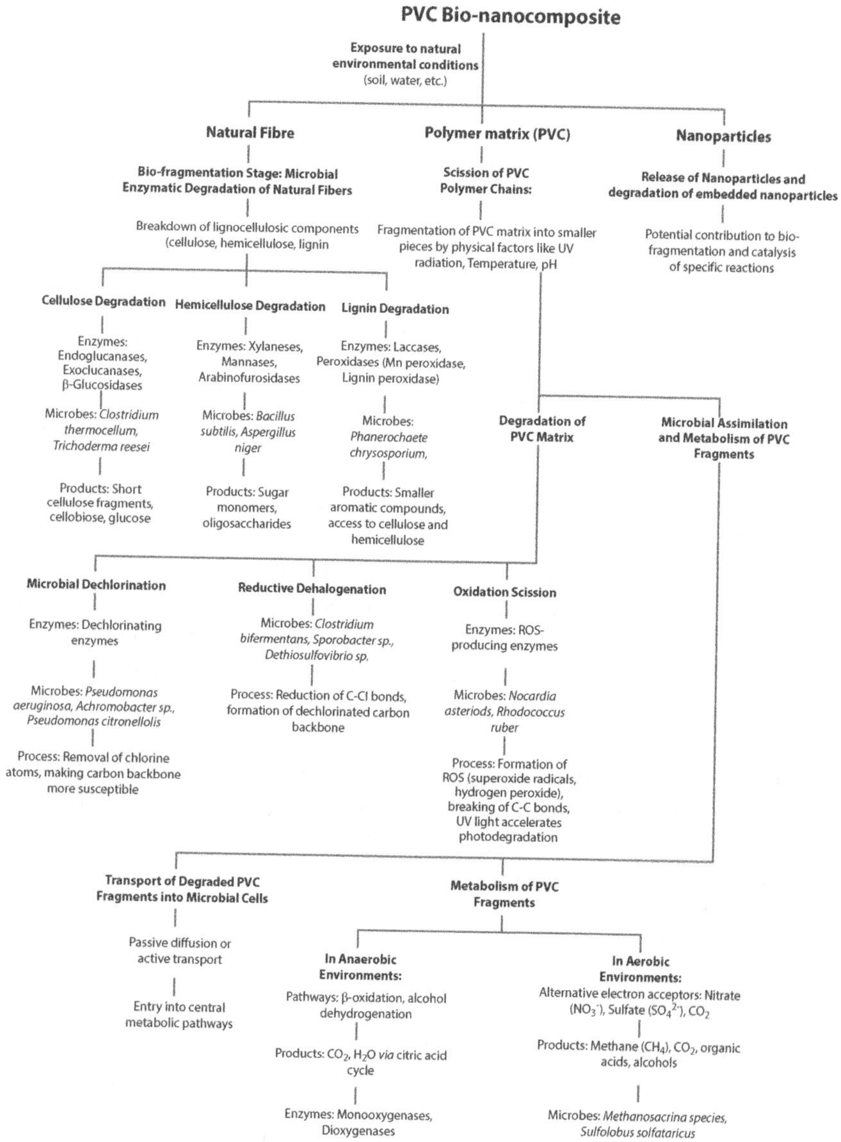

Figure 8.5 Involvement of microbes and enzymes in the metabolism of PVC-based natural fiber–reinforced bionanocomposites.

full potential of these composites. In addition, incorporating extremely stable engineered PVC degrading enzymes into PVC bionanocomposites would allow them to be degraded in the environment in an astonishingly short period of time. As global environmental concerns continue to mount, the utility of such biodegradable materials will play an increasingly crucial role in creating a more sustainable, circular economy across various industries.

8.5 Conclusion

In conclusion, the study of biodegradation in polyvinyl chloride (PVC)-based fibers within bionanocomposites demonstrates substantial advances in resolving PVC's environmental concerns. The chapter demonstrates how including natural fibers and particular nanoparticles can improve the biodegradability of PVC composites, encouraging more effective microbial and enzymatic breakdown. This development has significant implications for decreasing plastic pollution and improving trash management. However, obstacles remain, such as discovering new potent microbes and enzymes, optimizing composite formulations, comprehending long-term environmental implications, and resolving economic and regulatory issues. Future research should concentrate on improving enzyme activity through genetic engineering, analyzing materials to balance performance and sustainability, and eventually contributing to more environmentally acceptable solutions in the battle against plastic pollution.

AI disclosure statement: The author acknowledges the use of AI-based tools for minor language editing or grammatical correction. However, all intellectual contributions, analyses, and conclusions presented in this chapter are entirely the author's own.

References

1. Gilbert, M. and Patrick, S., Poly(Vinyl Chloride), in: *Brydson's Plastics Materials: Eighth Edition*, pp. 329–388, Elsevier Ltd, 2017, http://dx.doi.org/10.1016/B978-0-323-35824-8.00013-X.
2. Visakh, P. and Darie-Nita, R.N., Preparation, characterization and application, in: *Polyvinylchloride-Based Blends*, V.P.M. and R.N. Darie-Nita (Eds.), pp. 1–8, Springer International Publishing, Cham, Springer Series on Polymer and Composite Materials, Cham, 2022, https://link.springer.com/10.1007/978-3-030-78455-3.
3. Visakh, P.M. and Darie-Nita, R.N., Polyvinylchloride (PVC)-Based Blends: State of Art, New Challenges and Opportunities, in: *Polyvinylchloride-Based Blends*, pp. 1–17, 2021.

4. Pal, S.N., Ramani, A.V., Subramanian, N., Gas permeability studies on poly(vinyl chloride) based polymer blends intended for medical applications. *J. Appl. Polym. Sci.*, 46, 981–990, 1992.
5. Nizzetto, L., Futter, M., Langaas, S., Are Agricultural Soils Dumps for Microplastics of Urban Origin? 2016. https://doi.org/10.1021/acs.est.6b04140.
6. Tang, K.H.D., Climate Change and Plastic Pollution: A Review of Their Connections. *Trop. Environ. Biol. Technol.*, 1, 110–120, 2023.
7. Ru, J., Huo, Y., Yang, Y., Microbial Degradation and Valorization of Plastic Wastes. 2020. https://doi.org/10.3389/fmicb.2020.00442.
8. Lu, L., Li, W., Cheng, Y., Liu, M., Chemical recycling technologies for PVC waste and PVC-containing plastic waste: A review. *Waste Manag.*, 166, 245–258, 2023.
9. Braun, D., Poly(vinyl chloride) on the Way from the 19th Century to the 21st Century. *J. Polym. Sci. Part A Polym. Chem.*, 42, 578–586, 2004.
10. Puiggalí, J. and Katsarava, R., Bionanocomposites, in: *Clay-Polymer Nanocomposites*, pp. 239–272, Elsevier Inc, 2017, http://dx.doi.org/10.1016/B978-0-323-46153-5.00007-0.
11. Amobonye, A., Bhagwat, P., Singh, S., Pillai, S., *Plastic biodegradation: Frontline microbes and their enzymes, Science of the Total Environment,* 759, p. 143536, Elsevier B.V., 2021, https://doi.org/10.1016/j.scitotenv.2020.143536.
12. Shchipunov, Y., Bionanocomposites: Green sustainable materials for the near future. 2012. https://doi.org/10.1351/PAC-CON-12-05-04.
13. Khalil, H.P.S.A., Tehrani, M.A., Davoudpour, Y., Bhat, A.H., Jawaid, M., Hassan, A., Natural fiber reinforced poly(vinyl chloride) composites: A review. 2013. https://doi.org/10.1177/0731684412458553.
14. Chang, B.P., Mohanty, A.K., Misra, M., Studies on durability of sustainable biobased composites: a review, 10, 31, pp.17955–17999, Royal Society of Chemistry, 2020, doi: 10.1039/c9ra09554c.
15. Das, S. and Natarajan, G., Silk fiber composites in biomedical applications, in: *Materials for Biomedical Engineering: Biopolymer Fibers*, pp. 309–338, Elsevier Inc, 2019, http://dx.doi.org/10.1016/B978-0-12-816872-1.00011-X.
16. Thapliyal, D., Verma, S., Sen, P., Kumar, R., Thakur, A., Natural Fibers Composites: Origin, Importance, Consumption. *J. Compos. Sci.*, 7, 1–44, 2023.
17. Kumar, R.S., Devi, V.V., Nivedhitha, D.M., Satya, S.S., Visakan, S., Sharma, S.B., Kumar, S.B.S., Effect of nanoparticles in natural fiber reinforced polymer composites. *Mater. Today Proc.*, 1–7, 2023.
18. Hakamy, A., Shaikh, F.U.A., Low, I.M., Characteristics of hemp fabric reinforced nanoclay-cement nanocomposites. *Cem. Concr. Compos.*, 50, 27–35, 2014.
19. Ismail, H., Shuhelmy, S., Edyham, M.R., The effects of a silane coupling agent on curing characteristics and mechanical properties of bamboo fibre filled natural rubber composites. *Eur. Polym. J.*, 38, 39–47, 2002.

20. Shakir, M.H. and Singh, A.K., Mechanical properties, surface treatments, and applications of jute fiber reinforced composites—A review. *Polym. Adv. Technol.*, 35, e6268, 2024.
21. Yadav, P., Srivastava, C.M., Vaya, D., Plant Fibers - Based Sustainable Biocomposites, in: *Handbook of Nanomaterials and Nanocomposites for Nnergy and Environmental Applications*, pp. 513–547, 2021.
22. Djidjelli, H., Boukerrou, A., Founas, R., Rabouhi, A., Kaci, M., Farenc, J., Martinez-Vega, J.J., Benachour, D., Preparation and characterization of poly(vinyl chloride)/virgin and treated sisal fiber composites. *J. Appl. Polym. Sci.*, 103, 3630–3636, 2007.
23. Boraei, A.S.B., Bakhshandeh, B., Mohammadzadeh, F., Haghighi, D.M., Mohammadpour, Z., Clay-reinforced PVC composites and nanocomposites. *Heliyon*, 10, e29196, 2024.
24. Adeyemo, A.A., Adeoye, I.O., Bello, O.S., Adsorption of dyes using different types of clay: a review. *Appl. Water Sci.*, 7, 543–568, 2017.
25. Das, P., Manna, S., Behera, A.K., Shee, M., Basak, P., Sharma, A.K., Current synthesis and characterization techniques for clay-based polymer nano-composites and its biomedical applications: A review. *Environ. Res.*, 212, 1–13, 2022.
26. Balkanloo, P.G., Marjani, A.P., Zanbili, F., Mahmoudian, M., Clay mineral/polymer composite: characteristics, synthesis, and application in Li-ion batteries: A review. 2022, https://doi.org/10.1016/j.clay.2022.106632.
27. Pareek, P. and Ledwani, L., Synthesis and Applications of Polymer–Nano Clay Composites in Wastewater Treatment: A Review, in: *Industrial Wastewater Treatment*, pp. 237–256, 2022.
28. Rani, P., Ahamed, M.B., Deshmukh, K., Significantly enhanced electromagnetic interference shielding effectiveness of montmorillonite nanoclay and copper oxide nanoparticles based polyvinylchloride nanocomposites. *Polym. Test.*, 91, 106744, 2020.
29. Tomaszewska, J., Wieczorek, M., Skórczewska, K., Klapiszewska, I., Lewandowski, K., Klapiszewski, L., Preparation, Characterization and Tailoring Properties of Poly(Vinyl Chloride) Composites with the Addition of Functional Halloysite–Lignin Hybrid Materials. *Mater. (Basel)*, 15, 2022.
30. Bouchareb, B., Haddaoui, N., Zahra, S., Effect of montmorillonite addition (modified and unmodified) on the structural, mechanical, thermal, and morphological properties of nanocomposites based poly (vinyl chloride). *Mater. Tech.*, 110, 103, 2022.
31. Pan, M., Shi, X., Li, X., Hu, H., Zhang, L., Morphology and properties of PVC/clay nanocomposites *via in situ* emulsion polymerization. *J. Appl. Polym. Sci.*, 94, 277–286, 2004.
32. Gong, F., Feng, M., Zhao, C., Zhang, S., Yang, M., Thermal properties of poly(vinyl chloride)/montmorillonite nanocomposites. *Polym. Degrad. Stab.*, 84, 289–294, 2004.

33. Yang, D.F., Liu, Q.X., Xie, X.L., Zeng, F.D., Structure and thermal properties of exfoliated PVC/layered silicate nanocomposites *via in situ* polymerization. *J. Therm. Anal. Calorim.*, 84, 355–359, 2006.
34. Haiyan, H., Mingwang, P., Xiucuo, L., Xudong, S., Liucheng, Z., Preparation and characterization of poly(vinyl chloride)/organoclay nanocomposites by *in situ* intercalation. *Polym. Int.*, 53, 225–231, 2004.
35. Klapiszewski, L., Tomaszewska, J., Skórczewska, K., Jesionowski, T., Preparation and characterization of eco-friendly Mg(OH)2/lignin hybrid material and its use as a functional filler for poly(vinyl chloride). *Polym. (Basel)*, 9, 2017.
36. Ping, Q.W., Xiao, J., Zhao, J., The preparation and property of organic solvent lignin and PVC composite materials. *Adv. Mater. Res.*, 236–238, 1195–1198, 2011.
37. Zhang, W., Wu, H., Zhou, N., Cai, X., Zhang, Y., Hu, H., Feng, Z., Huang, Z., Liang, J., Enhanced Thermal Stability and Flame Retardancy of Poly(Vinyl Chloride) Based Composites by Magnesium Borate Hydrate-Mechanically Activated Lignin. *J. Inorg. Organomet. Polym. Mater.*, 31, 3842–3856, 2021.
38. Feldman, D. and Banu, D., Contribution to the study of rigid PVC polyblends with different lignins. *J. Appl. Polym. Sci.*, 66, 1731–1744, 1997.
39. Klapiszewski, L., Pawlak, F., Tomaszewska, J., Jesionowski, T., Preparation and characterization of novel pvc/silica-lignin composites. *Polym. (Basel)*, 7, 1767–1788, 2015.
40. Ali, M., II, Perveen, Q., Ahmad, B., Javed, I., Razi-Ul-Hussnain, R., Andleeb, S., Atique, N., Ghumro, P.B., Ahmed, S., Hameed, A., Studies on biodegradation of cellulose blended polyvinyl chloride films. *Int. J. Agric. Biol.*, 11, 577–580, 2009.
41. Taurino, R., Sciancalepore, C., Collini, L., Bondi, M., Bondioli, F., Functionalization of PVC by chitosan addition: Compound stability and tensile properties. *Compos. Part B Eng.*, 149, 240–247, 2018.
42. Rosa, D.S., Rios, A.R., Viana, H.M., Tavares, M., II, Characterization of poly(vinyl chloride) compounds containing different levels of starch. *J. Vinyl Addit. Technol.*, 22, 396–404, 2016.
43. Bigot, S., Louarn, G., Kébir, N., Burel, F., Click grafting of seaweed polysaccharides onto PVC surfaces using an ionic liquid as solvent and catalyst. *Carbohydr. Polym.*, 98, 1644–1649, 2013.
44. Sobahi, T.R., Makki, M.S., II, Abdelaal, M.Y., Carrier-mediated blends of Chitosan with polyvinyl chloride for different applications. *J. Saudi Chem. Soc.*, 17, 245–250, 2013.
45. Badr, I.H.A., Gouda, M., Abdel-Sattar, R., Sayour, H.E.M., Reduction of thrombogenicity of PVC-based sodium selective membrane electrodes using heparin-modified chitosan. *Carbohydr. Polym.*, 99, 783–790, 2014.
46. Xu, Y., Wu, Q., Lei, Y., Yao, F., Zhang, Q., Natural Fiber Reinforced Poly (vinyl chloride) Composites: Effect of Fiber Type and Impact Modifier. *J. Polym. Environ.*, 16, 250–257, 2008.

47. Mengeloglu, F. and Matuana, L.M., J. A. King, of Rigid PVC/Wood-Fiber Composites *. *J. Vinyl Addit. Technol.*, 6, 153–157, 2000.
48. Rocha, N., Kazlauciunas, A., Gil, M.H., Gonçalves, P.M., Guthrie, J.T., Composites: Part A Poly (vinyl chloride)–wood flour press mould composites: The influence of raw materials on performance properties. *Compos. Part A*, 40, 653–661, 2009.
49. Abdellah, S.F., El, M., Abdelhamed, M., Rafey, E.E., Formulation and characterization of new ternary stable composites: Polyvinyl chloride-wood flour-calcium carbonate of promising physicochemical properties. *Integr. Med. Res.*, 9, 12840–12854, 2020.
50. Fasihi, M. and Garmabi, H., Evaluation and Optimization of the Mechanical Properties of Highly Filled PVC/(Wood Flour) Composites by Using Experimental Design. *J. Vinyl Addit. Technol.*, 17, 112–119, 2011.
51. Zheng, Y., Wang, D., Chen, J., Study on the interface modification of bagasse fibre and the mechanical properties of its composite with PVC. *Compos. Part A*, 38, 20–25, 2007.
52. Wirawan, R. and Sapuan, S.M., Natural Fibre Reinforced Vinyl Ester and Vinyl 7 - Sugarcane Bagasse-Filled Poly (Vinyl Chloride) Composites: A Review *. *Nat. Fibre Reinf. Vinyl Ester Vinyl Polym. Compos.*, 157–168, 2018.
53. Shcbani, A. and Algoul, S.M., Impact strength and surface hardness properties: virgin PVC versus recycled PVC composites filled with two different natural fibers Impact strength and surface hardness properties: virgin PVC versus recycled PVC composites filled with two different natu. *2nd Conf. Eng. Sci. Technol.*, 1–10, 2019.
54. Hamida, B., Mazouzi, B., Eddine, D., Effect of Chemicals Treatments on the Morphological, Mechanical, Thermal and Water Uptake Properties of Polyvinyl Chloride/Palm Fibers Composites. *J. Compos. Adv. Mater.*, 29, 3–5, 2019.
55. Maou, S., Meghezzi, A., Grohens, Y., Meftah, Y., Kervoelen, A., Magueresse, A., Effect of various chemical modifications of date palm fibers (DPFs) on the thermo-physical properties of polyvinyl chloride (PVC) – high-density polyethylene (HDPE) composites. *Ind. Crop. Prod.*, 171, 113974, 2021.
56. Sombatsompop, N. and Chaochanchaikul, K., Average Mixing Torque, Tensile and Impact Properties, and Thermal Stability of Poly (vinyl chloride)/Sawdust Composites with Different Silane Coupling Agents. *J. Appl. Polym. Sci.*, 96, 213–221, 2005.
57. Bakar, A.A. and Baharulrazi, N., Mechanical Properties of Benzoylated Oil Palm Empty Fruit Bunch Short Fiber Reinforced Poly (vinyl chloride) Composites Mechanical Properties of Benzoylated Oil Palm Empty Fruit Bunch Short Fiber Reinforced Poly (vinyl chloride) Composites. *Polym. Plast. Technol. Eng.*, 47, 1072–1079, 2008.
58. Bakar, A.A., Hassan, A., Fuad, A., Yusof, M., Bakar, A.A., Hassan, A., Fuad, A., Yusof, M., Mechanical Properties of Silane and Zirconate Coupling Agent-Treated Oil Palm Empty Fruit Bunch Fiber-Filled Acrylic-Impact

Modified Poly (Vinyl Chloride) Composites Mechanical Properties of Silane and Zirconate Coupling Agent-Treated Oil Palm Empty Fru. *Polym. Plast. Technol. Eng.*, 49, 1563–1570, 2010.
59. Dutta, N., Bhadra, B., Gogoi, G., Maji, T.K., Development of polyvinyl chloride/waste rice husk ash/modified montmorillonite nanocomposite using epoxidized soybean oil as green additive substituting synthetic plasticizer and compatibiliser. *Clean. Mater.*, 2, 100033, 2021.
60. Petchwattana, N. and Sanetuntikul, J., Static and Dynamic Mechanical Properties of Poly (vinyl chloride) and Waste Static and Dynamic Mechanical Properties of Poly (vinyl chloride) and Waste Rice Husk Ash Composites Compatibilized with γ -aminopropyltrimethoxysilane. *Silicon*, 2018. doi: 10.1007/s12633-016-9440-x.
61. Allahbakhsh, A., PVC/rice straw/SDBS-modified graphene oxide sustainable Nanocomposites : Melt mixing process and electrical insulation characteristics. *Compos. Part A*, 134, 105902, 2020.
62. Kim, J., Peck, J.H., Hwang, S., Hong, J., Hong, S.C., Huh, W., Lee, S., Preparation and Mechanical Properties of Poly (vinyl chloride)/Bamboo Flour Composites with a Novel Block Copolymer as a Coupling Agent. *J. Appl. Polym. Sci.*, 108, 2654–2659, 2008.
63. Bahari, S.A., Grigsby, W.J., Krause, A., Flexural Properties of PVC/Bamboo Composites under Static and Dynamic-Thermal Conditions: Effects of Composition and Water Absorption. *Int. J. Polym. Sci.*, 2017, 2717848, 2017.
64. Wang, H., Sheng, K.C., Lan, T., Adl, M., Qian, X.Q., Zhu, S.M., Role of surface treatment on water absorption of poly (vinyl chloride) composites reinforced by Phyllostachys pubescens particles. *Compos. Sci. Technol.*, 70, 847–853, 2010.
65. Abdrahman, M.F. and Zainudin, E.S., Properties of Kenaf Filled Unplasticized Polyvinyl Chloride Composites. *Key Eng. Mater.*, 472, 507–512, 2011.
66. El-Shekeil, Y.A., Sapuan, S.M., Jawaid, M., Al-shuja'a, O.M., Influence of Fiber Content on Mechanical, Morphological and Thermal Properties of Kenaf Fibers Reinforced Poly (vinyl chloride)/Thermoplastic Polyurethane Poly-blend Composites. *Mater. Des.*, 14, 2014.
67. Mousa, A., Heinrich, G., Wagenknecht, U., Thermoplastic Composites Based on Renewable Natural Resources : Unplasticized PVC / Olive Husk. *Int. J. Polym. Mater. Polym. Biomater.*, 59, 843–853, 2010.
68. Khan, R.A., K.M.A., Zaman, H.U., Parvin, F., Fabrication and Characterization of Jute Fabric-Reinforced PVC-based Composite. *J. Thermoplast. Compos.*, 25, 45–58, 2012.
69. Li, Y., Wang, B., Wang, B., Ma, M., The enhancement performances of cotton stalk fiber/PVC composites by sequential two steps modification. *J. Appl. Polym. Sci.*, 135, 1–8, 2018.
70. Nayak, S., Mohanty, J.R., Samal, P.R., Polyvinyl Chloride Reinforced with Areca Sheath Fiber Composites — An Experimental Study. *J. Nat. Fibers*, 00, 1–12, 2018.

71. Sah, M.H.M., Noor, A.M., Abbas, M.R., Mechanical Properties of Coconut Shell Powder Reinforced PVC Composites in Mechanical Properties of Coconut Shell Powder Reinforced PVC Composites in Automotive Applications. *J. Mech. Eng.*, 14, 49–61, 2017.
72. Joglekar, J.J., Munde, Y.S., Jadhav, A.L., Bhutada, D.S., Radhakrishnan, S., Kulkarni, M.B., Studies on effective utilization of Citrus Maxima fibers based PVC composites. *Mater. Today Proc.*, 42, 578–583, 2020.
73. Shah, A.A., Hasan, F., Hameed, A., Ahmed, S., Biological degradation of plastics: A comprehensive review. 2008. https://doi.org/10.1016/j.biotechadv.2007.12.005.
74. Silva, R.R.A., Marques, C.S., Arruda, T.R., Teixeira, S.C., de Oliveira, T.V., Biodegradation of Polymers: Stages, Measurement, Standards and Prospects. *Macromol.*, 3, 2, 371–399, 2023.
75. Yu, J., Sun, L., Ma, C., Qiao, Y., Yao, H., Thermal degradation of PVC: A review, Elsevier Ltd, 2016, https://doi.org/10.1016/j.wasman.2015.11.041.
76. Liu, K., Wang, X., Wei, N., Song, Z., Li, D., Accurate quantification and transport estimation of suspended atmospheric microplastics in megacities: Implications for human health. *Environ. Int.*, 132, 105127, 2019.
77. Law, K.L., Plastics in the Marine Environment. 2017. https://doi.org/10.1146/annurev-marine-010816-060409.
78. Wang, C., Zhao, L., Lim, M.K., Chen, W.Q., Sutherland, J.W., Structure of the global plastic waste trade network and the impact of China's import Ban. *Resour. Conserv. Recycl.*, 153, 104591, 2020.
79. Yang, H., Ma, M., Thompson, J.R., Flower, R.J., Waste management, informal recycling, environmental pollution and public health. *J. Epidemiol. Community Health*, 72, 237–243, 2018.
80. Crawford, C.B. and Quinn, B., *Microplastic Pollutants*, 2016.
81. Wadsö, L. and Karlsson, O.J., Alkaline hydrolysis of polymers with ester groups studied by isothermal calorimetry. *Polym. Degrad. Stab.*, 98, 73–78, 2013.
82. Zhang, K., Hamidian, A.H., Tubić, A., Zhang, Y., Fang, J.K.H., Wu, C., Lam, P.K.S., Understanding plastic degradation and microplastic formation in the environment: A review. 2021. https://doi.org/10.1016/j.envpol.2021.116554.
83. Yang, X.G., Wen, P.P., Yang, Y.F., Jia, P.P., Li, W.G., Pei, D.S., Plastic biodegradation by *in vitro* environmental microorganisms and *in vivo* gut microorganisms of insects. 2023. https://doi.org/10.3389/fmicb.2022.1001750.
84. Danso, D., Chow, J., Streita, W.R., Plastics: Environmental and biotechnological perspectives on microbial degradation. 2019. https://doi.org/10.1128/AEM.01095-19.
85. Miri, S., Saini, R., Davoodi, S.M., Pulicharla, R., Brar, S.K., Magdouli, S., *Biodegradation of microplastics: Better late than never*, 286, p.131670, Elsevier Ltd, Chemosphere, 2022, https://doi.org/10.1016/j.chemosphere.2021.131670.

86. Zhang, Z., Peng, H., Yang, D., Zhang, G., Zhang, J., Ju, F., Polyvinyl chloride degradation by a bacterium isolated from the gut of insect larvae. *Nat. Commun.*, 13, 5360, 2022.
87. Sumathi, T., Viswanath, B., Sri Lakshmi, A., Saigopal, D.V.R., Production of Laccase by Cochliobolus sp. Isolated from Plastic Dumped Soils and Their Ability to Degrade Low Molecular Weight PVC. *Biochem. Res. Int.*, 2016, 9519527, 2016.
88. Novotný, C., Fojtík, J., Mucha, M., Malachová, K., Biodeterioration of Compost-Pretreated Polyvinyl Chloride Films by Microorganisms Isolated From Weathered Plastics. *Front. Bioeng. Biotechnol.*, 10, 832413, 2022.
89. Karimi-Avargani, M., Bazooyar, F., Biria, D., Zamani, A., Skrifvars, M., The promiscuous potential of cellulase in degradation of polylactic acid and its jute composite. *Chemosphere*, 278, 130443, 2021.
90. Temporiti, M.E.E., Nicola, L., Nielsen, E., Tosi, S., Fungal Enzymes Involved in Plastics Biodegradation. 2022. https://doi.org/10.3390/microorganisms10061180.
91. Wong, D.W.S., Structure and action mechanism of ligninolytic enzymes. 2009. https://doi.org/10.1007/s12010-008-8279-z.
92. Karimi, M. and Biria, D., The promiscuous activity of alpha-amylase in biodegradation of low-density polyethylene in a polymer-starch blend. *Sci. Rep.*, 9, 2612, 2019.
93. Le, B. and Yang, S.H., Microbial chitinases: properties, current state and biotechnological applications. *World J. Microbiol. Biotechnol.*, 35, 1–12, 2019.
94. Khatoon, N., Jamal, A., Ali, M., II, Lignin peroxidase isoenzyme: a novel approach to biodegrade the toxic synthetic polymer waste. *Environ. Technol. (United Kingdom)*, 40, 1366–1375, 2019.
95. Brebu, M., Environmental degradation of plastic composites with natural fillers-a review. *Polym. (Basel)*, 12, 2020.
96. Saba, N., Jawaid, M., Alothman, O.Y., Paridah, M.T., A review on dynamic mechanical properties of natural fibre reinforced polymer composites. *Constr. Build. Mater.*, 106, 149–159, 2016.
97. Gu, J.D., Microbiological deterioration and degradation of synthetic polymeric materials: Recent research advances. *Int. Biodeterior. Biodegrad.*, 52, 69–91, 2003.
98. John, M.J. and Thomas, S., Biofibres and biocomposites. *Carbohydr. Polym.*, 71, 343–364, 2008.
99. A. Bledzki, J., Gassan, Composites reinforced with cellulose.pdf. *Prog. Polym. Sci.*, 24, 221–274, 1999.
100. Li, M., Pu, Y., Thomas, V.M., Yoo, C.G., Ozcan, S., Deng, Y., Nelson, K., Ragauskas, A.J., Recent advancements of plant-based natural fiber–reinforced composites and their applications. *Compos. Part B Eng*, 200, 2020.
101. Girijappa, Y.T.G., Rangappa, S.M., Parameswaranpillai, J., Siengchin, S., Natural Fibers as Sustainable and Renewable Resource for Development of

Eco-Friendly Composites: A Comprehensive Review. *Front. Mater.*, 6, 1–14, 2019.
102. Oliveira, J., Belchior, A., da Silva, V.D., Rotter, A., Petrovski, Z., Almeida, P.L., Lourenço, N.D., Gaudêncio, S.P., Marine Environmental Plastic Pollution: Mitigation by Microorganism Degradation and Recycling Valorization. 2020. https://doi.org/10.3389/fmars.2020.567126.
103. Tokiwa, Y., Calabia, B.P., Ugwu, C.U., Aiba, S., Biodegradability of plastics. 2009. https://doi.org/10.3390/ijms10093722.
104. Chiellini, E., Corti, A., Del Sarto, G., D'Antone, S., Oxo-biodegradable polymers - Effect of hydrolysis degree on biodegradation behaviour of poly(vinyl alcohol). *Polym. Degrad. Stab.*, 91, 3397–3406, 2006.
105. Otsuka, Y., Sonoki, T., Ikeda, S., Kajita, S., Nakamura, M., Katayama, Y., Detection and characterization of a novel extracellular fungal enzyme that catalyzes the specific and hydrolytic cleavage of lignin guaiacylglycerol β-aryl ether linkages. *Eur. J. Biochem.*, 270, 2353–2362, 2003.
106. Banerjee, A., Chatterjee, K., Madras, G., Enzymatic degradation of polymers: A brief review. 2014. https://doi.org/10.1179/1743284713Y.0000000503.
107. Sánchez, C., Lignocellulosic residues : Biodegradation and bioconversion by fungi. *Biotechnol. Adv.*, 27, 185–194, 2009.
108. Andrady, A.L., Microplastics in the marine environment. *Mar. Pollut. Bull.*, 62, 1596–1605, 2011.
109. Laycock, B., Pratt, S., Halley, P., A perspective on biodegradable polymer biocomposites - from processing to degradation. *Funct. Compos. Mater.*, 4, 1–12, 2023.
110. Zeenat, A.E., Bukhari, D.A., Shamim, S., Rehman, A., Plastics degradation by microbes: A sustainable approach. The Authors, 2021. https://doi.org/10.1016/j.jksus.2021.101538.
111. Kawai, F., Biodegradation of Polymers (Bioassimilation, Biomineralization, Biodisintegration, Compost), Overview, in: *Encyclopedia of Polymeric Nanomaterials*, pp. 1–6, 2015.
112. Kirkwood, D., Microbial corrosion of metals in seawater, in: *Uhlig's Corrosi Handbook*, pp. 915–927, 1982.
113. Kalogerakis, N., Arff, J., Banat, I.M., Broch, O.J., Daffonchio, D., Edvardsen, T., Eguiraun, H., Giuliano, L., Handå, A., López-de-Ipiña, K., Marigomez, I., Martinez, I., Øie, G., Rojo, F., Skjermo, J., Zanaroli, G., Fava, F., The role of environmental biotechnology in exploring, exploiting, monitoring, preserving, protecting and decontaminating the marine environment. *N. Biotechnol.*, 32, 157–167, 2015.
114. Du, J., Zhang, X., Li, X., Zhao, J., Liu, G., Gao, B., Qu, Y., The cellulose binding region in Trichoderma reesei cellobiohydrolase I has a higher capacity in improving crystalline cellulose degradation than that of Penicillium oxalicum. *Bioresour. Technol.*, 266, 19–25, 2018.
115. Hirano, K., Kurosaki, M., Nihei, S., Hasegawa, H., Shinoda, S., Haruki, M., Hirano, N., Enzymatic diversity of the Clostridium thermocellum

cellulosome is crucial for the degradation of crystalline cellulose and plant biomass. *Sci. Rep.*, 6, 1–9, 2016.
116. Qu, F., Cheng, H., Han, Z., Wei, Z., Song, C., Identification of driving factors of lignocellulose degrading enzyme genes in different microbial communities during rice straw composting. *Bioresour. Technol.*, 381, 1–9, 2023.
117. Kumar, A. and Chandra, R., *Ligninolytic enzymes and its mechanisms for degradation of lignocellulosic waste in environment*, 6, 2, Elsevier Ltd, Heliyon, 2020, https://doi.org/10.1016/j.heliyon.2020.e03170.
118. Patil, R., Isolation of polyvinyl chloride degrading bacterial strains from environmental samples using enrichment culture technique. *African J. Biotechnol.*, 11, 7947–7956, 2012.
119. Rozana, K., Prabaningtyas, S., Widyatama, D.R., Distribution and biodegradation analysis of polyvinyl chloride microplastic by indigenous bacteria isolated from Supit Urang Landfill, Malang, Indonesia. *Biodiversitas*, 24, 3853–3859, 2023.
120. Nyamjav, I., Jang, Y., Lee, Y.E., Lee, S., Biodegradation of polyvinyl chloride by Citrobacter koseri isolated from superworms (Zophobas atratus larvae). *Front. Microbiol.*, 14, 1–12, 2023.
121. Peng, B.Y., Chen, Z., Chen, J., Yu, H., Zhou, X., Criddle, C.S., Wu, W.M., Zhang, Y., Biodegradation of Polyvinyl Chloride (PVC) in Tenebrio molitor (Coleoptera: Tenebrionidae) larvae. *Environ. Int.*, 145, 106106, 2020.
122. Das, G., Bordoloi, N.K., Rai, S.K., Mukherjee, A.K., Karak, N., Biodegradable and biocompatible epoxidized vegetable oil modified thermostable poly(vinyl chloride): Thermal and performance characteristics post biodegradation with Pseudomonas aeruginosa and Achromobacter sp. *J. Hazard. Mater.*, 209–210, 434–442, 2012.
123. Giacomucci, L., Raddadi, N., Soccio, M., Lotti, N., Fava, F., Polyvinyl chloride biodegradation by Pseudomonas citronellolis and Bacillus flexus. *N. Biotechnol.*, 52, 35–41, 2019.
124. Giacomucci, L., Raddadi, N., Soccio, M., Lotti, N., Fava, F., Biodegradation of polyvinyl chloride plastic films by enriched anaerobic marine consortia. *Mar. Environ. Res.*, 158, 104949, 2020.
125. Rosner, B.M., McCarty, P.L., Spormann, A.M., *In vitro* studies on reductive vinyl chloride dehalogenation by an anaerobic mixed culture. *Appl. Environ. Microbiol.*, 63, 4139–4144, 1997.
126. Baldera-Moreno, Y., Pino, V., Farres, A., Banerjee, A., Gordillo, F., Andler, R., Biotechnological Aspects and Mathematical Modeling of the Biodegradation of Plastics under Controlled Conditions. 2022. https://doi.org/10.3390/polym14030375.
127. Mohanan, N., Montazer, Z., Sharma, P.K., Levin, D.B., Microbial and Enzymatic Degradation of Synthetic Plastics. 2020. https://doi.org/10.3389/fmicb.2020.580709.

128. Lv, S., Li, Y., Zhao, S., Shao, Z., Biodegradation of Typical Plastics: From Microbial Diversity to Metabolic Mechanisms. 2024. https://doi.org/10.3390/ijms25010593.
129. Pivato, A.F., Miranda, G.M., Prichula, J., Lima, J.E.A., Ligabue, R.A., Seixas, A., Trentin, D.S., Hydrocarbon-based plastics: Progress and perspectives on consumption and biodegradation by insect larvae. 2022. https://doi.org/10.1016/j.chemosphere.2022.133600.
130. Malakhova, D.V., Egorova, M.A., Leontieva, M.R., Elcheninov, A.G., Panova, T.V., Aleksandrov, Y.D., Tsavkelova, E.A., Anaerobic Microbial Degradation of Polypropylene and Polyvinyl Chloride Samples. *Microbiol. Russian Fed.*, 92, 83–93, 2023.

9

Applications of Nano-ZnO-Blended Polymeric Membranes in Environmental Remediation and Sustainable Development

Narendra B. Patil[1*], Madhavi Vemula[2†], Subash C.B. Gopinath[3,4], Hussaini Adam[3], Thangavel Lakshmipriya[3,4] and Paresh Patil[5]

[1]*Department of Pharmacology, H. R. Patel Institute of Pharmaceutical Education and Research, Shirpur (Maharashtra), India*
[2]*Department of Chemistry, BVRIT Hyderabad, College of Engineering for Women, Hyderabad, India*
[3]*Department of Nanoelectronics, Institute of Nano Electronic Engineering, Universiti Malaysia Perlis, Kangar, Perlis, Malaysia*
[4]*Department of Chemistry, Faculty of Chemical Engineering & Technology, Universiti Malaysia Perlis, Arau, Perlis, Malaysia*
[5]*Department of Pharmacology, Ahinsa College of Pharmacy, Dondaicha (Maharashtra), India*

Abstract

The integration of nano-zinc oxide (ZnO) into polymeric membranes has emerged as a promising strategy to enhance membrane performance. The incorporation of nano-ZnO enhances membrane performance, providing improved mechanical strength, thermal stability, and antimicrobial properties. The synergy between nano-ZnO and polymeric materials is explored, revealing enhanced membrane properties and expanded applications. This chapter delves into the fabrication techniques, structural characterization, and property evaluation of nano-ZnO-blended polymeric membranes. Emerging trends, such as nanostructured membrane design and surface modification, are discussed. The chapter highlights the potential of nano-ZnO-blended polymeric membranes to address pressing environmental and energy concerns. The chapter concludes by identifying challenges

**Corresponding author*: rxnbpatil@gmail.com
†*Corresponding author*: madhuchem9@gmail.com

Shiv Kumari Panda and Visakh P. M (eds.) Polyvinyl Chloride-Based Biocomposites and Bionanocomposites, (271–296) © 2026 Scrivener Publishing LLC

and opportunities for further research, paving the way for the development of next-generation nano-ZnO-blended polymeric membranes.

***Keywords*:** Nano-zinc oxide, polymer, membranes, synergic properties, surface modifications

9.1 Introduction

Population growth, urbanization, and improper water resource management and climate change have exacerbated global challenges in environmental and biomedical domains [1]. Environmentally, the demand for clean water continues to rise with population expansion, placing immense pressure on water supplies. The World Health Organization predicts that nearly half of the world's population will experience water stress by 2025 [2], and over 15% of the population already faces deficiency in access to adequate fresh water for healthy living. Protecting existing freshwater reserves and advancing water supply technologies are crucial [3]. Industrial, municipal, and agricultural activities significantly contribute to water pollution, introducing contaminants such as pesticides [4], hazardous heavy metals [5], organic acids [6], fertilizers [7], dyes, and microbes [8], all found in wastewater. Heavy metals are particularly concerning due to their non-biodegradable nature and toxicity [9–11], leading to environmental persistence and bioaccumulation in the food chain [12]. Contaminated water is linked to severe health issues including cancer, skin disorders, organ failure, diarrhea, and more [13]. Therefore, eliminating these pollutants is vital to ensure clean water for all living beings [14]. Spoilage remains a major concern for fresh fruits and vegetables, affecting them throughout the logistics process, from harvest to consumer use [15]. Nearly 50% of fruits and vegetables are lost due to severe post-harvest diseases, highlighting a major agricultural issue [16]. The root causes of these problems include physical harm, chemical transformations, and microbial contamination. Most infections are attributed to fungal plant pathogens that release mycotoxins, posing threats to human and animal health. Packaging stands out as a common post-harvest practice, organizing fruits and vegetables into manageable units to simplify handling and shield them from the risks associated with transport and storage [17]. The prevalence of foodborne pathogens is a primary cause of illness, hospitalization, and death annually. These pathogens thrive on the surfaces and packaging of food products, resulting in substantial economic losses as the food industry invests millions to combat microbial contamination [18]. Pathogens like *Toxoplasma gondii*, *Staphylococcus aureus*,

Campylobacter spp., *Escherichia coli*, and *Salmonella species* are among the top contributors to foodborne illnesses and deaths in humans [19]. With escalating environmental concerns over plastic waste in recent decades and a global effort to enhance living standards, biodegradable and antimicrobial polymeric films have become a key focus in the packaging industry [20, 21]. The challenges discussed above have directed to the progress of innovative techniques, with nanotechnology emerging as a promising solution. Nanoparticles, characterized by their unique physicochemical properties distinct from bulk materials, have garnered significant attention in membrane manufacturing [22, 23]. They offer enhanced control and enable the creation of tailored structures and functionalities. Membranes, known for their operational flexibility, high functionality, and cost-effectiveness, have become a pivotal technology across various fields. Their applications span water treatment, energy storage, environmental protection, and biomedical advancements, showcasing their versatility and impact [24]. Table 9.1 lists common additives and solvents used in nano-ZnO blended polymeric membranes.

Zinc oxide nanoparticles (ZnO NPs), recognized for their affordability, environmental friendliness, and hydrophilic nature, have been extensively studied by researchers worldwide [19]. According to scientific literature,

Table 9.1 Shows the common additives and solvents used for nano-ZnO-blended polymeric membrane.

Sr. no.	Headings	Examples	Ref.
1	Solvents	Water, dimethyl sulfoxide, tetrahydrofuran, N,N-dimethylformamide, N,-dimethylacetamide, N-methyl-2-pyrrolidone	[24]
2	Crosslinking agents	Borax, urea, tetraethoxysilane, glutaraldehyde	
3	Fillers	Silica, titanium dioxide, halloysite, ZnO-SiO_2, ZnO-GO nanocomposites, alumina	
4	Polymers	Polyethylene glycol (PEG), poly vinyl pyrrolidone, PEG-2000, PEG-400, and PEG-600	
5	Surface modifiers	Zwitterionic copolymers [sulfobetaine methacrylate, sulfobetaine-2-vinylpyridine]	

ZnO NPs enhance polymeric membranes, making them suitable for various applications [25]. Consequently, polymer–ZnO composite membranes have become a prominent topic in membrane technology, with significant advancements documented over the past decade [26]. In this chapter, the synthetic methods for ZnO NPs and the diverse polymers used in membrane development has been briefly outlined. Furthermore, we shed light on the potential role of nano-ZnO-blended polymeric membranes in water purification, membrane fouling mitigation, biomedical application, food packaging, photocatalytic degradation of dyes, and energy storage. The chapter concludes by identifying key challenges and opportunities, providing insights for advancing next-generation nano-ZnO-blended polymeric membranes.

9.2 Synthesis Approaches of ZnO NPs

The synthesis of zinc oxide nanoparticles has received significant attention in recent years due to their unique optical, electrical and catalytic properties, making them suitable for a wide range of applications including optoelectronics, energy storage and biomedical devices. Various synthetic approaches have been developed to prepare ZnO nanoparticles with controlled size, shape, and surface properties. These methods can be broadly categorized into chemical, physical and biological approaches as illustrated in Figure 9.1.

9.2.1 Conventional Synthesis Methods

9.2.1.1 Hydrothermal Method

In this approach, precursor chemicals are subjected to high-pressure and high-temperature conditions in an aqueous solution, leading to the formation of nanoparticles. Furthermore, as a highly efficient alternative synthesis method, it has garnered considerable attention [27]. Notably, by adjusting the temperature, reaction time, and precursor concentration during the hydrothermal process, the shape and size of the particles can be effectively controlled [28]. Notably, this technology operates at a low process temperature, making it an eco-friendly option. Additionally, the production process is cost-effective and scalable, allowing for widespread adoption. Furthermore, the simplicity of the equipment and handling required makes it an attractive solution, eliminating the need for complex and specialized machinery.

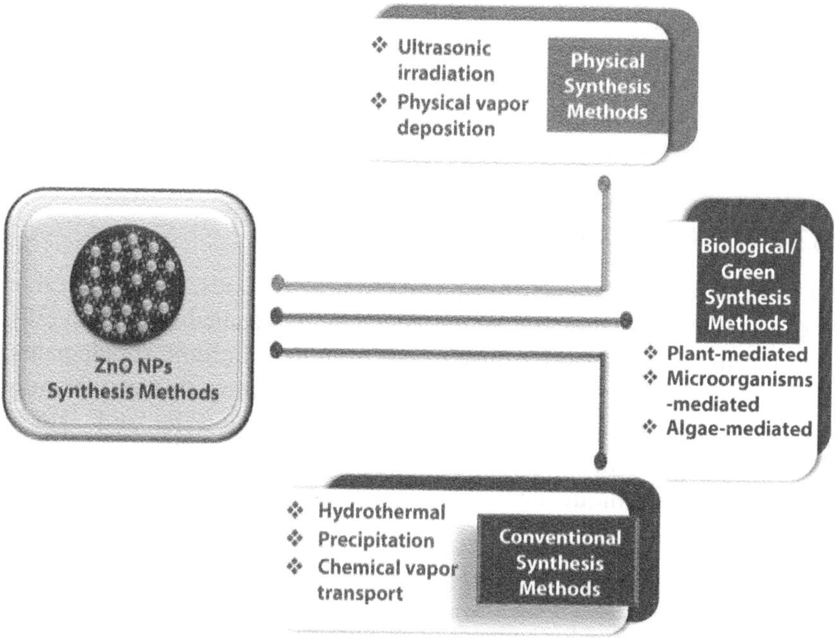

Figure 9.1 Synthesis approaches of ZnO NPs.

9.2.1.2 *Precipitation Method*

This process involves combining a zinc salt solution with a suitable reagent, such as an acid or base, under carefully controlled conditions, thereby creating ZnO nanoparticles [28]. Additionally, zinc nitrate and urea are frequently used as precursors in chemical methods that under extreme pressure and temperature within sealed aqueous solutions result in the formation of ZnO-NPs [29]. This innovative technology is remarkably time-saving, cost-efficient, and user-friendly, making it an attractive solution for various applications.

9.2.1.3 *Chemical Vapor Transport Method*

In this approach, the process begins with the powder being subjected to high temperatures in the presence of a transport agent, such as graphite or iodine. Subsequently, it is moved to a cooler zone, where condensation occurs, consequently in the formation of nanoparticles [30]. Remarkably, this process was pioneered by Bunsen in 1852 [31]. The chemical vapor deposition (CVD) method offers several advantages in the production of nanoparticles. Notably, this technique enables the synthesis of nanoparticles with precisely controlled sizes and high efficiency. Additionally, the CVD method yields nanoparticles of exceptionally high purity, making it a viable possibility for various applications. Furthermore, the process is highly versatile and scalable, allowing for seamless adaptation to different requirements and large-scale production.

9.2.2 Biological/Green Synthesis Methods

In contrast to traditional chemical and physical methods, the biological or green synthesis of nanoparticles provides a more sustainable solution [32]. Specifically, ZnO, which has historically been used in food, supplements, and medications, is now increasingly produced using this eco-friendly approach [33]. By employing safe reagents such as water and natural extracts, the green method avoids the use of hazardous ingredients, making it an ideal technique for the creation of metal nanoparticles [34].

9.2.2.1 Plant-Mediated Synthesis

A promising alternate for conventional chemical techniques is the fabrication of ZnO-NPs using plants and plant extracts [35]. Moreover, it is a safe, effective, and environmentally friendly method. Notably, this process for producing ZnO-NPs eliminates the need for harmful and hazardous substances, making it a more sustainable option [36]. Consequently, it offers a greener solution compared to conventional approaches. Plant-mediated synthesis offers numerous advantages, making it a highly desirable approach. This eco-friendly process is not only cost-effective and scalable but also takes place under mild conditions, ensuring efficiency and sustainability. By minimizing environmental impact, plant-mediated synthesis provides a responsible and green alternative for nanoparticle synthesis, aligning with the principles of sustainable development.

9.2.2.2 Microorganism-Mediated Synthesis

The concept of using microbes for nanoparticle synthesis has sparked considerable interest, and numerous studies have utilized different microorganism models [37]. Despite this, maintaining processes such as cell cultures, intracellular synthesis, and extensive purification steps has made microbial-assisted ZnO NPs synthesis challenging [38]. Furthermore, the mechanism behind the biological synthesis of ZnO-NPs using microbes is still not fully understood [39]. The selection of microorganisms, appropriate growth conditions, and the decision between intracellular or extracellular biosynthesis are critical to producing ZnO-NPs [40]. Frequently used microorganisms, including *Sphingobacterium thalpophilum*, *Staphylococcus aureus*, and *Bacillus megaterium* NCIM2326 and *Asperigillus fumigstus JCF* are capable of producing ZnO-NPs with various shapes, such as rod, cubic, triangular, and acicular, with particle sizes ranging from 10 to 95 nm [41, 42]. Microbial-mediated synthesis of ZnO-NPs presents several significant advantages. Firstly, this method enables the precise control of nanoparticle size and shape, resulting in uniform and tailored particles. Additionally, the synthesis process occurs under mild conditions, eliminating the need for extreme temperatures or hazardous chemicals. Moreover, microbial-mediated synthesis is an eco-friendly, cost-effective, and scalable approach, offering three-fold benefits that make it an attractive and sustainable option for large-scale industrial production.

9.2.2.3 Algae-Mediated Synthesis

Algae are considered one of the safest and most established natural materials utilized in the eco-friendly synthesis of nanoparticles, as they reduce the toxicity of chemical reagents in the production process [43]. Various marine macro- and microalgae, including *Chlorella*, brown *Sargassum muticum*, green *Caulerpa peltata*, and red *Gracilaria gracilis*, have been reported to be effective in the production of ZnO-NPs [44, 45]. The process occurs under mild conditions, eliminating the need for extreme temperatures, pressures, or hazardous chemicals. Moreover, this method is environmentally friendly, cost-effective, and scalable, making it a sustainable and viable option for the large-scale production of nanoparticles.

9.2.3 Physical Synthesis Methods

These methods are highly valued for their reliance on chemically pure reagents and straightforward technology, rendering them ideal for large-scale industrial applications that demand high-volume production.

9.2.3.1 Physical Vapor Deposition

This process offered a simple and efficient way to fabricate ZnO NPs at a relatively low temperature of 450 °C [46]. The catalyst-free nature of the growth process was demonstrated by the fact that the nanoparticle diameter expanded as the temperature increased. As a result, these findings provide valuable insight for constructing ZnO-based nanoscale devices on a variety of substrates with low-temperature endurance [47]. The physical vapor deposition (PVD) method offers several significant advantages for the synthesis of ZnO nanoparticles. This technique enables the production of ZnO nanoparticles at relatively low temperatures, reducing energy consumption and thermal damage. Additionally, PVD is a catalyst-free process, simplifying the synthesis and minimizing potential contamination. The resulting nanoparticles exhibit exceptionally high purity, making them suitable for various applications. Furthermore, the PVD method is an environmentally friendly option, as it avoids the use of hazardous chemicals and minimizes waste generation.

9.2.3.2 *Ultrasonic Irradiation*

This approach is commonly used in solution-phase operations to produce nanoparticles [49]. It distinguishes itself from surfactant-based techniques by being a natural method that retains the particles' original characteristics [50]. Ultrasonic irradiation is regarded as the best and most efficient way to fabricate pure materials with a wide array of controlled features [51]. The ultrasonication method offers several advantages for nanoparticle synthesis. Notably, this technique enables the production of nanoparticles with precisely controlled sizes and high efficiency. Additionally, the ultrasonication method is remarkably fast, yielding high-purity nanoparticles in a relatively short period. Furthermore, this approach is environmentally friendly, cost-effective, and scalable, making it an attractive option for large-scale industrial applications.

9.3 Application of Nano-ZnO-Blended Polymeric Membrane

Nanoparticle-embedded polymeric membranes are a good choice for the efficient elimination of hazardous pollutants from wastewater [52]. ZnO nanoparticles are of great interest as an emerging class of photocatalysts due to their lower bandgap, high photocatalytic efficiency, and high biological carbon and chemical oxygen demand removal efficiency [53]. Blending these nanoparticles in polymeric membranes can lead to the preparation of multifunctional membranes with efficient removal of dyes, bacteria, heavy metals, phenolic compounds, and inorganic oxyanions. Various polymers such as chitosan, Nafion, halloysite, polymethyl methacrylate, and polysulfone have been incorporated with ZnO and have been developed.

In this study, the potential of ZnO-polymeric blended membranes is explored comprehensively with a detailed overview of membranes prepared with various polymeric materials including synthetic and biopolymers. This chapter explores membranes processed in the form of nanofibers and emphasizes the novel preparation process in the form of a post-treated electrospun organic–inorganic blended membrane. The properties of thin films, active layers, pores, and surface chemical properties of ZnO-polymeric blended membranes are discussed. Consequently, the challenges in ZnO-polymeric membranes and existing knowledge gaps related to their preparation are presented. The production of ZnO nanoparticles with various morphologies, shapes, and sizes has been achieved with different methods [54]. The reaction precursors, type of reaction, solvents, pH, ageing temperatures, and other physical conditions are critical factors for synthesizing ZnO nanoparticles. Furthermore, the hydrothermal method, sol–gel method, solvothermal method, chemical vapor deposition, precipitation method, and flux method are used for synthesizing ZnO nanoparticles [55]. Polymeric membranes are used for the selective retention of small organic pollutants like dyes, heavy metals and salts [56]. Over the past two decades, numerous polymeric membranes have been developed for several pressure-driven water purification techniques, especially for reverse osmosis [57]. Despite several advantages of these membranes, their applications are limited primarily by insufficient antifouling properties and the cost of manufacture. Research in inorganic nanoparticle–polymer-blended RO membranes has been performed aggressively, and it has been indicated that the embedment of inorganic nanoparticles can effectively enhance the properties of the resultant mixed amorphous materials [56]. Among the numerous inorganic nanoparticles, ZnO nanoparticles have been known to possess intrinsic UV-responsive properties, promoting the dissociation of water molecules to generate active hydroxyl radicals, which induce the mineralization of the foulants [53]. Recent advancements in growth techniques of nano-ZnO have allowed many innovations in various forms and applications [58]. Despite the advantages of incorporating nano-ZnO in the polymeric RO membrane, concerns about the nanoparticulate interactions, membrane fabrication processes, and control of the leakages of nanoparticles have hindered the practical applications of many existing nano-ZnO-polymeric membranes [59]. Promising strategies such as coating on commercial polysulfone membranes, supercritical CO_2-antisolvent rapid depressurization method, sponge-assisted method, and triple-layer skin layer–integrated ZnO@poly(m-phenylenediamine) nanoparticles are recommended for the guarantee of a benefit–risk ratio [60]. Niche examples are also presented in the context of temperature-resistant, scalable, and fouling-resistant nano-ZnO-blended polyamide membranes for organic solvents or biomolecule filtration, casting on bottle-grade PET substrates and scalable pi-stacked and three-layered composite membrane biofouling-resilient properties. The combination of ZnO, Porifera-style aggregate analogous properties, anisotropic molecular separation chain, or space-holed nanoparticles embedding methylene blue that enhanced osmotic pressure and quick-healable networks in the new nano-ZnO structures is proposed [61]. Furthermore, to resolve the hydrogel shortage issue in gas and vapor separation, the usage of a pH-responsive interlayer to fabricate a magnetically aerocontaminants purged hybrid membrane is introduced. From a future perspective, the concern is on structural and structural–functional correlations, implementation stability, and challenges of a breakthrough with lower energy and a clearer viewpoint collaboration of the combination strategy of multivergent nano-ZnO/polymeric-blended water treatment membrane superior performances. Figure 9.2 illustrates methods to regulate membrane properties such as reducing fouling, increasing solvent flux, enhancing pollutant rejection, and controlling swelling and deswelling.

9.3.1 Food Packaging

The current study delves into biodegradable packaging materials designed to maintain high food quality, addressing key factors like UV degradation and oxidation that often result in poor flavor and nutritional loss [62]. Among these materials, ZnO-blended polymeric membranes have increased considerable attention due to their exceptional antibacterial properties and high sensitivity [63]. As a result, the development of functional packaging materials based on ZnO-blended polymeric membranes is progressing rapidly in the food industry [64]. This active packaging not only aims to reduce the environmental encounters posed by non-biodegradable packaging waste but also ensures food safety, enhanced

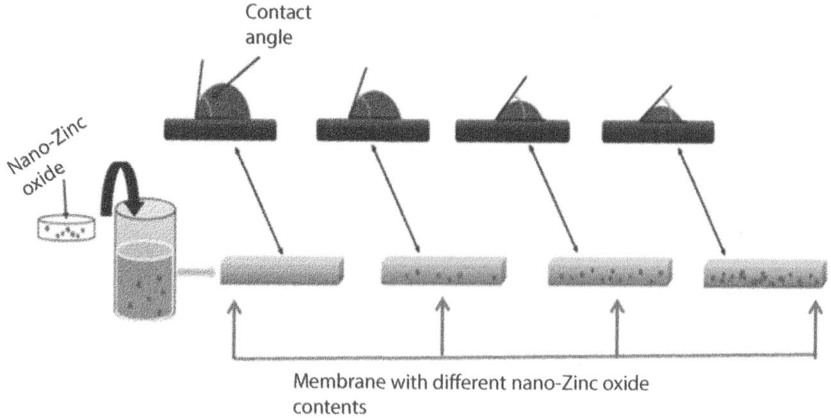

Figure 9.2 Methods to regulate membrane properties, including reducing fouling and biofouling, increasing solvent flux, enhancing pollutant rejection, and controlling swelling and deswelling.

quality, and extended shelf life. Moreover, this innovative packaging goes beyond serving as a protective barrier; it also acts as an effective preservation technique [65]. In this manner, Zhang and colleagues fabricated an antimicrobial fibrous membrane using polylactic acid (PLA) and ZnO NPs through the electrospinning technique. The composite membranes exhibited a tensile strength of 6.85 ± 0.19 MPa and an elongation at a break of 74.90 ± 0.59%. Additionally, zones of inhibition against *Escherichia coli* and *Staphylococcus aureus* measured 15.02 ± 0.18 mm and 14.74 ± 0.06 mm, singly. After 45 minutes of ultrasonication, these inhibition zones increased to 18.11 ± 0.15 mm for *E. coli* and 15.80 ± 0.11 mm for *S. aureus*. Overall, the membrane displayed promising antimicrobial efficacy and is a potential candidate for food packaging applications to prevent microbial contamination [66]. In another study, researchers developed a PLA-based composite membrane incorporating ZnO NPs using the solvent casting method. To create an active membrane, varying amounts of PEE (0.5, 1, 1.5, and 2 wt%) and 3 wt% ZnONPs were added to the PLA matrix. Notably, the membrane demonstrated excellent antioxidant properties, achieving maximum DPPH and ABTS scavenging activities of 96.2 ± 0.8% and 93.1 ± 0.5%, respectively. Furthermore, the membrane effectively inhibited *S. aureus* by 1.4 ± 0.05 Log CFU/mL and *E. coli* by 8.2 ± 0.35 Log CFU/mL, in comparison to the blank group. Consequently, this composite film exhibits significant prospective for applications in food packaging, particularly for extending shelf life and enhancing food safety [67]. Giannakas *et al.* synthesized hybrid nanostructures by growing zinc oxide nanorods on nanoclay surfaces. Subsequently, these nanostructures were incorporated into chitosan (CS)/polyvinyl alcohol (PVA), resulting in films that exhibited antimicrobial activity against *E. coli*, *S. aureus*, *S. enterica*, and *L. monocytogenes*. Therefore, these findings underscore their potential as innovative active food packaging materials [68]. In this context, Xuemei Sun *et al.* developed a novel food packaging material by loading ZnO NPs onto lecithin-montmorillonite (LEC-Mt) and incorporating the composite into CS. The resulting CS@ZnONPs/LEC-Mt films exhibited significant antibacterial properties against *E. coli* and *S. aureus*. These findings underscore the potential of CS@ZnONPs/LEC-Mt films for designing innovative food packaging materials [69]. Nano-food packaging was developed by synthesizing ZnNPs within a PLA polymer. Specifically, ZnNPs at concentrations of 0.4% (w/w) and 4% (w/w) were incorporated into PLA using the solution casting method. Furthermore, *Vitis vinifera* fruits were packed in the resulting biofilm and monitored for appearance, taste, and freshness during ambient storage. As a result, the biofilm-packed fruit samples maintained their original taste and freshness for up to 15 days [70]. Tilak Gasti and his colleagues synthesized clove essential oil (CEO)-loaded CS-ZnO hybrid nanoparticles (CS–ZnO@CEO or CPCZC) and incorporated them into CS/pullulan nanocomposite films. Notably, the CPCZC films exhibited enhanced antioxidant activity. Additionally, these films demonstrated strong antibacterial activity against *P. aeruginosa*, *S. aureus*, and *E. coli*. Furthermore, the CPCZC films effectively extended the shelf life of chicken meat by up to 5 days when stored at 8 ± 2 °C [71]. Besides that, Mallikarjunagouda *et al.* developed microwave-assisted poly(acrylamide-co-2-hydroxyethyl methacrylate)/CS (P(AM-co-HEMA)/CS)@ZnO-loaded blend membranes. Furthermore, the antimicrobial tests revealed that these ZnO NP-loaded membranes are ideal for the antimicrobial packaging of fruits and vegetables. As a conclusion, the flexibility, oxygen barrier efficiency, and biodegradability of these membranes position them as promising substrates for advanced packaging materials, suitable for both vegetables and processed food products [72]. In 2022, Huifang Wang *et al.* developed nano-ZnO@Xylan *via* the nanoprecipitation method. Furthermore, the nano-ZnO@Xylan showcased impressive antibacterial properties, achieving 99% inhibition zones against *E. coli* and *S. aureus*. Notably, ZnO@Xylan/QX/PVA films improved the shelf life of cherry tomatoes to a minimum of 21 days. Collectively, these findings underscore the immense potential of the fabricated composite films for innovative applications in food packaging [73]. Seyedeh *et al.* developed carboxymethyl cellulose-poly(ethylene terephthalate) (PET) bilayer nanocomposite films loaded with ZnO NPs. Subsequently, these films demonstrated strong antimicrobial activity against *E. coli* and *S. aureus*. Moreover, they showcased remarkable versatility for food packaging applications due to their antimicrobial properties, effectively extending the shelf life of packaged foods as active packaging materials [74]. In this study, a green and easy method was employed to develop polycaprolactone films loaded with ZnO NPs using ionic liquids. Notably, the nanocomposite films demonstrated commendable antioxidant activity and effective UV-light barrier properties. As a result, these films show significant potential for application in the food packaging industry [75].

9.3.2 Photocatalytic Degradation of Dyes

Water pollution caused by industrial activities has reached critical levels, posing a major environmental concern [76]. In response, researchers have made significant strides in developing efficient techniques and advanced materials for removing highly toxic dyes. Among these, photocatalytic removal stands out as a widely adopted approach due to its high efficiency, affordability, and simplicity [77, 78]. As a result, recent research has concentrated on advancing technologies to produce catalysts that are both high-performing and cost-effective [79]. In this context, ZnO nano-blended polymeric membranes have emerged as a prominent solution, offering benefits such as design versatility, enhanced physical and chemical attributes, and enhanced stability. Figure 9.3 illustrates the photocatalytic degradation of dye solution using ZnO blended polymer under light source. In 2023, Oranso *et al.* fabricated a polyethersulfone (PES) membrane embedded with graphene oxide-ZnO (GO-ZnO) and polyvinylpyrrolidone (PVP)-loaded GO-ZnO, which significantly enhanced the photocatalytic degradation of dyes. Specifically, the incorporation of GO-ZnO enhanced the degradation process. Furthermore, this membrane displayed superior performance, particularly in degrading carbamazepine and brilliant black. In conclusion, the membrane achieved the maximum dye degradation within a shorter duration [80]. Similarly, Muna *et al.* fabricated a cellulose acetate (CA)-ZnO NPs membrane for the photocatalytic degradation of methylene blue (MB). Notably, the membrane exhibited faster degradation of MB under UV irradiation. The study also evinced that UV-light irradiation was more efficient in achieving the desired photocatalytic degradation of MB compared to other methods [81].

Subsequently, in 2023, Kusworo *et al.* developed an innovative polysulfone (PSF)/nickel@ZnO (PSF/Ni@ZnO) membrane for the photocatalytic degradation of dyes. In this study, the photocatalytic activity was shown to increase due to a reduction in band-gap energy from 3.31 to 3.07 eV.

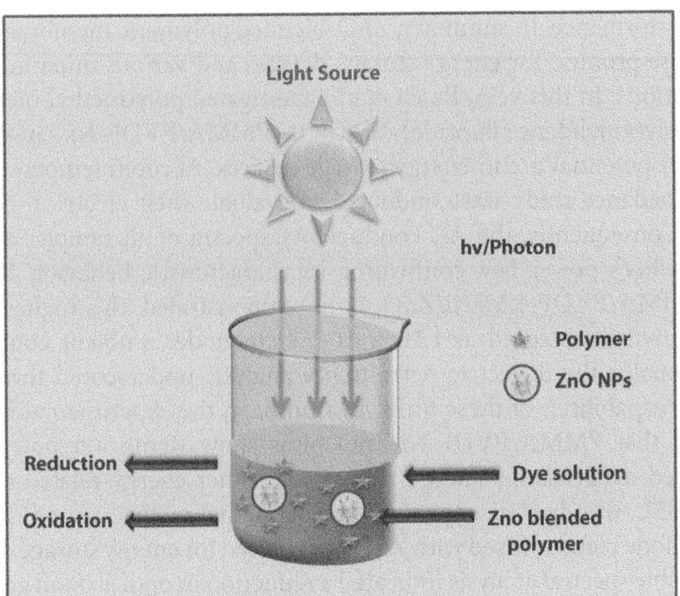

Figure 9.3 Nano-ZnO-blended polymer as photocatalysts for dye degradation.

Notably, the PSF/Ni@ZnO membrane with 1% (w/w) loading exhibited the highest efficiency in dye photodegradation [82]. Dana *et al.* employed the phase inversion method to synthesize polyvinylidene fluoride (PVDF) membranes blended with MWCNT-ZnO for RhB dye removal. As a result, under optimal conditions, 85% of the RhB dye was successfully degraded using UV light radiation [83]. In another study, Liliya *et al.* demonstrated the use of copper(II) oxide/ZnO@PET (Cu_2O/ZnO@PET) composite membranes for degrading carbendazim (Czm) in an aqueous system. Furthermore, the study achieved maximum Czm degradation efficiency at an optimal pH of 6 and a contact time of 140 minutes. Consequently, the Cu_2O/ZnO@PET composite exhibited enhanced photocatalytic activity, with a degradation rate of 98% [84].

9.3.3 Energy Storage

A polymeric matrix with suitable physicochemical and electrical properties is essential for the development of flexible devices. Notably, polymer matrices are remarkable materials widely utilized in electrochemical applications, particularly for energy storage. Among these, ZnO stands out as a non-toxic material with exceptional photocatalytic activity. Moreover, it has garnered significant attention due to its unique features, including leak-proof operation, excellent flexibility, superior thermal stability, and high performance. In summary, ZnO-blended polymeric membranes hold immense promise for energy storage devices and various other advanced applications. In this vein, Rajeh *et al.* investigated poly(methyl methacrylate)/poly(vinylidene fluoride)-Ni/ZnO (PMMA/PVDF-Ni/ZnO) films for their potential use in energy storage devices. At room temperature, an AC impedance study was conducted to evaluate their electrical conductivity. Consequently, the AC conductivity spectra of all samples adhered to Jonscher's power law, confirming its characteristic behavior. Notably, the PMMA/PVDF-1.5%Ni/ZnO films demonstrated the highest ionic conductivity, measured at 1.10×10^{-5} S/cm under ambient conditions. Additionally, the dielectric permittivity analysis underscored the charge storage capabilities of these films. In summary, the experimental findings suggest that PMMA/PVDF-Ni/ZnO films show significant potential as advanced materials for energy storage and other energy-related applications [85]. Alzaidy *et al.* engineered a polymethyl methacrylate (PMMA)/polysulfone blend infused with ZnO NP tailored for energy storage devices. UV-visible spectral analysis indicated a reduction in optical band gap energies, with the direct energy gap decreasing from 4.45 to 3.12 eV and the indirect gap declining from 3.33 to 1.82 eV. Significant improvements were noted in dielectric permittivity, dielectric loss, and AC conductivity with higher concentrations. Additionally, impedance measurements of Z′ and Z″ components were used to construct an equivalent electrical circuit [86]. In 2023, researchers developed a ZnO-doped CS/PVP-blended polymeric membrane, used in energy storage devices. This study demonstrated that the membrane enhances the optical band gap, Urbach energy, extinction coefficient, and optical conductivity. At normal temperature, the membrane exhibits an ionic conductivity of 4.2×10^{-5} S/cm. Furthermore, the electric modulus and dielectric permittivity were used to evaluate conductivity relaxation and charge storage characteristics. As a conclusion, the Cs/PVP-ZnO membrane is particularly suitable for use in thin-film solar cell applications [87]. Laila S. Alqarni and colleagues developed nanocomposite films using gelatin/chitosan doped with ZnO NPs for energy storage applications. These films demonstrated a significant influence on AC conductivity, dielectric constant, and loss tangent values. Notably, an ionic conductivity of 1.089×10^{-7} S/cm was observed at CS/Ge-3.5 wt% Co-doped ZnO NPs [88]. Another study focused on the development of a rGO/PANI/ZnO ternary nanocomposite for energy storage devices. Notably, the rGO/PANI/ZnO nanocomposite exhibited excellent capacitive behavior, achieving a specific capacity of 1546 F/g at a scan rate of 2 mV/s on an eggshell membrane electrode. Furthermore, it demonstrated outstanding cycling stability, retaining 97% of its capacity even after 3000 cycles [89]. Omar and his team employed the solution casting technique to create PVDF/ZnO/rGO membranes. These membranes demonstrated notable increases in total dipole moments, electrostatic potential, and bandgap energy, underscoring their strong potential for use in energy storage devices [54].

9.3.4 Biomedical Application

9.3.4.1 Wound Healing

Wounds, often described as a "silent epidemic," pose a significant healthcare issue worldwide, given their high prevalence and detrimental effects on global health, patient lifestyles, and mental health [90]. Furthermore, as their prevalence increases with age and related health conditions, a significant rise in cases is anticipated in the future [91]. To address this, antimicrobial bionanocomposites emerge as a cutting-edge solution, offering the biocompatibility and tissue-like features of biomacromolecules alongside the antimicrobial capabilities of metal or metal oxide nanoparticles [92]. The blending of ZnO NPs with polymers has demonstrated significant potential for applications in wound healing. Gutha *et al.* explored the wound-healing capabilities of CS/PVA/ZnO blends. The blends demonstrated antibacterial activity against *E. coli* and *S. aureus*. Furthermore, *in vitro* assessments validated their hemocompatibility and biocompatibility. Additionally, the *in vivo* wound-healing activity exhibited enhanced recovery of the injury [93]. In their work, Rifat *et al.* created a gentamicin-loaded CS-ZnO NP-based bionanocomposite, cross-linked with poly(ethylene glycol) (PEG) and sodium tripolyphosphate, for wound healing applications. The *in vivo* results indicated that the drug-loaded bionanocomposites provided better healing properties than commercial hydrogel dressings, without any scar formation. Additionally, cytotoxicity tests conducted on Vero and BHK 21 cells affirmed the biocompatibility of the bionanocomposite. Furthermore, *in vitro* analysis showed enhanced antibacterial activity against *E. coli* and *S. enterica* [94]. In another pioneering study, a ZnO-impregnated mesoporous silica (ZnO-MCM-41) was incorporated into a carboxymethyl cellulose (CMC) hydrogel to enhance its wound healing properties. Notably, citric acid was used as a cross-linking agent. Furthermore, the hydrogel demonstrated excellent cytocompatibility with adipose tissue–derived stem cells. These findings confirmed the hydrogels promising potential as a wound dressing material [95]. According to the literature, PVA, starch (St), and CS hydrogels incorporated with nano-zinc oxide (nZnO) (PVA/St/Cs/nZnO) were developed. As a result, the MTT assay, antibacterial tests, *in vivo* wound healing studies, and histopathological analysis revealed that the hydrogel exhibited significantly enhanced wound healing properties [96]. Similarly, another study investigated the use of heparinized PVA/CS/nZnO as a potential wound dressing material. The results showed that the bionanocomposite samples exhibited no toxicity in cell viability tests. Moreover, the hydrogel samples displayed excellent antibacterial properties, ensuring effective wound protection [97]. Likewise, the researchers designed a PVA/Chi/ZnO-Cur patch using chitosan (Chi) and PVA for its antimicrobial and wound-healing properties. This patch demonstrated excellent antibacterial effects against wound pathogens. Moreover, *in vitro* assessments verified its biocompatibility and hemocompatibility, while *in vivo* tests on Wistar rats highlighted its ability to promote accelerated wound healing [98]. Researchers synthesized CS–gelatin/ZnO nanocomposite hydrogel scaffolds (CS-GEL/nZnO) to examine their wound healing properties. Moreover, the scaffolds exhibited enhanced features such as swelling, biodegradation, and antibacterial activity. Thus, the *in situ* synthesized CS–GEL/nZnO is highly recommended for biomedical applications, particularly in skin tissue engineering [99].

9.3.4.2 Drug Delivery

Herein, we will summarize the recent progress on the use of nano-ZnO-blended polymeric membrane for drug delivery. In this vein, the therapeutic efficacy of curcumin in targeting cancer cells is limited by its poor water solubility and low bioavailability. In this context, curcumin-loaded PMMA-PEG/ZnO nanocomposites have been explored for delivering curcumin to cancer cells. These nanocomposites can trigger cancer cell apoptosis through a cell cycle–mediated pathway, increasing their potential to treat gastric cancer [100]. Iman *et al.* utilized CMC, starch, and ZnO-NPs for the delivery of doxorubicin. As a result, the ZnO NP-loaded CMC/starch hydrogel demonstrated prolonged and more controlled drug release. Furthermore, the cytotoxicity of the samples was confirmed through testing on human colon cancer cells (SW480) [101]. Similarly, Raman *et al.* introduced ZnO NPs encapsulated with the copolymer PMMA-AA for the delivery of curcumin. Consequently, the pH-responsive release mechanism of curcumin ensures safer and more controlled delivery at physiological pH. In addition, *in vitro* evaluations revealed that the Cur/PMMA-AA/ZnO NPs exhibited significant anticancer activity against various cancer cell lines [102]. Quercetin (QUR), a flavonoid found in plants and fruits, is known for its anticancer and medicinal effects on brain tumors. However, its low stability, bioavailability, and inability to cross the blood–brain barrier (BBB) hinder its effectiveness. In this study, the researchers developed CS/CMC/GQDs/ZnO@QUR nanocomposites for more efficient QUR delivery. The incorporation of GQDs improved BBB penetration. The release of QUR from the nanocomposites was more controlled, with half of the drug being released over 72 hours, offering a slower release compared to other carriers. Furthermore, MTT assays on U-87 MG and L929 cell lines confirmed that the nanocomposite possesses excellent anticancer properties [103]. Likewise, this study focused on the preparation of CS/PVA blended with ZnO NPs for folic acid (FA) delivery. The *in vitro* release study revealed a cumulative drug release of 95%. Additionally, *in vivo* toxicity tests demonstrated no toxicity associated with the CH/PVA/ZnO-FA sample. These findings suggest enhanced bioavailability and successful targeting of cancer tumors [104]. Sun and colleagues designed an eco-friendly colon-specific drug delivery system. Specifically, 5-fluorouracil was encapsulated within ZnO/CMC/CS bionanocomposite beads. Under optimal pH conditions, *in vitro* studies revealed sustained drug release behavior [105].

9.3.4.3 *Antibacterial Activity*

The antibacterial properties of ZnO nanoparticles blended with polymers have been evaluated for their efficacy against bacteria such as *Staphylococcus aureus, Klebsiella pneumoniae,* and *Pseudomonas aeruginosa.* In the study of Aswathy *et al.*, they developed a PVA-based nanocomposite blended with ZnO NPs using the solvent casting method to assess antimicrobial activity. Consequently, the nanocomposite exhibited outstanding antimicrobial effects against *Staphylococcus aureus, Klebsiella pneumonia,* and *Pseudomonas aeruginosa* [106]. Behzad and his colleagues developed a PVA/CMC-ZnO-loaded erythromycin (EM) nanocomposite for antibacterial applications. The nanocomposite demonstrated excellent antibacterial activity against *S. aureus* and *E. coli.* Additionally, the *in vitro* release profile revealed that the release of EM from the PVA-CMC/ZnO-EM nanofibers gradually increased over time, demonstrating a controlled-release behavior [107]. Similarly, Ying *et al.* designed a PLA/GO@ZnO blend for antibacterial properties. Notably, antibacterial tests revealed that the nanocomposite films displayed remarkable antibacterial effectiveness even at minimal GO–ZnO concentrations [108].

9.4 Conclusions and Future Outlook

We have explored the interesting potential applications of integrating nano-ZnO into polymer membranes that have emerged as promising strategy for environmental remediation, energy concern, and sustainable development. This chapter has highlighted the diverse applications of nano-ZnO-blended polymeric membranes including water purification, membrane fouling mitigation, biomedical applications, food packaging, photocatalytic degradation of dyes, and energy storage. The unique properties of nano-ZnO, such as its antimicrobial activity, photocatalytic activity, and mechanical strength, have been leveraged to enhance the performance and functionality of polymeric membranes. Looking ahead, though nano-ZnO-blended polymer membranes have developed and performed significantly, upcoming research should focus on their long-term stability and durability while maintaining their quality. Furthermore, developing multifunctional membranes that can simultaneously perform multiple tasks is an exciting area of research. By addressing these challenges and opportunities, nano-ZnO-blended polymeric membranes can play a vital role in promoting environmental sustainability and sustainable development.

AI disclosure statement: The author acknowledges the use of AI-based tools for minor language editing or grammatical correction. However, all intellectual contributions, analyses, and conclusions presented in this chapter are entirely the author's own.

References

1. Abraham, E., *et al.*, *Green nanocomposites of natural rubber/nanocellulose: Membrane transport, rheological and thermal degradation characterisations*, vol. 51, pp. 415–424, 2013.
2. Wen, Y., *et al.*, *Polymeric nanocomposite membranes for water treatment: a review*, vol. 17, pp. 1539–1551, 2019.
3. Adhi, T.P., *et al.*, *Sustainable Management of Water Resources for Drinking Water Supply by Exploring Nanotechnology*, vol. 16, p. 1896, 2024.
4. Arias-Estévez, M., *et al.*, *The mobility and degradation of pesticides in soils and the pollution of groundwater resources*, vol. 123, pp. 247–260, 2008.
5. Lin WenChi, L.W., Li ZhongRui, L.Z., Burns, M., A drinking water sensor for lead and other heavy metals. 2017.
6. Hunt, H. and Briscoe, H.J.T.J.P.C., *The Electrical Conductivity of Organic Acids in Water, Alcohols, and Acetone, and the Electronic Structures, of the Acids*, vol. 33, pp. 1495–1513, 2002.
7. Kremser, U. and Schnug, E.J.L.V., *Impact of fertilizers on aquatic ecosystems and protection of water bodies from mineral nutrients*, vol. 52, pp. 81–90, 2002.
8. Naim, M.M., *et al.*, *Macro-reticular ion exchange resins for recovery of direct dyes from spent dyeing and soaping liquors*, vol. 27, p. 1593, 2022.
9. Farhat Nabeela, F.N., *et al.*, Microbial contamination of drinking water in Pakistan-a review. 2014.
10. El Batouti, M., Al-Harby, N.F., Elewa, M.M.J.W., *A review on promising membrane technology approaches for heavy metal removal from water and wastewater to solve water crisis*, vol. 13, p. 3241, 2021.
11. Ali, S., *et al.*, *Efficient removal of zinc from water and wastewater effluents by hydroxylated and carboxylated carbon nanotube membranes: Behaviors and mechanisms of dynamic filtration*, vol. 365, pp. 64–73, 2019.
12. Noshin Masood, N.M., Abida Farooqi, A.F., Zafar, M., Health risk assessment of arsenic and other potentially toxic elements in drinking water from an industrial zone of Gujrat, Pakistan: a case study. 2019.
13. Pandey, N., Shukla, S., Singh, N.J.N., *Water purification by polymer nanocomposites: an overview*, vol. 3, pp. 47–66, 2017.
14. Sharma, A., *et al.*, *A comparative study of chromium (VI) removal using sawdust and eucalyptus bark*, vol. 9, pp. 343–347, 2009.

15. Guo, H., *et al.*, *Development of active and intelligent pH food packaging composite films incorporated with litchi shell extract as an indicator*, vol. 226, pp. 77–89, 2023.
16. Tamayo, L., *et al.*, *Copper-polymer nanocomposites: An excellent and cost-effective biocide for use on antibacterial surfaces*, vol. 69, pp. 1391–1409, 2016.
17. Al-Tayyar, N.A., Youssef, A.M., Al-Hindi, R.J.F., *Antimicrobial food packaging based on sustainable Bio-based materials for reducing foodborne Pathogens: A review*, vol. 310, p. 125915, 2020.
18. Avella, M., *et al.*, *Biodegradable starch/clay nanocomposite films for food packaging applications*, vol. 93, pp. 467–474, 2005.
19. Olmos, D., *et al.*, *Preparation and characterization of antimicrobial films based on LDPE/Ag nanoparticles with potential uses in food and health industries*, vol. 8, p. 60, 2018.
20. Mamo, F.T., *et al.*, *Mycotoxins in Ethiopia: a review on prevalence, economic and health impacts*, vol. 12, p. 648, 2020.
21. De Lucca, A.J.J.R., *Harmful fungi in both agriculture and medicine*, vol. 24, p. 3, 2007.
22. Li, J.-F., *et al.*, *Effect of TiO2 nanoparticles on the surface morphology and performance of microporous PES membrane*, vol. 255, pp. 4725–4732, 2009.
23. Cortalezzi, M.M., *et al.*, *Characteristics of ultrafiltration ceramic membranes derived from alumoxane nanoparticles*, vol. 205, pp. 33–43, 2002.
24. Shen, L., *et al.*, *Polymeric membranes incorporated with ZnO nanoparticles for membrane fouling mitigation: A brief review*, vol. 8, p. 224, 2020.
25. Liu, F., *et al.*, *Progress in the production and modification of PVDF membranes*, vol. 375, pp. 1–27, 2011.
26. Mahlangu, O., *et al.*, *Development of hydrophilic GO-ZnO/PES membranes for treatment of pharmaceutical wastewater*, vol. 76, pp. 501–514, 2017.
27. Zhang, D.-D., *et al.*, *Construction of ZnO@ mSiO2 antibacterial nanocomposite for inhibition of microorganisms during Zea mays storage and improving the germination*, vol. 168, p. 113907, 2022.
28. Wojnarowicz, J., *et al.*, *Effect of microwave radiation power on the size of aggregates of ZnO NPs prepared using microwave solvothermal synthesis*, vol. 8, p. 343, 2018.
29. Adam, R.E., *et al.*, *Synthesis of ZnO nanoparticles by co-precipitation method for solar driven photodegradation of Congo red dye at different pH*, vol. 32, pp. 11–18, 2018.
30. Xue, X., *et al.*, Review on nanomaterials synthesized by vapor transport method: growth and their related applications. 5, 97, 79249–79263, 2015.
31. Güell, F., *et al.*, Influence of colloidal Au on the growth of ZnO nanostructures. 11, 4, 870, 2021.
32. Barani, M., *et al.*, *Cell-free extract assisted synthesis of ZnO nanoparticles using aquatic bacterial strains: Biological activities and toxicological evaluation*, vol. 606, p. 120878, 2021.

33. Maťátková, O., *et al.*, *Antimicrobial properties and applications of metal nanoparticles biosynthesized by green methods*, vol. 58, p. 107905, 2022.
34. Ogunyemi, S.O., *et al.*, *Green synthesis of zinc oxide nanoparticles using different plant extracts and their antibacterial activity against Xanthomonas oryzae pv. oryzae*, vol. 47, pp. 341–352, 2019.
35. Bandeira, M., *et al.*, *Green synthesis of zinc oxide nanoparticles: A review of the synthesis methodology and mechanism of formation*, vol. 15, p. 100223, 2020.
36. Zheng, Y., *et al.*, *Green biosynthesis and characterization of zinc oxide nanoparticles using Corymbia citriodora leaf extract and their photocatalytic activity*, vol. 8, pp. 59–63, 2015.
37. Chen, C., *et al.*, *Investigation of nano-sized ZnO particles fabricated by various synthesis routes*, vol. 12, pp. 420–425, 2011.
38. Reghioua, A., *et al.*, *Synthesis of Schiff's base magnetic crosslinked chitosan-glyoxal/ZnO/Fe3O4 nanoparticles for enhanced adsorption of organic dye: modeling and mechanism study*, vol. 20, p. 100379, 2021.
39. Jin, S.-E. and Jin, H.-E.J.P., *Synthesis, characterization, and three-dimensional structure generation of zinc oxide-based nanomedicine for biomedical applications*, vol. 11, p. 575, 2019.
40. Hefny, M.E., *et al.*, *Biosynthesis of zinc nanoparticles using culture filtrates of Aspergillus, Fusarium and Penicillium fungal species and their antibacterial properties against gram-positive and gram-negative bacteria*, vol. 46, pp. 2009–2021, 2019.
41. Muhammad, W., *et al.*, *Optical, morphological and biological analysis of zinc oxide nanoparticles (ZnO NPs) using Papaver somniferum L*, vol. 9, pp. 29541–29548, 2019.
42. Rajan, A., Cherian, E., Baskar, G.J.I.J.M.S.T., *Biosynthesis of zinc oxide nanoparticles using Aspergillus fumigatus JCF and its antibacterial activity*, vol. 1, pp. 52–57, 2016.
43. Rudramurthy, G.R., *et al.*, *Nanoparticles: alternatives against drug-resistant pathogenic microbes*, vol. 21, p. 836, 2016.
44. Sanaeimehr, Z., Javadi, I., Namvar, FJC., *Antiangiogenic and antiapoptotic effects of green-synthesized zinc oxide nanoparticles using Sargassum muticum algae extraction*, vol. 9, pp. 1–16, 2018.
45. Khalafi, T., Buazar, F., Ghanemi, KJS., *Phycosynthesis and enhanced photocatalytic activity of zinc oxide nanoparticles toward organosulfur pollutants*, vol. 9, p. 6866, 2019.
46. Lyu, S.C., *et al.*, *Low-temperature growth of ZnO nanowire array by a simple physical vapor-deposition method*, vol. 15, pp. 3294–3299, 2003.
47. Eremina, A., *et al.*, *Mesoporous silicon nanoparticles covered with PEG molecules by mechanical grinding in aqueous suspensions*, vol. 331, p. 111641, 2022.
48. Shahidi, S., Moazzenchi, B., Ghoranneviss, M.J.T.E.P.J.A.P., *A review-application of physical vapor deposition (PVD) and related methods in the textile industry*, vol. 71, p. 31302, 2015.

49. Hajnorouzi, A., Afzalzadeh, R., Ghanati, FJU., *Ultrasonic irradiation effects on electrochemical synthesis of ZnO nanostructures*, vol. 21, pp. 1435–1440, 2014.
50. Xu, J., *et al.*, *Preparation of ZnS nanoparticles by ultrasonic radiation method*, vol. 66, pp. 639–641, 1998.
51. Čech Barabaszová, K., *et al.*, *Effects of ultrasound on zinc oxide/vermiculite/chlorhexidine nanocomposite preparation and their antibacterial activity*, vol. 9, p. 1309, 2019.
52. Nath, M.R., *et al.*, Innovations in nanomaterial-embedded polymer membranes for water and wastewater treatment: a comprehensive review. 2024.
53. Hussain, R.T., Hossain, M.S., Shariffuddin, J.H.J.M.T.S., *Green synthesis and photocatalytic insights: A review of zinc oxide nanoparticles in wastewater treatment*, p. 100764, 2024.
54. Omar, A., *et al.*, *Investigation of morphological, structural and electronic transformation of PVDF and ZnO/rGO/PVDF hybrid membranes*, vol. 55, p. 381, 2023.
55. Strachowski, T., *et al.*, *Hydrothermal synthesis of zinc oxide nanoparticles using different chemical reaction stimulation methods and their influence on process kinetics*, vol. 15, p. 7661, 2022.
56. Birniwa, A.H., *et al.*, *Membrane technologies for heavy metals removal from water and wastewater: A mini review*, vol. 9, p. 100538, 2024.
57. Aziz, S., *et al.*, *A comprehensive review of membrane-based water filtration techniques*, vol. 14, p. 169, 2024.
58. Gulab, H., *et al.*, *Advancements in zinc oxide nanomaterials: synthesis, properties, and diverse applications*, vol. 39, p. 101271, 2024.
59. Vevers, R., *et al.*, *Photocatalytic Zinc Oxide Nanoparticles in Antibacterial Ultrafiltration Membranes for Biofouling Control*, vol. 29, p. 1274, 2024.
60. Li, Z., Jia, Y., Bai, SJR., *Polysulfone foam with high expansion ratio prepared by supercritical carbon dioxide assisted molding foaming method*, vol. 8, pp. 2880–2886, 2018.
61. Rogers, N., *et al.*, *Mechanical and solubility properties of bio-nanocomposite film of semi refined kappa carrageenan/ZnO nanoparticles*, vol. 3, pp. 7–8, 2007.
62. Hussain, S., Akhter, R., Maktedar, S.S.J.S.F.T., *Advancements in sustainable food packaging: from eco-friendly materials to innovative technologies*, vol. 2, pp. 1297–1364, 2024.
63. Alp, E., *et al.*, *The effect of ZnO nanoparticles morphology on the barrier and antibacterial properties of hybrid ZnO/graphene oxide/montmorillonite coatings for flexible packaging*, vol. 55, p. 105307, 2024.
64. Smaoui, S., *et al.*, *Zinc oxide nanoparticles in meat packaging: A systematic review of recent literature*, vol. 36, p. 101045, 2023.
65. Amin, U., *et al.*, *Biodegradable active, intelligent, and smart packaging materials for food applications*, vol. 33, p. 100903, 2022.

66. Zhang, R., *et al.*, *Development of polylactic acid/ZnO composite membranes prepared by ultrasonication and electrospinning for food packaging*, vol. 135, p. 110072, 2021.
67. Dai, L., *et al.*, *Development of Pomegranate Peel Extract and Nano ZnO Co-Reinforced Polylactic Acid Film for Active Food Packaging*, vol. 12, p. 1108, 2022.
68. Giannakas, A.E., *et al.*, *Nanocomposite Film Development Based on Chitosan/Polyvinyl Alcohol Using ZnO@Montmorillonite and ZnO@Halloysite Hybrid Nanostructures for Active Food Packaging Applications*, vol. 12, p. 1843, 2022.
69. Sun, X., *et al.*, *Fabrication of antimicrobial chitosan/ZnO nanoparticles/lecithin-montmorillonite films for food packaging application*, vol. 159, p. 110686, 2025.
70. Akshaykranth, A., *et al.*, *Antibacterial activity study of ZnO incorporated biodegradable poly (lactic acid) films for food packaging applications*, vol. 80, pp. 1369–1384, 2023.
71. Gasti, T., *et al.*, *Chitosan/pullulan based films incorporated with clove essential oil loaded chitosan-ZnO hybrid nanoparticles for active food packaging*, vol. 277, p. 118866, 2022.
72. Patil, M.B., *et al.*, *Microwave-assisted synthesis of poly (acrylamide-co-2-hydroxyethyl methacrylate)/chitosan semi-IPN ZnO nanocomposite membranes for food packaging applications*, vol. 20, pp. 3537–3548, 2022.
73. Wang, H., *et al.*, *Preparation, characterization and food packaging application of nano ZnO@ Xylan/quaternized xylan/polyvinyl alcohol composite films*, vol. 215, pp. 635–645, 2022.
74. Nasiri, S.L., *et al.*, *Potential perspectives of CMC-PET/ZnO bilayer nanocomposite films for food packaging applications: physical, mechanical and antimicrobial properties*, vol. 15, pp. 3731–3740, 2021.
75. Amini, E., *et al.*, *Promising nanocomposites for food packaging based on cellulose–PCL films reinforced by using ZnO nanoparticles in an ionic liquid*, vol. 193, p. 116246, 2023.
76. Vemula, M., *Exploring the potential of Coriandrum sativum seed powder for the adsorptive removal of malachite green dye*, Vol 14, 252–260, 2024.
77. Zhang, Y., *et al.*, Visible light response ZnO–C 3 N 4 thin film photocatalyst. 40, 96–104, 2021.
78. Giziński, D., *et al.*, *Nanostructured anodic copper oxides as catalysts in electrochemical and photoelectrochemical reactions*, vol. 10, p. 1338, 2020.
79. Mohadesi, M., Sanavi Fard, M., Shokri, A.J.I.J.E.A.C., *The application of modified nano-TiO2 photocatalyst for wastewater treatment: A review*, vol. 104, pp. 2571–2592, 2024.
80. Mahlangu, O.T., Mamba, G., Mamba, B.B.J.J.E.C.E., *A facile synthesis approach for GO-ZnO/PES ultrafiltration mixed matrix photocatalytic membranes for dye removal in water: Leveraging the synergy between photocatalysis and membrane filtration*, vol. 11, p. 110065, 2023.

81. Abu-Dalo, M.A., Al-Rosan, S.A., Albiss, B.A.J.P., *Photocatalytic degradation of methylene blue using polymeric membranes based on cellulose acetate impregnated with ZnO nanostructures*, vol. 13, p. 3451, 2021.
82. Kusworo, T., *et al.*, *Fabrication, characterization, and application of PSf/Ni@ ZnO amalgamated membrane for photocatalytic degradation of dyeing wastewater from batik industry*, vol. 30, p. 101493, 2023.
83. Toloman, D., *et al.*, *Photocatalytic self-cleaning PVDF membrane blended with MWCNT-ZnO nanocomposites for RhB removal*, vol. 13, p. 594, 2023.
84. Altynbaeva, L.S., *et al.*, *A novel Cu2O/ZnO@ PET composite membrane for the photocatalytic degradation of carbendazim*, vol. 12, p. 1724, 2022.
85. Rajeh, A., *et al.*, *Alteration in the structural, optical, thermal, electrical, and dielectric properties of PMMA/PVDF blend by incorporation of Ni/ZnO nanohybrid for optoelectronic and energy storage devices*, vol. 34, pp. 1221–1231, 2024.
86. AlZaidy, G.A.J.J.I., *O. Polymers, and Materials, Boosting of the optical properties, and electrical conductivity of polymethyl methacrylate (PMMA)/ Polystyrene (PS) Blend with zinc oxide nanoparticles for high-performance energy storage devices*, pp. 1–12, 2024.
87. Alghamdi, H.M. and Rajeh, A.J.P.T., *Study of the photoluminescence, optical, thermal, and electrical parameters of the Cs/PVP blend/zinc oxide nanorods films for energy storage devices*, vol. 124, p. 108093, 2023.
88. Alqarni, L.S., *et al.*, *Enhancing the optical, electrical, dielectric properties and antimicrobial activity of chitosan/gelatin incorporated with Co-doped ZnO nanoparticles: Nanocomposites for use in energy storage and food packaging*, vol. 1297, p. 137011, 2024.
89. Abbas, S., *et al.*, *Facile synthesis of rGO/PANI/ZnO ternary nanocomposites for energy storage devices*, .
90. Järbrink, K., *et al.*, *Prevalence and incidence of chronic wounds and related complications: a protocol for a systematic review.*
91. Tottoli, E.M., *et al.*, *Skin wound healing process and new emerging technologies for skin wound care and regeneration*, vol. 12, p. 735, 2020.
92. Boateng, J. and Catanzano, O.J.J., *Advanced therapeutic dressings for effective wound healing—a review*, vol. 104, pp. 3653–3680, 2015.
93. Gutha, Y., *et al.*, *Antibacterial and wound healing properties of chitosan/poly (vinyl alcohol)/zinc oxide beads (CS/PVA/ZnO)*, vol. 103, pp. 234–241, 2017.
94. Masud, R.A., *et al.*, *Preparation of novel chitosan/poly (ethylene glycol)/ZnO bionanocomposite for wound healing application: Effect of gentamicin loading*, vol. 12, p. 100785, 2020.
95. Rakhshaei, R., Namazi, H.J.M.S., E.C., *A potential bioactive wound dressing based on carboxymethyl cellulose/ZnO impregnated MCM-41 nanocomposite hydrogel*, vol. 73, pp. 456–464, 2017.
96. Baghaie, S., *et al.*, *Wound healing properties of PVA/starch/chitosan hydrogel membranes with nano Zinc oxide as antibacterial wound dressing material*, vol. 28, pp. 2220–2241, 2017.

97. Khorasani, M.T., *et al.*, *Incorporation of ZnO nanoparticles into heparinised polyvinyl alcohol/chitosan hydrogels for wound dressing application*, vol. 114, pp. 1203–1215, 2018.
98. Niranjan, R., *et al.*, *Chitosan based wound dressing patch loaded with curcumin tagged ZnO nanoparticles for potential wound healing application*, vol. 154, p. 110885, 2023.
99. Rakhshaei, R., *et al.*, *In situ synthesized chitosan–gelatin/ZnO nanocomposite scaffold with drug delivery properties: Higher antibacterial and lower cytotoxicity effects*, p. 136(22)47590, 2019.
100. Dhivya, R., *et al.*, *Biocompatible curcumin loaded PMMA-PEG/ZnO nanocomposite induce apoptosis and cytotoxicity in human gastric cancer cells*, vol. 80, pp. 59–68, 2017.
101. Gholamali, I. and Yadollahi, M.J.I., *Doxorubicin-loaded carboxymethyl cellulose/Starch/ZnO nanocomposite hydrogel beads as an anticancer drug carrier agent*, vol. 160, pp. 724–735, 2020.
102. Dhivya, R., *et al.*, *Enhancing the anti-gastric cancer activity of curcumin with biocompatible and pH sensitive PMMA-AA/ZnO nanoparticles*, vol. 82, pp. 182–189, 2018.
103. Ostovar, S., Pourmadadi, M., Zaker, M.A.J.I., *Co-biopolymer of chitosan/carboxymethyl cellulose hydrogel improved by zinc oxide and graphene quantum dots nanoparticles as pH-sensitive nanocomposite for quercetin delivery to brain cancer treatment*, vol. 253, p. 127091, 2023.
104. Al-Tememe, E., Malk, F.H., Hraishawi, R.M., Synthesis, characterization and in vitro drug release studies of polymeric drug coated zinc oxide nanoparticles, in: *AIP Conference Proceedings*, AIP Publishing, 2021.
105. Sun, X., *et al.*, *pH-sensitive ZnO/carboxymethyl cellulose/chitosan bionanocomposite beads for colon-specific release of 5-fluorouracil*, vol. 128, pp. 468–479, 2019.
106. Jayakumar, A., *et al.*, *Lipopeptide and zinc oxide nanoparticles blended polyvinyl alcohol-based nanocomposite films as antimicrobial coating for biomedical applications*, vol. 102, pp. 220–228, 2021.
107. Darbasizadeh, B., *et al.*, *Crosslinked-polyvinyl alcohol-carboxymethyl cellulose/ZnO nanocomposite fibrous mats containing erythromycin (PVA-CMC/ZnO-EM): Fabrication, characterization and in-vitro release and anti-bacterial properties*, vol. 2019, pp. 1137–1146, p. 141.
108. Huang, Y., *et al.*, *Poly (lactic acid)/graphene oxide–ZnO nanocomposite films with good mechanical, dynamic mechanical, anti-UV and antibacterial properties*, vol. 90, pp. 1677–1684, 2015.

10

Polyvinyl Chloride (PVC)-Based Hybrid Biocomposites and Bionanocomposites

S. Tejas Krishna[1] and G.Vinitha[2*]

[1]School of Computer Science and Engineering, Vellore Institute of Chennai, Chennai, Tamilnadu, India
[2]Department of Physics, School of Advanced Sciences, Vellore Institute of Technology, Chennai, India

Abstract
They combine the desirable traits of polyvinyl chloride with the ecological benefits of bio-based and nanoscale reinforcements. While challenges remain, ongoing research and technological advances are rapidly expanding the frontiers of this domain. Their development not only reflects innovation in materials science but also resonates with the global imperative for climate-conscious, sustainable manufacturing. From lightweight automotive parts to biomedical devices, smart packaging, and flame-retardant panels, these materials are shaping the future of responsible materials engineering. The PVC-based hybrid biocomposites and bionanocomposites present a highly promising class of materials that bridge the gap between performance engineering and sustainability.

***Keywords*:** Polyvinyl chloride, hybrid materials, biocomposites and bionanocomposites, nano, bionano, composites, nanocomposites, biomaterials

10.1 Introduction

In the evolving landscape of advanced materials, there is a growing demand for the development of environmentally sustainable and high-performance composites. The 21st century has ushered in a heightened awareness of ecological concerns, including climate change, plastic pollution, and the

**Corresponding author*: vinitha.g@vit.ac.in

Shiv Kumari Panda and Visakh P. M (eds.) Polyvinyl Chloride-Based Biocomposites and Bionanocomposites, (297–338) © 2026 Scrivener Publishing LLC

depletion of non-renewable resources. As a response, the scientific community has increasingly turned toward the design and implementation of bio-based and hybrid composite materials that not only match but in some cases surpass the mechanical and functional properties of traditional petroleum-based plastics. One such material that has garnered significant attention in this context is polyvinyl chloride (PVC), particularly in the development of hybrid biocomposites and bionanocomposites.

PVC is a widely used thermoplastic polymer known for its affordability, excellent chemical resistance, flame retardancy, and versatility. Its broad range of applications spans from construction materials (such as pipes and window frames) to healthcare devices and electrical insulation. However, like many synthetic polymers, traditional PVC exhibits poor biodegradability and poses environmental challenges. This has led researchers to explore ways to enhance its sustainability profile without compromising its utility. The incorporation of bio-based fillers, fibers, and nanomaterials into PVC matrices presents a promising pathway to achieving this balance. When properly engineered, such composites not only reduce the polymer's environmental footprint but can also deliver improved mechanical, thermal, and barrier properties.

The terms *biocomposite* and *bionanocomposite* refer to advanced material systems that incorporate natural components or nanostructured biomaterials into a polymer matrix. In the case of biocomposites, natural fibers such as jute, kenaf, hemp, sisal, flax, or agricultural residues like rice husk and coconut coir are used as reinforcement agents. These materials are biodegradable, renewable, and often abundantly available, making them suitable candidates for sustainable composites. Bionanocomposites go a step further by incorporating nanomaterials derived from biological sources or those that are biocompatible, such as cellulose nanocrystals, chitin nanowhiskers, starch nanoparticles, or nano-clays. These nanomaterials offer a high aspect ratio, excellent surface area, and unique interfacial properties that can significantly enhance the performance of the base polymer matrix.

A hybrid composite, in this context, refers to a material system that combines two or more types of reinforcement—for example, a blend of natural fiber and synthetic fiber or natural fiber and nanoparticles—to achieve synergistic effects. When these reinforcements are integrated into a PVC matrix, the result is a hybrid biocomposite or bionanocomposite that leverages the strengths of multiple materials to overcome the individual limitations of each component. Such composites are at the forefront of material research due to their ability to be tailored for specific applications in industries such as automotive, aerospace, packaging, biomedical, and construction.

Several challenges must be addressed to effectively develop and commercialize PVC-based hybrid biocomposites and bionanocomposites. These include ensuring compatibility between the hydrophobic PVC matrix and hydrophilic natural fillers, improving dispersion of nanomaterials, maintaining mechanical integrity under stress, and ensuring cost-effectiveness at scale. To overcome these challenges, various modification techniques—including chemical treatments, compatibilizers, surface functionalization, and nanotechnology-based dispersion methods—have been explored. These strategies help to improve the interfacial bonding between the PVC matrix and the reinforcement phases, which is critical for load transfer and overall composite performance.

The synthesis and characterization of such hybrid composites also involve a multidisciplinary approach, combining principles from polymer chemistry, nanotechnology, materials engineering, and environmental science. A comprehensive understanding of processing techniques such as extrusion, injection molding, compression molding, and solution casting is essential. Additionally, characterization tools like scanning electron microscopy (SEM), Fourier transform infrared spectroscopy (FTIR), X-ray diffraction (XRD), thermogravimetric analysis (TGA), and differential scanning calorimetry (DSC) are routinely employed to analyze the morphology, chemical composition, thermal stability, and crystallinity of these materials.

From an environmental standpoint, PVC-based hybrid biocomposites offer a partial solution to the global plastic waste problem. The use of biofillers not only displaces a portion of the petroleum-based polymer but also enables valorization of agricultural waste and natural resources. Furthermore, these composites can be engineered to be recyclable, biodegradable, or biocompatible, depending on the selection of reinforcements and processing techniques. This aligns with the global shift toward a circular economy, where materials are designed for reuse, recycling, or safe degradation at the end of their life cycle.

In recent years, research on PVC-based hybrid biocomposites has accelerated, supported by a growing body of literature and successful pilot-scale implementations. Studies have reported notable improvements in tensile strength, thermal resistance, flame retardancy, and moisture resistance of such composites compared to pure PVC or traditional biocomposites. The incorporation of nanomaterials, in particular, has shown exceptional promise in enhancing the barrier properties and durability of PVC composites, making them suitable for packaging and biomedical applications where material integrity is critical.

Looking ahead, the integration of machine learning, AI-based design and green chemistry principles in the development of PVC-based hybrid biocomposites opens new frontiers for innovation. These emerging technologies can accelerate material discovery, optimize formulations, and ensure sustainability throughout the product life cycle.

In conclusion, the development of PVC-based hybrid biocomposites and bionanocomposites represents a significant leap forward in the quest for high-performance, eco-friendly materials. By intelligently combining the advantages of synthetic polymers, natural fibers, and nanotechnology, these composites offer a pathway toward materials that are not only functionally superior but also environmentally responsible. This chapter delves deeper into the fundamental concepts, classifications, processing techniques, performance attributes, and potential applications of these advanced composites, beginning with an exploration of hybrid materials in the following section.

10.2 Hybrid

The term "hybrid" in material science denotes the strategic combination of two or more different materials to produce a new material system with synergistic or improved properties. In the realm of polymer composites, a hybrid typically refers to the blending of two or more types of reinforcing agents within a polymer matrix to achieve a balance of mechanical strength, thermal stability, cost-effectiveness, and environmental compatibility. These reinforcements can include a mixture of natural fibers, synthetic fibers, fillers, and nanomaterials, each contributing distinct advantages to the final composite material.

Hybrid materials bridge the gap between single-filler composites and complex engineered systems, making them especially relevant in fields where multifunctionality and material performance under varying conditions are critical. The concept of hybridization is not limited to fiber reinforcements; it can extend to the matrix phase itself (e.g., mixing two different polymers) or to hybrid additives that combine different structural, chemical, or morphological features. However, within the scope of this chapter, we focus primarily on hybrid reinforcements within a PVC matrix, particularly in biocomposites and bionanocomposites.

10.2.1 Definition and Types of Hybrid Composites

A hybrid composite is a composite material made by combining two or more types of fibers or fillers within a single matrix to improve the composite's overall performance. The rationale behind hybridization is to exploit the positive attributes of each reinforcement while mitigating their individual weaknesses. For example, natural fibers offer sustainability and low density but often suffer from low moisture resistance and lower mechanical strength compared to synthetic fibers. By combining them with synthetic reinforcements or nanomaterials, a hybrid system can be engineered to provide enhanced functionality.

Hybrid composites are broadly categorized into the following:

- **Fiber–fiber hybrids**: involving two or more fiber types (e.g., glass fiber and jute).
- **Fiber–filler hybrids**: using fibers along with particulate fillers (e.g., hemp fiber + fly ash).
- **Filler–filler hybrids**: employing a mix of micro- and nano-fillers (e.g., rice husk ash + nanoclay).
- **Nanohybrids**: integrating nano-scale reinforcements with micro-scale fillers or fibers (e.g., cellulose nanocrystals + natural fiber).

Depending on the arrangement and distribution, hybrid composites can also be classified into the following:

- **Interply hybrids**: where different reinforcements are layered alternately (common in laminates).
- **Intraply hybrids**: where two types of fibers are woven together within the same layer.
- **Random hybrids**: where fillers or fibers are randomly distributed within the matrix.

10.2.2 Significance of Hybridization

The primary motivation for developing hybrid composites is performance tailoring—achieving a desired set of mechanical, thermal, or barrier properties through strategic material selection and arrangement. In single-fiber or mono-filler systems, performance is often limited by the intrinsic properties of the chosen reinforcement. However, hybrid composites enable a multiscale reinforcement strategy, enhancing strength, toughness, modulus, and fatigue life simultaneously.

Some key benefits of hybrid materials include the following:

- **Optimized mechanical properties**: combining stiff fibers with ductile ones to balance strength and toughness.
- **Improved thermal performance**: leveraging thermally stable reinforcements for high-heat applications.
- **Enhanced environmental resistance**: using nanofillers to reduce water absorption in natural fiber composites.
- **Cost efficiency**: using cheaper natural fillers to reduce the overall cost of high-performance composites.

In the context of PVC-based systems, hybridization is particularly useful due to PVC's relatively rigid structure and low impact resistance in its pure form. By introducing flexible reinforcements like natural fibers or impact modifiers, the composite can become more durable and suitable for demanding applications like automotive interiors or external building panels.

10.2.3 Material Selection Criteria

The design of hybrid composites depends heavily on the compatibility and performance synergy between the selected reinforcements. Several criteria must be considered during material selection:

- **Mechanical compatibility**: matching the modulus and elongation of reinforcements to prevent premature failure.
- **Thermal behavior**: ensuring similar degradation temperatures during processing.
- **Surface chemistry**: compatibility with the matrix, or surface treatments to improve adhesion.
- **Density and weight**: especially important in automotive and aerospace applications.
- **Cost and availability**: local or abundant natural fibers can be economically beneficial.
- **Environmental footprint**: preference for renewable or recyclable components for green composites.

Natural fibers such as jute, kenaf, hemp, flax, and sisal are often combined with inorganic fillers like calcium carbonate, talc, or nanoclays.

In high-performance applications, carbon fibers, aramid fibers, or graphene may be introduced alongside biodegradable reinforcements to create functional hybrids.

10.2.4 Processing and Fabrication Techniques

The fabrication of hybrid composites is inherently more complex than that of single-reinforcement systems. Proper dispersion, alignment, and interfacial adhesion are critical to achieving the desired performance. Common processing techniques include the following:

- **Compression molding**
- **Injection molding**
- **Extrusion**
- **Solution casting**
- **Pultrusion**
- **Resin transfer molding (RTM)**

Each technique has implications on the final microstructure and property distribution of the composite. For PVC matrices, extrusion and injection molding are the most widely used, offering compatibility with commercial processing lines. Surface treatments such as alkali treatment, silane coupling agents, or maleic anhydride compatibilizers are often applied to natural fibers to improve their bonding with the hydrophobic PVC matrix.

10.2.5 Challenges in Hybrid Composite Design

Despite their advantages, hybrid composites also present several technical challenges:

- **Incompatibility between materials**: poor interfacial bonding leads to delamination or void formation.
- **Processing complexity**: multiple reinforcement types may require different processing conditions.
- **Uneven dispersion**: particularly in nanohybrids, where agglomeration of nanoparticles is common.
- **Moisture absorption**: natural fibers can introduce hygroscopic behavior, affecting dimensional stability.
- **Aging and durability**: degradation due to UV, humidity, or thermal cycles needs to be addressed.

Overcoming these challenges requires not only advanced processing technologies but also material modeling and simulation tools to predict composite behavior under operational conditions.

10.2.6 Applications of Hybrid Composites

Hybrid composites have made significant inroads across various industries due to their tailorability:

- **Automotive**: door panels, dashboards, seat backs, and trim using natural/synthetic fiber hybrids.
- **Construction**: cladding, window profiles, and decorative panels.
- **Packaging**: biodegradable containers and barrier films.
- **Biomedical**: biocompatible implants and prosthetics using natural fiber/nanofiller hybrids.
- **Aerospace**: lightweight hybrid laminates for secondary structures.

PVC-based hybrid composites are particularly popular in building and construction due to their flame resistance, dimensional stability, and cost-effectiveness. With added reinforcements, they find use in load-bearing structures, thermal insulation panels, and protective sheaths.

10.2.7 Future Trends in Hybrid Composite Development

With the advent of nanotechnology and sustainable material engineering, the future of hybrid composites lies in the creation of multifunctional smart materials. These include the following:

- **Self-healing composites** using microcapsules.
- **Sensor-integrated composites** for health monitoring.
- **Green hybrids** based entirely on biodegradable components.
- **3D-printed hybrid materials** using additive manufacturing.

The synergy between biocompatibility, sustainability, and high performance is expected to guide future research. The combination of PVC with bio-based nanofillers, advanced compatibilizers, and low-energy processing techniques could further position hybrid composites as a staple in the materials ecosystem of the future.

10.2.8 Interfacial Bonding in Hybrid Composites

A critical factor determining the performance of any composite material—especially hybrids—is the quality of the interface between the reinforcing phases and the polymer matrix. Interfacial bonding governs the load transfer efficiency, energy absorption during impact, and the overall durability of the material under mechanical stress. In hybrid composites, where two or more distinct types of reinforcements are present, achieving uniform and strong interfacial adhesion becomes a multifaceted challenge.

PVC is a polar polymer due to the presence of chlorine atoms along its backbone. This polarity can enhance interactions with polar fillers but poses compatibility issues with non-polar or highly hydrophilic reinforcements. When incorporating natural fibers, which are inherently hydrophilic due to their cellulose content, it becomes essential to use surface modification techniques such as the following:

- **Alkali treatment (mercerization)**: removes lignin and hemicellulose, increases surface roughness.
- **Acetylation**: reduces moisture uptake, enhances thermal stability.
- **Silane coupling agents**: form covalent bonds between fiber and matrix.
- **Maleic anhydride grafted PVC (PVC-g-MA)**: acts as a compatibilizer that enhances adhesion.

Additionally, nanoparticles like clays, graphene oxide, and carbon nanotubes are frequently functionalized with carboxyl, hydroxyl, or amine groups to improve dispersion and bonding with the matrix.

10.2.9 Role of Hybridization in Enhancing PVC Performance

Pure PVC exhibits certain limitations—such as brittleness at low temperatures, limited impact resistance, and relatively low elongation at break—which restrict its application in demanding environments. Hybrid reinforcement techniques can help address these drawbacks by modifying the mechanical and thermal profile of the material.

Some notable enhancements achieved through hybridization in PVC include the following:

- **Increased tensile and flexural strength** due to better stress distribution among fillers

- **Improved impact resistance** by adding ductile components like rubber particles or natural fibers
- **Enhanced flame retardancy** when mineral fillers like aluminum hydroxide or magnesium hydroxide are added
- **Improved UV and weather resistance** through the inclusion of UV-absorbing nanoparticles

Research has shown that the combination of natural fibers (e.g., flax, kenaf) and nanoclay in PVC matrices can increase tensile strength by up to 40%, improve modulus by 30%, and reduce water absorption by 20–25% compared to unmodified PVC.

10.2.10 Environmental and Economic Aspects of Hybrid Composites

One of the most significant drivers for adopting hybrid composites is the environmental benefit of replacing petroleum-based materials with bio-derived or recycled alternatives. The use of agricultural residues (e.g., wheat husk, sugarcane bagasse, and banana fiber) helps in waste valorization—converting waste into high-value materials. Additionally, biofillers tend to be lighter, renewable, and biodegradable, reducing the overall carbon footprint of the final product.

On the economic side, natural fibers are generally lower in cost than synthetic fibers or engineered fillers. By incorporating them into hybrid systems, manufacturers can produce cost-effective composites that do not compromise on essential properties. Moreover, innovations in processing technologies and fiber extraction methods continue to reduce the cost of hybrid materials, enabling their adoption in price-sensitive industries like construction and consumer goods.

10.2.11 Standards, Testing, and Quality Control

To ensure the commercial viability of hybrid composites, especially in regulated industries, compliance with national and international standards is critical. These standards help establish benchmarks for the following:

- **Mechanical properties** (tensile, flexural, impact strength)
- **Thermal properties** (softening point, thermal degradation)
- **Water absorption and permeability**
- **Fire resistance**
- **Aging and weathering**

Testing is typically performed according to standards such as ASTM D638, ISO 527, ASTM D256, and UL 94 for flammability. These standardized protocols ensure consistent quality and performance of hybrid composites across different applications.

10.2.12 Case Studies and Industrial Examples

Several real-world examples demonstrate the success of hybrid composite systems based on PVC:

- In automotive door trims, manufacturers use PVC reinforced with a hybrid of kenaf fiber and talc to achieve lightweight and sound-dampening properties.
- In construction, PVC sheets reinforced with a hybrid of fly ash and coconut coir have shown promise in roofing and wall panel applications due to improved thermal insulation and reduced material cost.
- In electrical cable insulation, PVC/nanoclay composites hybridized with rubber particles have demonstrated better flexibility, electrical resistance, and flame retardancy.

These case studies reinforce the idea that hybrid materials are not just theoretical solutions but practical innovations that deliver tangible benefits.

10.2.13 Summary

In conclusion, the development and application of hybrid materials represent a pivotal advancement in the field of composite materials, particularly within the context of PVC-based systems. Hybrid composites offer a unique platform for property enhancement, sustainability, and application-specific customization. By strategically selecting and combining reinforcements such as natural fibers, mineral fillers, and nanomaterials, engineers can tailor the mechanical, thermal, and barrier properties of the final material to suit diverse industrial requirements.

Moreover, the hybrid approach enables the integration of green chemistry principles and circular economy goals into materials engineering. As industries move toward more sustainable practices, hybrid composites—particularly those involving bio-based reinforcements—are expected to play a central role in replacing conventional, environmentally harmful plastics.

The next section explores hybrid-based biocomposites, a sub-class of these materials that exclusively incorporates bio-sourced reinforcements, offering further environmental benefits and unique property combinations.

10.3 Hybrid-Based Biocomposites

10.3.1 Introduction to Biocomposites

Biocomposites are a class of composite materials wherein at least one of the components (matrix or reinforcement) is derived from biological, renewable, or biodegradable sources. In the context of hybrid-based biocomposites, these materials are formulated by combining two or more types of reinforcements, typically involving natural fibers, agricultural waste, or biofillers, to enhance their performance characteristics in a synergistic manner.

The growing emphasis on sustainable development, coupled with regulatory pressures to reduce the use of fossil-based and non-degradable materials, has pushed industries and researchers toward exploring eco-friendly alternatives. Hybrid-based biocomposites not only address the performance limitations of single-fiber systems but also contribute significantly to waste minimization, resource conservation, and carbon footprint reduction.

10.3.2 Composition and Structure

The typical structure of a hybrid-based biocomposite comprises:

- A matrix, which can be bio-based (like polylactic acid, starch-based polymers, or bio-polyesters) or conventional (like PVC, polypropylene, or polyethylene)
- Multiple reinforcements, such as the following:
  - **Natural fibers**: jute, kenaf, sisal, hemp, flax, banana and coir
  - **Mineral fillers**: calcium carbonate, talc, mica
  - **Nanoreinforcements**: cellulose nanocrystals, nano-clay, or biochar

Hybridization refers to the simultaneous incorporation of two or more of these reinforcement types to optimize the trade-offs between mechanical strength, durability, biodegradability, and processing cost.

10.3.3 Objectives and Benefits of Hybridization in Biocomposites

The primary objectives of hybridizing biocomposites include the following:

- **Mechanical reinforcement**: combining stiff fibers with ductile ones to balance strength and toughness
- **Moisture resistance**: adding hydrophobic fillers to reduce water uptake typical of natural fibers
- **Dimensional stability**: improving resistance to thermal expansion and contraction
- **Cost optimization**: using low-cost agricultural waste in conjunction with performance fillers
- **Biodegradability tuning**: slowing or accelerating degradation depending on the application

For example, hybrid composites made of flax and basalt fibers have shown superior impact resistance compared to either fiber used alone. Similarly, combining sisal fiber with nanoclay results in biocomposites with improved stiffness and reduced water absorption, making them ideal for semi-structural applications.

10.3.4 Common Natural Fiber Combinations

Hybrid-based biocomposites are often tailored using specific natural fiber blends to optimize desired properties:

- **Flax–Hemp hybrids**: good tensile strength and thermal insulation
- **Jute–Coir hybrids**: economical with improved toughness and biodegradability
- **Kenaf–Sisal hybrids**: balanced performance for automotive interiors
- **Banana–Pineapple leaf fiber (PALF)**: high elongation and tensile modulus

These fibers can be used in different configurations (randomly oriented, woven, chopped, or layered), and their volume fractions play a crucial role in determining the final properties.

10.3.5 Processing Methods

The manufacturing of hybrid-based biocomposites is influenced by the type of matrix and reinforcement and typically includes the following processes:

- **Compression molding**: preferred for thermoplastics and fiber mats
- **Injection molding**: used when fibers are chopped and dispersed in the matrix
- **Extrusion**: often applied in profile and sheet manufacturing
- **Resin transfer molding (RTM)**: suitable for producing components with high fiber content and complex geometries
- **Pultrusion and filament winding**: for continuous fiber reinforcement in structural components

Processing parameters like temperature, pressure, fiber content, and compounding techniques must be optimized to prevent degradation of natural fibers and ensure good fiber–matrix adhesion.

10.3.6 Mechanical Properties

Hybridization generally results in improved mechanical performance when compared to mono-fiber biocomposites. Some observed trends include the following:

- **Tensile strength and modulus**: enhanced through inclusion of stiff fillers like basalt or nano-cellulose
- **Flexural properties**: improved due to better interfacial bonding and stress transfer
- **Impact resistance**: benefitted by ductile natural fibers like coir or banana
- **Fatigue resistance**: increased due to reduction in micro-crack formation

An example is a PLA-based biocomposite reinforced with kenaf and coconut shell powder, which demonstrated a 30–50% increase in tensile strength and better thermal stability over unreinforced PLA.

10.3.7 Thermal and Barrier Properties

Hybrid biocomposites also exhibit notable enhancements in thermal and barrier performance:

- **Thermal conductivity**: reduced due to the porous nature of many natural fibers
- **Thermal degradation resistance**: enhanced by introducing thermally stable fillers like biochar
- **Moisture uptake**: minimized using hybrid fillers with hydrophobic properties (e.g., hybridizing jute with talc or graphene oxide)
- **UV resistance**: improved using nanotitanium dioxide or clay nanoparticles

These improvements allow the use of hybrid-based biocomposites in outdoor applications, packaging and building insulation materials.

10.3.8 Environmental and Sustainability Aspects

Hybrid-based biocomposites contribute significantly to environmental sustainability due to the following:

- **Use of renewable resources**: fibers and fillers derived from plants or waste
- **Reduction in petrochemical usage**: especially when combined with bio-based matrices
- **Lower embodied energy**: as compared to conventional composites
- **End-of-life options**: such as composting, biodegradation, or recycling

Life cycle assessments (LCAs) have shown that biocomposites made from flax and wood flour embedded in biopolyester matrices have 40–60% lower CO_2 emissions than glass fiber–reinforced plastics (GFRPs) [1].

10.3.9 Applications of Hybrid-Based Biocomposites

These materials are increasingly used across diverse sectors:

- **Automotive**: door panels, dashboards, trunk liners (e.g., jute–glass/PVC or PLA hybrids)
- **Construction**: wall panels, ceiling boards, flooring tiles (e.g., rice husk–wood fiber hybrids)
- **Consumer goods**: furniture, kitchenware, casing materials
- **Electronics**: housings for small devices due to their low conductivity and insulation
- **Packaging**: biodegradable trays and containers with barrier properties

Hybrid biocomposites meet performance demands while aligning with eco-friendly product trends.

10.3.10 Challenges in Hybrid Biocomposite Development

Despite their potential, hybrid biocomposites face several challenges:

- **Incompatibility between matrix and reinforcement**: leading to poor adhesion
- **Moisture sensitivity of fibers**: resulting in dimensional instability
- **Processing difficulty**: due to non-uniform dispersion or fiber degradation
- **Lack of standardization**: particularly in testing and characterization protocols
- **End-of-life complexity**: especially when combining biodegradable and non-biodegradable components

These challenges can be addressed through advanced surface treatments, compatibilizers, and design optimization techniques.

10.3.11 Future Prospects

Research is actively focused on enhancing the performance and processability of hybrid-based biocomposites through the following:

- **Nanotechnology integration**: adding nano-cellulose, nanosilica, or bionanoclay
- **Advanced fiber treatments**: using plasma or enzymatic processes for better adhesion

- **Smart biocomposites**: embedded with sensors or stimuli-responsive materials
- **Hybrid green matrices**: combining biodegradable polymers with functional bio-based resins

These innovations aim to create next-generation biocomposites suitable for high-performance, sustainable applications.

10.3.12 Summary

Hybrid-based biocomposites represent a powerful convergence of ecological responsibility and materials innovation. By strategically blending different bio-based reinforcements, it is possible to overcome the shortcomings of traditional biocomposites while enhancing mechanical, thermal, and functional properties. With applications spanning industries and a growing demand for green materials, these hybrids are poised to play a transformative role in the future of sustainable materials science.

10.4 Hybrid-Based Bionanocomposites

10.4.1 Introduction to Bionanocomposites

Bionanocomposites are an emerging class of composite materials that integrate biodegradable matrices with nanoscale reinforcements, including both natural and synthetic nanomaterials. These systems are developed with the goal of enhancing mechanical, thermal, and barrier properties while maintaining a low environmental impact. Hybrid-based bionanocomposites take this approach a step further by incorporating multiple nanoscale and/or microscale reinforcements—either of similar or distinct origin—into a single biopolymeric matrix, thereby creating a synergistic enhancement in performance.

The growing field of bionanotechnology has significantly influenced material science by enabling the design of functional and sustainable composites. Hybridization in bionanocomposites allows for multifunctionality, improved dispersion of nanofillers, and balanced mechanical–biodegradability trade-offs. These materials find applications across various industries such as biomedicine, packaging, automotive, construction, and environmental engineering.

10.4.2 Fundamental Principles of Hybrid-Based Bionanocomposites

At the core of hybrid bionanocomposites lies the principle of synergistic reinforcement—achieved by combining different nano and micro-reinforcements into a biodegradable polymer. Typical components include the following:

- **Biopolymer matrix**:
  - Natural: starch, chitosan, cellulose, gelatin, alginate [2]
  - Synthetic biodegradable: polylactic acid (PLA), polycaprolactone (PCL), polyhydroxyalkanoates (PHAs), polyethylene succinate (PES)
- **Hybrid nanofillers**:
  - Organic: nanocellulose, chitin nanofibers, lignin nanoparticles
  - Inorganic: nanoclay, graphene oxide (GO), carbon nanotubes (CNTs), metal oxides (ZnO, TiO_2)
  - Biowaste-based: nanosilica from rice husk, nano-hydroxyapatite from eggshells

The interfacial adhesion, dispersion, and alignment of these fillers within the matrix are critical in determining the final composite's properties. Advanced techniques like surface functionalization, chemical coupling, and *in situ* polymerization are used to improve matrix–filler interaction.

10.4.3 Importance of Hybridization at the Nanoscale

Hybridizing nanomaterials in biocomposites addresses several inherent limitations:

- **Mechanical property optimization**: Combining rigid (e.g., CNTs) and flexible (e.g., nanocellulose) nanofillers leads to a balance of strength and ductility.
- **Barrier enhancement**: Incorporating both nanoclays and biochar improves oxygen and moisture resistance in packaging films.
- **Thermal stability**: Metal oxide nanoparticles like ZnO or TiO_2 boost decomposition temperatures and slow thermal degradation.

- **Antimicrobial action**: Hybrid nanocomposites infused with AgNPs and chitosan nanoparticles can impart powerful antibacterial properties.
- **Electrical conductivity**: Incorporation of graphene and CNTs introduces conductive pathways suitable for sensors and bioelectronics.

Thus, hybrid-based bionanocomposites allow material scientists to fine-tune and multifunctionalize biodegradable polymers for a wide spectrum of uses.

10.4.4 Preparation and Processing Techniques

The development of hybrid bionanocomposites requires specialized processing techniques to ensure even dispersion and strong bonding of multiple nanofillers. Common methods include the following:

1. **Solution casting**:
   - Suitable for water-soluble polymers like starch or chitosan
   - Nanoparticles are dispersed using ultrasonication or magnetic stirring before being cast into films.
2. **Melt blending**:
   - Applicable to thermoplastics like PLA and PCL
   - Nanofillers are blended during extrusion at high temperatures.
3. ***In situ* polymerization**:
   - Used to generate a polymer matrix around well-dispersed nanofillers
   - Often used with CNTs or graphene derivatives for better interfacial strength
4. **Electrospinning**:
   - For creating fibrous mats used in biomedical and filtration applications
   - Nanoparticles are incorporated into the polymeric jet during spinning.
5. **Layer-by-layer assembly**:
   - Ideal for building multi-layered films or coatings with controlled thickness and composition
   - Enables precise placement of each nanocomponent.

Maintaining nanoparticle dispersion stability, avoiding agglomeration, and optimizing fiber/matrix compatibility are essential throughout these processes.

10.4.5 Mechanical Properties and Structural Behavior

Hybrid-based bionanocomposites generally outperform single-filler systems due to the complementary action of their components:

- **Tensile strength**: improved significantly by using high aspect ratio reinforcements like graphene oxide and cellulose nanocrystals (CNCs)
- **Young's modulus**: enhanced due to uniform stress transfer and rigid particle loading
- **Elongation at break**: optimized by adding flexible fillers such as gelatin nanoparticles
- **Impact strength**: increased when nano-reinforcements help absorb and redistribute energy

For instance, PLA reinforced with chitin nanofibers and nanoclay has demonstrated 80–100% higher tensile strength compared to neat PLA, with notable gains in modulus and thermal stability.

10.4.6 Thermal and Barrier Properties

The inclusion of multiple nanoscale reinforcements provides superior thermal resistance and barrier performance:

- **Thermal stability**: Hybrid systems delay onset of degradation and increase char formation.
- **Glass transition and melting temperature**: can be tuned based on the type of nanofiller used.
- **Moisture/oxygen barrier**: improved due to tortuous pathways formed by platelet-like fillers such as nanoclay or graphene oxide.
- **UV resistance**: enhanced using titanium dioxide or lignin nanoparticles.

Such enhancements make hybrid bionanocomposites ideal for biopackaging applications requiring shelf-life extension and protection from environmental stressors.

10.4.7 Biodegradability and Environmental Performance

Despite their advanced functionalities, hybrid-based bionanocomposites retain their biodegradable nature, especially when both matrix and reinforcements are bio-derived. Key factors influencing biodegradability include the following:

- **Matrix composition**: Natural polymers degrade faster than synthetic biodegradable ones.
- **Filler type**: Organic fillers support microbial activity, whereas inorganic fillers slow degradation.
- **Environmental conditions**: Moisture, temperature, pH, and microbial diversity play a role.

Studies show that PLA-based nanocomposites reinforced with nanochitin and rice husk silica degraded up to 60–70% within 6 months in composting environments, with minimal eco-toxicity.

Life cycle assessments (LCAs) of such materials reveal reduced greenhouse gas emissions, lower energy consumption, and minimal end-of-life waste compared to petroleum-based counterparts.

10.4.8 Biomedical Applications of Hybrid Bionanocomposites

Hybrid-based bionanocomposites have made significant inroads into the biomedical field, particularly due to their biocompatibility, antimicrobial properties, and functionalization potential:

- **Drug delivery**: controlled release systems using PLA/chitosan/nanoclay hybrids
- **Tissue engineering**: scaffolds made from gelatin/alginate/CNTs with tailored porosity and strength
- **Wound dressings**: incorporation of silver nanoparticles into cellulose nanofiber films
- **Antimicrobial coatings**: chitosan and ZnO hybrids applied to surgical tools or devices

In these applications, material safety, cytotoxicity, and degradation kinetics are critical performance indicators.

10.4.9 Industrial and Engineering Applications

Beyond biomedical uses, these composites find adoption in the following:

- **Food packaging**: films with antibacterial and gas barrier functions
- **Automotive**: interior panels, dashboards using biodegradable PLA/graphene/flax systems
- **Construction**: lightweight panels with fire resistance and insulation (e.g., starch–CNT–nanosilica)
- **Electronics**: flexible printed substrates using nanocellulose and carbon-based fillers

Such applications demonstrate the adaptability and multifunctionality of hybrid bionanocomposites across sectors.

10.4.10 Limitations and Current Challenges

Despite their promise, hybrid-based bionanocomposites face several technological and scientific challenges:

- **Nanoparticle aggregation**: Poor dispersion reduces mechanical reinforcement potential.
- **Compatibility issues**: Lack of interfacial bonding leads to stress concentration and early failure.
- **Processing complexity**: Multi-filler systems often require precise mixing and parameter optimization.
- **Scalability**: High production costs and inconsistency in properties hinder industrial adoption.
- **Safety and toxicity**: Long-term health effects of engineered nanomaterials (ENMs) are still being evaluated.

Overcoming these challenges requires multidisciplinary collaboration, development of green synthesis methods and formulation of global safety regulations for nanomaterials.

10.4.11 Recent Advancements and Future Prospects

Current research in hybrid-based bionanocomposites focuses on several cutting-edge innovations:

- **Green synthesis routes**: utilizing plant extracts for nanoparticle synthesis (e.g., "green silver nanoparticles")
- **3D and 4D printing**: using bionanocomposite inks for programmable shape-changing materials
- **Smart packaging**: integration of sensors and indicators in packaging films
- **Self-healing materials**: systems that repair microcracks autonomously using embedded healing agents
- **Fully bio-based systems**: completely eliminating fossil-derived content through innovative polymer and nanofiller combinations

The future of hybrid bionanocomposites is poised to intersect with circular economy models, zero-waste production, and personalized medicine, where tailored materials can meet specific environmental, physiological, or industrial needs.

10.4.12 Summary

Hybrid-based bionanocomposites represent a frontier in sustainable materials engineering, offering high performance, multifunctionality, and biodegradability. By intelligently combining nanosized reinforcements of varied origins, researchers can achieve superior mechanical, thermal, and barrier properties while maintaining environmental compatibility. Although technological and economic challenges exist, the ongoing surge in research and innovation indicates that hybrid bionanocomposites will play a pivotal role in the development of next-generation eco-functional materials for a variety of applications.

10.5 Polyvinyl Chloride (PVC)/Hybrid-Based Biocomposites

10.5.1 Introduction to PVC in Composite Materials

Polyvinyl chloride (PVC) is a widely used thermoplastic polymer recognized for its chemical resistance, dimensional stability, flame retardancy, and cost-effectiveness. Despite being non-biodegradable, PVC continues to dominate several industrial sectors such as construction, automotive, packaging, and electronics. However, environmental concerns associated with its disposal and plasticizer leaching have led to renewed interest in modifying PVC with eco-friendly reinforcements.

In this context, hybrid-based biocomposites, where PVC is reinforced with a combination of natural fibers and other bio-derived or synthetic fillers, offer a compelling solution. These composites seek to bridge the gap between PVC's performance limitations and the need for more sustainable materials by incorporating biofillers, agricultural waste fibers, and hybrid nanoparticles to improve mechanical, thermal, and environmental characteristics.

10.5.2 Fundamentals of Hybrid-Based PVC Biocomposites

Hybrid-based biocomposites are defined by their multi-filler reinforcement approach, where two or more fillers—often from different origins—are introduced into a single polymer matrix. In the case of PVC, these fillers may include the following:

- **Natural fibers**: kenaf, jute, hemp, flax, coir, rice husk, bagasse [3]
- **Agricultural waste fillers**: groundnut shells, wheat straw, banana fiber, pineapple leaf fiber
- **Inorganic reinforcements**: nanoclay, calcium carbonate ($CaCO_3$), aluminum oxide (Al_2O_3), titanium dioxide (TiO_2)
- **Biofillers**: lignin, cellulose microfibrils, starch-based derivatives, nanosilica from husk ash

The resulting synergistic effects of hybrid reinforcement enhance overall properties such as impact strength, stiffness, resistance to creep, and weatherability, making these materials suitable for a wide range of applications while lowering the carbon footprint.

10.5.3 Processing Techniques and Fabrication

PVC/hybrid biocomposites can be fabricated using several industrial-scale and lab-scale processing methods. The method chosen largely depends on the type of filler, particle size, and final application. Common techniques include the following:

1. **Twin-screw extrusion**:
    - The most prevalent method for blending PVC with natural and hybrid fillers

   - Ensures uniform mixing and thermal stability during processing
2. **Compression molding**:
   - Used for sheet or panel production
   - Offers excellent dimensional control and filler distribution
3. **Injection molding**:
   - Applicable for complex shapes in automotive or consumer goods
   - Provides fast production cycles and precise reinforcement alignment
4. **Solvent casting and *in situ* polymerization**:
   - For lab-scale or specialized composites with functional coatings or chemical grafting [4]

The filler content generally varies between 5 and 40% by weight, and the use of compatibilizers such as maleic anhydride-grafted PVC or silane-treated fibers improves interfacial bonding, leading to better mechanical performance.

10.5.4 Mechanical Properties and Performance Characteristics

The mechanical performance of PVC/hybrid biocomposites is significantly influenced by filler composition, morphology, orientation, and interfacial adhesion. Common trends observed in such composites include the following:

- **Tensile strength**: increases with the incorporation of stiff fillers such as nanoclay or cellulose fibers, especially when combined with chemical coupling agents.
- **Flexural modulus**: enhanced due to increased matrix rigidity and filler stiffness.
- **Impact resistance**: improved by hybridizing with flexible fibers (e.g., kenaf + nanoclay or coir + fly ash) which help in energy dissipation.
- **Hardness**: boosted through the presence of inorganic micro-fillers like calcium carbonate or nanosilica

For instance, PVC reinforced with kenaf fiber and nanoclay has shown a 50–70% increase in tensile and flexural strength compared to neat PVC, demonstrating how hybrid reinforcement outperforms single-fiber systems.

10.5.5 Thermal Stability and Flame Retardancy

One of the standout features of PVC is its inherent flame resistance, which is further improved through hybridization. Hybrid biocomposites can provide enhanced:

- **Thermal stability**: The onset decomposition temperature can be elevated by adding fillers like Al_2O_3, fly ash, or graphene oxide.
- **Char formation**: improved due to synergistic action between PVC and carbon-rich biofillers, helping form a protective layer
- **Flame retardancy**: enhanced using hybrid fillers such as lignin-nanoclay or metal oxide-natural fiber combinations
- **Thermogravimetric behavior**: shows reduced weight loss rates and better residual mass retention with hybrid systems

As an example, PVC/hemp fiber/nanoclay composites demonstrated significantly delayed degradation with a 15–20 °C increase in thermal stability due to barrier effects and intumescent behavior.

10.5.6 Morphological and Structural Insights

Understanding the internal morphology of PVC/hybrid composites is essential for optimizing performance. Techniques such as scanning electron microscopy (SEM) and Fourier-transform infrared spectroscopy (FTIR) are used to analyze the following:

- **Fiber dispersion**: Even distribution reduces stress concentrations and improves overall integrity.
- **Interfacial adhesion**: assessed by examining gaps, voids, and debonding zones
- **Crosslinking**: presence of functional groups or coupling agents that bond matrix to filler
- **Porosity and density**: critical in structural applications where lightweight yet strong materials are needed

Hybrid systems with surface-modified lignocellulosic fillers and nanoclay have shown dense, void-free matrices, contributing to superior durability and mechanical behavior.

10.5.7 Environmental Performance and Biodegradability Aspects

Although PVC is not biodegradable, incorporating natural and biowaste-based fillers reduces the overall environmental burden of the material. Key benefits include the following:

- **Material substitution**: Up to 40% of petroleum-based content can be replaced with biowaste.
- **Lower carbon footprint**: due to less energy-intensive filler production
- **End-of-life management**: Enhanced recyclability and reduced toxic gas release when natural fibers are used.
- **Improved eco-toxicological profile**: especially when non-toxic plasticizers and additives are employed.

Life cycle assessments have shown that PVC/rice husk/nanosilica composites produce 30–45% fewer greenhouse gas emissions than conventional PVC-based panels.

10.5.8 Applications in Various Industries

PVC/hybrid-based biocomposites offer great versatility across several domains:

- **Building and construction**:
  - Wall claddings, roofing, window profiles
  - Replacing wood with PVC/wood fiber/nanoclay panels that resist moisture and UV exposure
- **Automotive sector**:
  - Door panels, dashboards, flooring
  - Composites using PVC/hemp fiber/fly ash provide lightweight and durable interior parts
- **Consumer goods**:
  - Footwear, packaging trays, garden equipment
  - Use of PVC/banana fiber/graphene oxide for robust and cost-effective tools
- **Electrical and cable industry**:
  - Insulators, conduit pipes, wiring jackets
  - Improved fire resistance using PVC/nano-MgO/natural fibers

These hybrid materials not only match performance criteria but also fulfill regulatory and sustainability standards in most industrial sectors.

10.5.9 Challenges and Limitations

Despite their potential, PVC/hybrid biocomposites [6] face several limitations:

- **Thermal degradation of natural fibers**: limits processing temperatures, affecting composite quality.
- **Compatibility issues**: poor interfacial bonding can reduce strength and durability.
- **Moisture absorption**: natural fillers may swell or degrade in humid environments without proper treatment.
- **Recycling difficulty**: hybrid materials may be harder to sort or reuse in conventional recycling systems.
- **Toxicity of additives**: some stabilizers or plasticizers may counteract the eco-friendly nature of the fillers.

Research continues on green coupling agents, nano-coating technologies, and bioplasticizer systems to overcome these hurdles.

10.5.10 Recent Developments and Innovations

Ongoing advancements are addressing key technical and sustainability concerns in PVC/hybrid biocomposites:

- **Nano-treated fibers**: improved bonding and thermal resistance.
- **3D printing-compatible PVC composites**: hybrid formulations suitable for additive manufacturing.
- **Green flame retardants**: replacing brominated systems with natural intumescent agents.
- **Surface functionalization**: using plasma, UV, or enzyme treatments for improved filler compatibility.
- **Smart composites**: incorporating sensing capabilities using carbon nanotubes or piezoelectric particles.

Furthermore, research in bio-PVC derived from bioethylene opens doors to truly green hybrid PVC composites in the future.

10.5.11 Summary

PVC/hybrid-based biocomposites represent a powerful evolution in composite materials technology—blending cost-effective, high-performance PVC with natural and inorganic reinforcements to create systems that are mechanically robust, thermally stable, and environmentally conscious. Through intelligent hybridization, these composites overcome the limitations of conventional PVC while reducing ecological footprints. Their growing role in structural, automotive, and consumer applications demonstrates their viability in a world increasingly focused on sustainable development.

10.6 Polyvinyl Chloride (PVC)/Hybrid-Based Bionanocomposites

10.6.1 Introduction to Bionanocomposites

Bionanocomposites represent a new frontier in materials science, where bio-based materials and nanotechnology converge to enhance mechanical, thermal, and barrier properties of polymer matrices. These materials incorporate nanoscale reinforcements (such as nanoclay, nanosilica, carbon nanotubes, graphene oxide, or metal oxides) along with biodegradable or bio-derived fillers (like cellulose nanocrystals, lignin, chitosan, and starch) into a conventional polymer such as PVC.

When applied to PVC systems, hybrid bionanocomposites exploit the flame retardancy, chemical resistance, and structural stability of PVC and augment them using green nanomaterials. This results in lightweight, high-strength composites that can meet the growing demand for eco-efficient and smart materials across multiple industries.

10.6.2 Components of PVC-Based Hybrid Bionanocomposites

PVC-based hybrid bionanocomposites generally consist of as follows:

1. **Polyvinyl chloride (matrix)**
    - Provides processability, structural rigidity, chemical and flame resistance
2. **Bio-based fillers**
    - Examples: lignocellulosic fiber, microcrystalline cellulose, starch, rice husk, banana fiber, chitosan

   - Role: improve biodegradability, reduce cost, and act as a green alternative to synthetic fillers
3. **Nanomaterials**
   - Examples: nanoclay, TiO_2 nanoparticles, carbon nanotubes (CNTs), graphene oxide (GO), nanosilica, nano-Al_2O_3
   - Role: enhance mechanical properties, thermal stability, flame retardancy, electrical conductivity
4. **Compatibilizers/coupling agents**
   - Necessary to ensure adhesion between the hydrophilic biofillers and the hydrophobic PVC matrix
   - Common agents: maleic anhydride-grafted PVC (PVC-g-MA), silanes, titanates, or polyethylene-grafted polymers

The hybrid effect in such systems enables multiple reinforcing mechanisms to operate concurrently, leading to synergistic performance improvements.

10.6.3 Fabrication Techniques

Processing PVC/hybrid-based bionanocomposites requires precise dispersion of nanofillers and strong fiber–matrix interaction. The common techniques include the following:

- **Melt compounding with extrusion**:
  - Twin-screw extruders ensure good mixing of nanoparticles and biofillers in PVC.
  - Challenges include avoiding nanoparticle agglomeration and thermal degradation of biofillers.
- **Solution casting**:
  - Ideal for thin films and membrane applications
  - Nanoparticles are dispersed in a solvent with PVC and biofillers, then cast and dried.
- ***In situ* polymerization or grafting**:
  - For stronger interfacial bonding, *in situ* polymerization is used to bond fillers with matrix chains.
- **Ultrasonic dispersion + extrusion**:
  - Ultrasonication helps to break up nanomaterial aggregates before compounding.

Proper filler content optimization is crucial—typically ranging from 1 to 10% nanofiller and 10 to 40% biofiller—to achieve balance between performance and processability.

10.6.4 Mechanical Properties

Hybrid bionanocomposites have demonstrated remarkable mechanical property enhancements over unmodified PVC:

- **Tensile strength**:
  - Increases by 30–60% with the addition of cellulose nanocrystals or nanoclay, due to increased matrix stiffness
- **Young's modulus**:
  - Reinforced significantly with nanosilica and natural fiber combinations [7]
- **Impact resistance**:
  - Hybrid fillers such as PVC/banana fiber/TiO_2 or PVC/jute fiber/nanoclay composites show higher energy absorption capacity
- **Flexural properties**:
  - Increased through the use of treated fibers and exfoliated clay platelets which improve stress transfer

Studies show that adding 2–5% nanoclay and 20–30% biofiber results in up to 75% improvement in tensile strength, depending on the compatibilizer used and fiber treatment method.

10.6.5 Thermal Behavior and Fire Retardancy

The integration of nanofillers significantly influences the thermal degradation behavior of the PVC matrix:

- **Thermal stability**:
  - Increased onset degradation temperature due to barrier effects from layered silicates or metal oxide nanoparticles.
- **Char formation**:
  - Nanoclay and GO facilitate the formation of a cohesive char layer, which shields the composite during combustion.

- **Flame retardancy**:
    - Enhanced by bio-nano synergy (e.g., lignin and nanoclay together), reducing heat release rate and smoke emission.
- **Thermal conductivity**:
    - Improved using fillers like carbon nanotubes or nano-graphite for applications requiring dissipation of heat.

Thermogravimetric analysis (TGA) and differential scanning calorimetry (DSC) are commonly used to confirm the improvements in thermal stability. In some cases, the thermal degradation temperature of PVC composites has increased by over 50 °C due to hybrid filler action.

10.6.6 Morphological and Structural Characterization

Microscopic and spectroscopic analyses are essential in confirming homogeneous distribution and bonding quality:

- **Scanning electron microscopy (SEM)**:
    - Reveals fiber pull-out, voids and nanoparticle dispersion
- **Transmission electron microscopy (TEM)**:
    - Especially useful for understanding nanoparticle dispersion at the nanometer scale
- **FTIR and XRD**:
    - Used to detect chemical interactions and crystallinity shifts after filler incorporation

Well-dispersed and chemically compatible hybrid fillers exhibit dense structures, minimal microcracking, and excellent bonding, resulting in high-performance composites.

10.6.7 Environmental Impact and Sustainability

The bionanocomposite approach aligns with the principles of green chemistry and sustainable materials engineering:

- **Biodegradability enhancement**:
    - While PVC is not biodegradable, the inclusion of starch, cellulose, or lignin increases surface erosion and microbial activity over time.

- **Reduction of PVC content**:
  - With filler loadings reaching up to 50%, the total use of petroleum-derived material is reduced.
- **Improved recyclability**:
  - Some hybrid systems allow for multiple processing cycles without significant property degradation.
- **Life cycle benefits**:
  - Lower CO_2 emissions, reduced fossil energy use, and better end-of-life disposal options

Hybrid PVC bionanocomposites therefore help mitigate the ecological footprint of traditional plastics while offering functionality required for advanced applications.

10.6.8 Industrial Applications

Applications of these composites span various high-value sectors:

- **Medical devices**:
  - Antibacterial and biocompatible properties using PVC/chitosan/nano-ZnO systems
- **Packaging films**:
  - Lightweight and durable films using PVC/starch/nanosilica for food packaging
- **Construction panels and pipes**:
  - Structural panels with fire-retardant properties using PVC/lignin/nanoclay
- **Automotive components**:
  - Door trims and dashboard panels using PVC/kenaf/TiO_2 combinations for weight reduction and durability
- **Electronics and insulation**:
  - Improved dielectric and thermal behavior with carbon-based nanofillers

PVC/hybrid bionanocomposites have shown cost-effective potential to replace conventional polymers in multiple sectors, particularly when performance and environmental regulations are stringent.

10.6.9 Limitations and Challenges

Despite their promise, these composites still face notable challenges:

- **Processing complexity**:
  - Dispersing nanofillers uniformly without agglomeration remains difficult.
- **Filler compatibility**:
  - Hydrophilic biofillers often have poor compatibility with hydrophobic PVC unless treated.
- **Moisture absorption**:
  - Natural fillers tend to absorb water, reducing long-term mechanical performance.
- **Cost of nanomaterials**:
  - High-purity nanofillers can be expensive, limiting economic feasibility in some cases.
- **Health concerns**:
  - Inhalation risks of nanoparticles require careful handling and safety protocols during processing.

Further research into bio-sourced nanomaterials [8], low-cost compatibilizers, and green processing technologies is essential to overcome these issues.

10.6.10 Recent Advances and Future Outlook

Emerging trends in PVC-based bionanocomposites are focused on the following:

- **Nano-coated biofibers**:
  - Using surface nanolayers to improve barrier and mechanical properties
- **Smart composites**:
  - Embedding sensors or conductive nanofillers for real-time monitoring or self-healing functions
- **3D printing and additive manufacturing**:
  - Developing printable PVC/nano-biocomposites for customized parts
- **Bio-PVC synthesis**:
  - From renewable feedstocks to make the entire matrix eco-friendly
- **Hybrid fire-retardant systems**:
  - Incorporating multifunctional nanomaterials with intumescent biofillers for enhanced safety

As research progresses, PVC-based hybrid bionanocomposites are poised to become key materials for the sustainable economy, balancing performance, and cost and environmental impact.

10.6.11 Summary

PVC/hybrid-based bionanocomposites combine the strengths of polyvinyl chloride with biodegradable fillers and advanced nanomaterials, creating composites with enhanced thermal, mechanical, and environmental characteristics [11]. These materials offer versatile solutions for construction, medical, automotive, and packaging industries, addressing both technical performance and sustainability goals. With continued innovation in processing and material design, they represent a crucial step toward the development of next-generation green composite materials.

10.7 Conclusion

10.7.1 Overview and Recapitulation

The field of polymer composites has undergone a remarkable transformation over the past few decades, particularly with the evolution of hybrid biocomposites and bionanocomposites that integrate polyvinyl chloride (PVC) with various organic and inorganic reinforcements. This chapter has systematically reviewed the development, structure, processing, properties, and applications of PVC-based hybrid biocomposites and bionanocomposites. These innovative materials reflect the ongoing shift towards eco-efficient, high-performance alternatives to conventional plastics, meeting both industrial performance requirements and environmental sustainability goals.

PVC, with its advantageous properties such as chemical inertness, flame retardancy, low cost, and processability, remains a cornerstone thermoplastic material. However, its inherent drawbacks—such as non-biodegradability, plasticizer leaching, and mechanical limitations—necessitate its modification or reinforcement using hybrid techniques. The incorporation of natural fibers, biodegradable fillers, and nanostructured materials has enabled the production of novel composite systems with tailored functionalities, expanding the application spectrum of PVC.

Each section of this chapter has revealed how combining biofillers and nanofillers into the PVC matrix provides synergistic effects that surpass the limitations of traditional monolithic or single-reinforced composites.

From mechanical strength enhancements to thermal stability, barrier properties, and sustainability benefits, these hybrid materials stand at the forefront of next-generation polymer composites.

10.7.2 Key Findings from Each Subsection

Introduction established the importance of developing environmentally friendly composites. It discussed the global push for sustainable materials and set the foundation for understanding the role of PVC in hybrid biocomposite innovation.

Subsection Hybrid Materials delved into the basic concepts of hybrid composites, defining their classification, design strategies, and the rationale behind dual reinforcement. It emphasized the importance of structural hierarchy and filler compatibility in ensuring performance enhancements.

Subsection Hybrid-Based Biocomposites focused on materials that integrate bio-sourced fillers such as lignin, cellulose, rice husk, jute, and kenaf fibers. It outlined their biodegradable nature, renewable sourcing, low density, and mechanical reinforcing potential, while highlighting processing challenges and fiber treatment techniques.

Subsection Hybrid-Based Bionanocomposites explored the role of nanotechnology in enhancing composite functionality. It introduced various nanofillers—such as nanoclays [10], metal oxides, graphene oxide, and carbon nanotubes—and their ability to improve barrier, thermal, and flame retardant properties. The combination of nano- and bio-reinforcements resulted in multifunctional, lightweight, and eco-compliant composites.

Subsection PVC/Hybrid-Based Biocomposites narrowed the focus to systems based on PVC matrices, highlighting their formulation, structural behavior, and application potential. It illustrated how adding natural fibers (e.g., hemp, flax, or bamboo) along with coupling agents enhances interfacial bonding and matrix compatibility.

Subsection PVC/Hybrid-Based Bionanocomposites discussed advanced formulations integrating PVC, biodegradable fillers, and nanomaterials. The section reviewed mechanical, thermal, and morphological improvements, along with industrial applications and environmental implications. It concluded that such systems embody the future of sustainable high-performance composites [12].

10.7.3 Significance of Hybrid Reinforcement in PVC Systems

The hybrid reinforcement approach, involving both bio-derived fillers and nanoscale particles, offers the best of both worlds: the environmental and cost advantages of renewable resources, combined with the strength, toughness, and thermal resilience of nanomaterials. In PVC-based matrices, this hybrid strategy mitigates key limitations such as brittleness, thermal degradation, and environmental non-degradability.

The unique advantages of hybridization include the following:

- **Property synergy**: simultaneous improvement of multiple properties—e.g., toughness, stiffness, and flame retardancy—through complementary filler behavior
- **Filler efficiency**: reduced filler loading due to the high aspect ratio and surface area of nanoparticles, avoiding composite densification
- **Design flexibility**: tailorable mechanical and thermal properties for specific applications
- **Cost optimization**: partial replacement of expensive nanofillers with low-cost biofillers without performance compromise

Such benefits are particularly crucial for automotive, aerospace, construction, packaging, and biomedical sectors, which require highly customizable materials that align with regulatory and environmental standards.

10.7.4 Challenges and Limitations

Despite the promising potential of PVC-based hybrid biocomposites and bionanocomposites, certain technical and economic challenges persist:

1. **Interfacial compatibility**:
   - Biofillers are inherently hydrophilic, while PVC is hydrophobic. Without proper surface treatments or compatibilizers, interfacial bonding remains weak.
2. **Nanofiller dispersion**:
   - Nanoparticles tend to agglomerate due to high surface energy, limiting their effectiveness unless dispersed uniformly through techniques like ultrasonication or surface functionalization.
3. **Processing constraints**:
   - PVC is thermally sensitive; overprocessing or improper filler loading can lead to degradation or loss of material integrity.

4. **Cost and scalability**:
   - While nanofillers offer great benefits, their cost and processing complexity pose barriers for large-scale commercial adoption.
5. **Long-term performance**:
   - Moisture absorption by natural fibers can compromise long-term durability, especially under outdoor or humid conditions.
6. **Recyclability and end-of-life management**:
   - Hybrid systems can complicate recycling due to multiphase compositions and possible degradation of biocomponents [14].

Addressing these challenges will require continued innovation in material design, green processing technologies, and multiscale modeling approaches to predict performance and lifecycle impacts accurately.

10.7.5 Environmental and Societal Impact

One of the central themes in this chapter is the role of PVC-based hybrid biocomposites and bionanocomposites in advancing sustainability. By reducing the reliance on petroleum-based polymers and minimizing environmental hazards, these materials align with global efforts to reduce carbon footprints, landfill waste, and resource depletion.

Benefits include the following:

- **Resource efficiency**: Use of agricultural and industrial waste products as fillers adds value to waste streams.
- **Lightweighting**: Reducing material mass in transport and construction industries contributes to energy savings.
- **Reduced toxicity**: Natural fillers often do not require toxic additives, improving safety for end-users and processors.
- **Biodegradability potential**: Partial biodegradability offers improved disposal and environmental assimilation compared to pure PVC.

Such contributions support the transition towards a circular economy, where materials are continuously reused, repurposed, and reintegrated into the production cycle with minimal ecological disruption.

10.7.6 Future Directions and Research Opportunities

There remains vast untapped potential in the field of PVC/hybrid-based bionanocomposites. Future research should focus on the following:

- **Developing biocompatible PVC derivatives**: synthesis of fully biodegradable PVC analogs that maintain mechanical integrity [15].
- **Green nanotechnology integration**: exploring bionanoparticles (e.g., cellulose nanocrystals, nanochitin) as functional nanofillers.
- **Smart composite systems**: Embedding responsive nanomaterials for sensing, actuation, or self-healing capabilities.
- **Advanced characterization techniques**: Nanoscale property mapping (e.g., AFM, XPS) to understand filler interactions and stress distribution [16–19].
- **Life cycle assessment (LCA)**: Holistic environmental impact studies for various PVC hybrid formulations across different industries [20].

Collaborative efforts between materials scientists, chemical engineers, environmental experts, and industry stakeholders will be essential to translate laboratory advances into market-ready solutions.

10.8 Conclusions

In conclusion, **PVC-based hybrid biocomposites and bionanocomposites** present a highly promising class of materials that bridge the gap between performance engineering and sustainability. They combine the desirable traits of polyvinyl chloride with the ecological benefits of bio-based and nanoscale reinforcements. While challenges remain, ongoing research and technological advances are rapidly expanding the frontiers of this domain.

From lightweight automotive parts to biomedical devices, smart packaging, and flame-retardant panels, these materials are shaping the future of responsible materials engineering. Their development not only reflects innovation in materials science but also resonates with the global imperative for climate-conscious, sustainable manufacturing.

AI disclosure statement: The author acknowledges the use of AI-based tools for minor language editing or grammatical correction. However, all intellectual contributions, analyses, and conclusions presented in this chapter are entirely the author's own.

Bibliography

1. Ashori, A., Wood–plastic composites as promising green-composites for automotive industries. *Bioresour. Technol.*, 99, 11, 4661–4667, 2008.
2. Bledzki, A.K. and Gassan, J., Composites reinforced with cellulose-based fibres. *Prog. Polym. Sci.*, 24, 2, 221–274, 1999.
3. Cao, Y., Shibata, S., Fukumoto, I., Mechanical properties of biodegradable composites reinforced with bagasse fibre before and after alkali treatments. *Composites, Part A*, 37, 3, 423–429, 2006.
4. Chand, N. and Hashmi, S.A., *Advanced Functional Materials*, Springer, 2015.
5. Faruk, O., Bledzki, A.K., Fink, H.P., Sain, M., Biocomposites reinforced with natural fibers: 2000–2010. *Prog. Polym. Sci.*, 37, 11, 1552–1596, 2012.
6. George, J., Sreekala, M.S., Thomas, S., A review on interface modification and characterization of natural fiber reinforced plastic composites. *Polym. Eng. Sci.*, 41, 9, 1471–1485, 2001.
7. Karger-Kocsis, J., Mahmood, H., Pegoretti, A., Recent advances in fiber/matrix interphase engineering for polymer composites. *Prog. Mater Sci.*, 73, 1–43, 2015.
8. Kausar, A., Polyvinyl chloride (PVC) reinforced with natural fillers and nanomaterials: A review of mechanical and thermal properties. *Polym. Plast. Technol. Eng.*, 56, 15, 1583–1600, 2017.
9. Kuo, S.W. and Chang, F.C., Synergistic effects in polymer/clay nanocomposites. *Macromolecules*, 34, 13, 4568–4571, 2001.
10. Lu, N. and Oza, S., A comparative study of the mechanical properties of hemp fiber with virgin and recycled high-density polyethylene matrix. *Composites, Part B*, 45, 1, 1651–1656, 2013.
11. Mohanty, A.K., Misra, M., Drzal, L.T., Sustainable bio-composites from renewable resources: Opportunities and challenges in the green materials world. *J. Polym. Environ.*, 10, 1–2, 19–26, 2002.
12. Mohanty, A.K., Vivekanandhan, S., Pin, J.M., Misra, M., Composites from renewable and sustainable resources: Challenges and innovations. *Science*, 362, 6414, 536–542, 2018.
13. Niska, K.O. and Sain, M. (Eds.), *Wood-Polymer Composites*, Woodhead Publishing, 2008.
14. Oksman, K., Skrifvars, M., Selin, J.F., Natural fibres as reinforcement in polylactic acid (PLA) composites. *Compos. Sci. Technol.*, 63, 9, 1317–1324, 2003.

15. Paul, S.A. and Alam, M.A., Hybrid natural fiber-reinforced composites and their applications: A review. *J. Reinf. Plast. Compos.*, 40, 7–8244–263, 2021.
16. Ramesh, M., Palanikumar, K., Reddy, K.H., Mechanical property evaluation of sisal–jute–glass fiber reinforced polyester composites. *Composites, Part B*, 48, 1–9, 2013.
17. Reddy, N. and Yang, Y., Biofibers from agricultural byproducts for industrial applications. *Trends Biotechnol.*, 23, 1, 22–27, 2005.
18. Sathishkumar, T.P., Navaneethakrishnan, P., Shankar, S., Hybrid fiber reinforced polymer composites - a review. *J. Reinf. Plast. Compos.*, 31, 5, 357–369, 2012.
19. Shubhra, Q.T.H., Alam, A.K.M.M., Quaiyyum, M.A., Mechanical and degradation characteristics of natural fiber reinforced plastic composites: A review. *Polym. Degrad. Stab.*, 98, 12, 2642–2658, 2013.
20. Thakur, V.K. and Thakur, M.K., Processing and characterization of natural cellulose fibers/thermoset polymer composites. *Carbohydr. Polymers*, 109, 102–117, 2014.

11

Preparation and PVC/Rubber Bio Nano-Blends

Shibin N. B.

Qualitas Material Testing Laboratory, Abu Dhabi, UAE

Abstract

In this chapter discussing about the Preparation and PVC/rubber Bio nano blends, in this chapter I have discussing with different topics such as preparation of PVC/rubber bio nano, characterisation of PVC/rubber nano, micro and macro blends, mechanical properties, dynamic mechanical analysis (DMA), thermo gravimetric analysis (TGA), differential scanning calorimetry (DSC), scanning electron microscopy (SEM), atomic force microscopy (AFM), transition electron microscopy (TEM), bheological measurements, X-ray diffraction (XRD), SAXS and WAXS analysis.

Keywords: PVC, rubber bio nano, characterisation, macro blends, mechanical properties, dynamic mechanical analysis, thermo gravimetric analysis, differential scanning calorimetry, scanning electron microscopy (SEM), atomic force microscopy

Email: shibinsivasakthi@gmail.com

Shiv Kumari Panda and Visakh P. M (eds.) Polyvinyl Chloride-Based Biocomposites and Bionanocomposites, (339–352) © 2026 Scrivener Publishing LLC

11.1 Introduction

Some of the most prominent characteristics of PVC are the relatively low price, high resistance to environmental factors, chemicals, high hardness, and extra tensile strength. The presence of chlorine atom at the molecular structure of the PVC causes the release of HCl and interconnects polymer chains when it is degraded. On the other hand, the presence of chlorine in the molecule helps to protect the polymer against environmental factors such as hydrocarbon fluids. Consequently, unlike some polymers such as polypropylene, degradation of the PVC is not accompanied by severe reduction of strength. PVC has a very high density, which makes it a hard and stiff plastic; therefore, it is mainly used in the construction industry to insulate electrical wires or to produce flooring for houses, schools, and hospitals; packaging; medical devices; food industries; and also all places where a sterilized environment is considered to be a top priority. Use of PVC in some applications is restricted by high glass transition temperatures and brittleness at low temperature. The PVC modification by blending is an inexpensive, above-board, and versatile method to combine the effective properties of different rubbers that can provide a significant increment of physical and mechanical properties [1].

The mentioned rubbers are recognized to indicate numerous properties such as oil resistance, low gas permeability, and high mechanical strength. A wide variety of nano additives are used in the PVC/rubber blend for several purposes, of which the most significant are reduction in material cost and improvement in reinforcement and processing. For example, nitrile rubbers are highly miscible with PVC because of near solubility parameters. They, in blending with PVC, exhibit excellent ozone, oil, and flame resistance. Kaushik *et al.* improved the toughness of PVC using blending with nitrile butadiene rubber (NBR) [2]. Yang investigated probability of toughening of PVC loaded with $CaCO_3$ nanoparticles. Here, silicone rubber was used as an interfacial modifier to reduce incompatibility between polymer–nanoparticle interfaces [3]. Some researches have been performed on PVC blend with natural rubber (NR) and synthetic rubbers such as epoxy nitrile rubber (ENR), styrene–butadiene rubber (SBR), and NBR, which were found miscible at all composition ranges.

On the other hand, incorporation of nano additives into PVC/rubber blend influences rheological, curing, elastomeric, and mechanical properties and attracted great interest of people. It is believed that the interfacial adhesion between polymer phases increased by reducing particle size. Use of small particles leads to internal cavitations of rubber particles, and the blend leads to increase in impact strength [4]. In this chapter, we review the current researches in the field of PVC/rubber nano-blend. The blending preparation methods were determined, and characterization of PVC/rubber nano-blends by different analysis is also discussed in this chapter.

11.2 Preparation of PVC/Rubber Bio Nano

a) Emulsion polymerization is a type of radical addition polymerization that is usually performed in a heterogeneous system containing monomer, surfactant, and water. In this process, monomer droplets are emulsified in continuous water phase called oil-in-water emulation, which is the most common type of emulsion. Also, researchers applied this method for preparation of PVC/rubber-based bionano-blend [5]. Modified PVC composite emulsion resins by rubber polyurethane (PU). In this method, PU emulsion was prepared from toluene di-isocyanate, polypropylene glycol, and dimethylol propionic acid in a four-necked flask equipped with a reflux condenser, a mechanical stirrer, a thermometer, and a nitrogen gas inlet. The continuous water phase comprises soluble polymer particles in water. The particles are in the size range 10- to 1000-nm diameter and are commonly spherical. Emulsion polymerization used to produce several commercial polymers such as PVC, polyvinyl acetate, and rubber neutralizes the carboxy groups to obtain the PU prepolymer, with triethylamine added to reaction mixture. After pouring the PU prepolymer, it dispersed in determined amount of deionized water under high-speed stirring.

b) In suspension polymerization, the monomer is polymerized instead of dissolved in an inert liquid. At the start of the reaction, the monomer-containing initiator is usually dispersed in the continuous phase, and eventually, hard and spherical polymer particles of 50 to 1000 μm are obtained. In the suspension polymerization, one or more insoluble monomers in water, a monomer soluble initiator, and a small amount of stabilizing agent are dispersed by a strong stirrer in a continuously aqueous phase. The mechanical mixing of the contents of the reaction container is necessary to maintain the suspension. For polymers that dissolve in their monomers, interruptions in stirring or mixing excessively or less than the required amount may cause droplets to accumulate during polymerization. If the amount of suspended agent is low, even polymers such as PVC that do not dissolve in its monomer will also adhere together [6, 7]. Uniform dispersion of graphene, $CaCO_3$, and other nanoparticles in a solution of vinyl chloride monomer by *in situ* suspension polymerization is a simple way of obtaining the PVC nanocomposite [8]. During the polymerization process, the droplets imperfectly polymerize and have adhesive state. Furthermore, suspension polymerization for polymers such as elastomers, which are inherently sticky, is not a suitable method, because they would unwittingly adhere together during polymerization.

c) In the mixing process first, the required amount of PVC is premixed using a high-speed mixer to homogeneous solution. In the following, the melt mixing process is carried out in a Brabender Plasti-Corder that has a mixing cam attachment. When, the temperature of melt mixing reaches the desired amount, rubber charges into the mixing chamber and is mixed for a short time. Then, the PVC compound and bionanoparticle are added and mixed to obtain the PVC/rubber nanocomposite.

Prepared PVC/ENR blend with and without CNT nanocomposite using this process. In a typical process, they have prepared a PVC homogeneous solution at high speed. PVC solution was added into the dispersed CNT in ENR solution and mixed in a mix chamber.

d) Polymer solution casting is a manufacturing process used to make thin film, which is typically utilized in membrane processes. The solution casting includes several polymers, which are dissolved in common solvent, coated onto a substrate, and then dried to remove water or solvent. The casting solution method is used to prepare the mixed matrix gas separation membrane filled with fillers. First, the polymers (PVC/SBR) with a ratio of 95:5 are dissolved into a tetrahydrofuran solvent, and the ratio of polymer to solvent is 10:1. The solution is mixed under a mechanical stirrer for more than 10 hours. Then, zeolite particles as inorganic filler additives are dispersed in the polymeric solution. It is worth mentioning that, for breaking up agglomeration of the zeolite particles, the solution was sonicated for about 2 hours using ultrasonic instrument.

e) High-shear mixing uses a high-speed machine that is designed for mixing the fluid with any of the other phases where materials would normally be immiscible. For emulsification, homogenization, particle size reduction, disintegration, and dispersion for many different solid and liquid materials, it is widely used in food, pharmaceutical industries, plastic, cosmetics, adhesion, and chemicals. The fluid speed difference between the outside diameters and the center of the rotor leads to create a tangent force, namely, shear. The use of high-shear mixing has many advantages, such as high mixing speed and usable immiscible materials, and, above all, the mixing time is much shorter than other standard mixer [9].

f) In *in situ* polymerization, it is possible to prepare blend PVC and rubber by melt mixing; it requires high temperature, which often leads to partial PVC degradation. In the synthesis of PVC/rubber blend, *in situ* polymerization is used and obtained by mixing a polymer with miscible monomers that are subsequently polymerized. Parnell *et al.* reported polymerization of rubber PU *in situ* with PVC to produce PVC/rubber PU blend. Then, poly(butylene adipate) (PBA) was added in proportions required for PVC and mixed for 10 minutes to obtain a homogeneous solution. Then, the stoichiometric quantities of 1,4-butanediol (BDO) were added to the PVC/PBA blend. After incorporating BDO into all PVC/PBA blend composition, stoichiometric quantities of 4,4-diphenylmethane di-isocyanate were added to it and mixed, which shows rubber PU formation [10]. Hajibaba and coworkers found that when two polymers are blended by high-shear mixing, the minor phase well is dispersed in major phase. When CNT nanoparticle is incorporated in the PVC phase, the viscosities of the NBR and PVC/SWNT system almost get close together. High-shear mixing is used less in laboratory scale; therefore, few studies have been devoted to prepare the PVC/rubber blend by this method.

g) Most mixing methods used in blending of polymers are shear flow mixing and elongation flow mixing; however, shear flow mixing is not very efficient in obtaining dispersive mixing, because the particles undergo rotation in addition to shearing that results in formation of large clusters. But in the elongation flow, there is no rotation, and the particles undergo a stretching type of deformation. Consequently, tensile force leads to creation of small clusters, in which case this could prove the effectiveness of this method [11]. The operation of the elongation flow is based on circulation flows between two opposite cylindrical chambers that are created by removable static mixing element. In this process, all components discharge into a mixer and are mixed at a controlled speed and presser for preparation of a dry blend of pellets. The determined speed and presser depend on the components' viscosity [12].

h) Extrusion is one of the volumetric methods for polymer deformation to reduce the thickness or cross-section of the material, which results in the production of flat and vertical cross-section. To perform the extrusion process, the primary material is compressed into an extrusion mold whose desired cross-section is formed, and thus, the extruded piece is produced with the molded cross-section. The advantages of this process, in comparison to other forming methods, are high capability in the production of components with a complex cross-section, as well as the formation of brittle materials, because the material in this method is subjected only to stress and shear stress [13].

i) Photo-induced polymerization process is individually used to preparation of PVC and type of rubber, although no studies have been found based on the use of this method for the preparation of PVC/rubber nanocomposites. This method is recognized as a surface modification, and because of the advantages of low cost, simple operation, and creating interconnected layers on the surface, it has been applied to different purposes. It may be attributed to the difficult control of reaction conditions in terms of radiation time and intensity. Someone used UV light for nanoparticle immobilization into a polymer matrix. Nawi *et al.* photoetched out TiO_2 nanoparticles on PVC/ENR50 blend surface and prepared TiO_2-ENR50-PVC composite by dip coating and then composite coated on glass plates. In the end, it exposed UV light to acceptable adhesion of the nanoparticles onto the composite for 45 minutes [14].

11.3 Characterization of PVC/Rubber Nano-, Micro-, and Macro-Blends

11.3.1 Mechanical Properties

In general, the interfacial adhesion between polymers matrix and nanoparticles results from blending of PVC/rubber, which has an effective role in the toughening of polymers. Mechanical properties include hardness, strength, modulus of elasticity and fracture toughness, fatigue behavior, and properties that determine the behavior of materials against the forces involved. The existence of nanoparticles in polymer matrix provides a higher surface for the interaction between nanoparticles and the polymer matrix; as a result, it highlights the reinforcement role of nanoparticle in tensile strength [15]. Therefore, the study of mechanical properties to identify the products produced in terms of strength and stability of materials helps to improve the properties of materials. The result of tensile strength can be interpreted by the molecular length of the organic modifier. Because the length of aliphatic group of the organic modifier increases, the gap size of the organoclay increases and intercalates into layers, hence reinforcing the properties of organoclay is improved [16].

11.3.2 Dynamic Mechanical Analysis

This technique can simply be defined as applying oscillatory force to a sample and analyzing the material response to force. In this technique, the sample is placed in the device, and the suitable frequency and temperature were given into the device as input data. Then, the device enters the sinusoidal stress into the sample at the selected temperature range, and the mechanical response of the material is measured in a controlled temperature. The results obtained were mostly loss tangent or the loss modulus in terms of temperature. The temperature at the maximum point of the tanδ or the peak of the loss modulus curve is used to define the T_g. Dynamic mechanical analysis is more sensitive than other methods, and it can detect small transmission areas that cannot be detected by differential scanning calorimetry (DSC). This method can be used to calculate properties such as viscosity and stiffness (modulus), respectively, from phase delay and sample return [17]. The effect of O-MMT on the dynamical mechanical properties of PVC/XNBR blend was evaluated in the temperature range between −80°C and 100°C. The PVC/XNBR/O-MMT nanocomposite showed higher storage modules in comparison with the blend without O-MMT. Also, the storage modules decreased with increasing temperature, it demonstrates that O-MMT has the ability to absorb heat and retard the softening of the polymers.

The presence of two damping peaks corresponds to the T_{gs} of individual polymers for immiscible polymer blends. For a miscible polymer blend, the curves show only a single peak between the T_g of the pure components. The tanδ curves of NBR and PVC show peaks at −15°C and 90°C, which correspond to the T_g of NBR and PVC, respectively. But in case of the PVC/NBR blend, the tanδ curve lies in between (24°C) the two component polymers showing miscibility [18].

11.3.3 Thermogravimetric Analysis

In this test, thermal stability applies to the sample according to the specified temperature schedule. When the sample decomposes during heating or reacted with gas environment, it provides useful information. In addition, thermal stability, degradation properties, and kinetics of the reaction can also be investigated [19]. Factors such as sample mass, volume, shape, physical properties, and heating velocity can all have an effect on the curve obtained. The effect of cross-link density on thermal stability of ENR/PVC blend–filled SiO_2 was studied using thermogravimetric analysis measurement; increasing the cross-link density increased the rigidity of the system, which in turn improved the thermal stability [2]. Incorporation of nanoparticles in rubber/PVC composite improves thermal stability. Jon *et al.* compared thermal stability of epoxidized NR/PVC with 5% SiO_2 nanoparticles and triethoxysilane (TEOS). Two distinctive degradation stages occurred at 50°C to 650°C. The first stage in the range of 250°C to 350°C is related to degradation of PVC, whereas another stage occurred in the range of 370°C to 500°C corresponding to the decomposition of polyene chains in ENR. It seems that the decomposition temperature of the ENR/PVC blend in the presence of TEOS and SiO_2 is higher than that of the unfilled ENR/PVC blend, which ascribed to the strong interaction of TEOS with epoxide groups in ENR [20].

11.3.4 Differential Scanning Calorimetry

This test provides useful information such as melting point, glass transitions, thermal capacity, crystallization temperature, and the enthalpy changes. In this test, the energy required to establish a zero-temperature difference between the material and reference sample is measured as a function of temperature or time. The specimen should be well condensed in the container and fitted to the container in a thermal contact to minimize the number of holes between the sample particles. Thus, the effect of low thermal conductivity on DSC results is negligible. High-purity N_2 gas is the most commonly used. The nature of the N_2 gas has an important impact on the shape of the DSC peak [21]. The existence of a single glass transition temperature in the DSC thermograms shows the criterion of component compatibility in polymer blends. Kaushik *et al.* utilized the DSC technique to study the thermal response of the nitrile PVC/NR blends containing different types of carbon black; samples show the single T_g at around −60°C, but in higher concentrations of fillers, T_g shifted to the lower side, which can be explained by the dilution effects [2]. The effect of zeolite loading on the thermal behavior of PVC/SBR blend was investigated by Moghadassi *et al.* The remarkable molecular interactions between the particle and polymer result from well dispersion of particles into the polymer matrix. Another reason is dissimilarity between organic and inorganic components properties [22].

11.3.5 Scanning Electron Microscopy

This technique is based on the scanning of the sample surface by interacting electron beam with atoms of the sample. When the electron beam meets with the surface, the interaction occurs between them; consequently, emitted beams received and identified using a detector that resulted in detection of sample characteristics. The type of obtained information from this method depends on the interaction of the beam and the sample. The most important beams used for the sample study include secondary electrons and back-scattering electrons. Secondary electrons resulted from atoms excited that were applied for investigating the surface topography and morphology, while chemical and structural analysis was performed by scattering electrons [23]. The scanning electron microscopy (SEM) observations showed that most NBR-ENP/unreacted Na-MMT were relieved due to its structure and higher crosslinking degree on surface dispersed into PVC matrix. Organic montmorillonite modified with quaternary ammonium (O-MMT) is incorporated into PVC/carboxylated NBR (PVC/XNBR) composite, and SEM analysis was used for the control sample in terms of dynamically cured filled O-MMT counterpart; it is clear that O-MMT filled all free spaces into the PVC matrix and increased interaction between the composite components, a positive influence of dynamic curing. However, it is difficult to separately detect the XNBR or PVC phase using the SEM images. Also, SEM analysis is used to compare tensile fracture surfaces of the different types of PVC/NBR composite by the change of the surface nature from rough to smooth. In this case, to improve the compatibility of PVC/NBR blends, maleic anhydride (MAH) was used. The morphology of the surface of PVC/NBR blends in the presence of MAH showed many tear lines with different sizes clearly different from those of commonly used composites. The existence of MAH resulted in increase of the interfacial interaction between PVC and NBR phases, thus improving compatibility and hence enhancing mechanical resistance [24].

11.3.6 Atomic Force Microscopy

Microscopes cannot be used to observe samples in the size of small molecules and atoms, because these samples have nano dimensions, and they are capable to show in the micrometer sizes. Therefore, the more accurate and advanced tools should be used to observe nano samples. One of these tools is the atomic force microscopy (AFM) technique, which gives information in an indirect way using tweezers and cubes made of a single diamond atom. This microscope, in addition to being used for nanolithography and nanostructures and nanomechanical production, can also be used to study mechanical properties, abrasion or scratches, and surface roughness. In a static state, the cantilever is located at a low distance from the sample surface. In this case, the force between the cantilever and the specimen is a repulsive force. Static mode works with two fixed height and constant force modes. In dynamic mode, the cantilever resonant frequency can be used as a force change criterion (or distance change needle to sample). In this case, the atomic force between the cantilever and the sample is a type of gravity. In dynamic mode, due to the lack of contact with soft samples, no damage is caused, but there is less scanning speed than the static mode [25, 26].

Utilized the AFM technique to evaluate the influence of dynamically vulcanized PVC/NBR blends on surface morphology. The dark and bright regions related to the valleys and hills of the surface helped to identify the phase-separated regions. It was observed that the particle size became higher with increasing the NBR content in the blends induced by aggregating of the NBR particles.

11.3.7 Transition Electron Microscopy

The basis of transition electron microscopy (TEM) performance is similar to optical electron microscope, with the difference that the light is used instead of electron beam. Therefore, TEM magnification is thousands times higher than this one. TEM images are formed first from the interaction of the electrons with the samples when the electron beams are transmitted through the ultrathin sample. Then, image is focused and magnified onto fluorescent screen or detected by spatial camera (charge-coupled device camera). On the other hand, this microscope can be used for elemental analysis, crystal structures, symmetry, orientation, and crystalline defects. These mentioned cases lead to TEM being recognized as the most important equipment in many advanced researches such as chemistry, physics, crystallography, material science, and biology [27, 28].

11.3.8 Rheological Measurements

Rheology science is the study of flow and deformation of fluids under the influence of the stress field applied to them. Depending on the viscosity changes against the applied stresses, fluids divide into Newton and non-Newton classes. One of the most prominent rheological properties of fluids is viscosity, which is a criterion of the resistance of fluids to flow or their deformation under the influence of the stress field applied to them. A rheometer is the laboratory test used to measure some of the increase parameters, including dynamic viscosity, shear rate, and shear stress. In a rheometer, the specimen in a chamber is placed on a small disk and baked at a certain temperature and pressure. Along with the changes made during the baking process in the rubber sample, the stiffness and resistance of the specimen will fluctuate until the baking is complete, and the force required moving the disc reaches the maximum value. The result of the rheometer test is the torque curve in terms of time, which with the help of this curve can analyze the baking features of a rubber compound [29].

The changes in stabilization torque values of PVC/NBR blend with clay content were studied by Mousa *et al.*; with further addition of clay, torque values sharply increased. The increase in torque values could be explained by uniform dispersion of the clay layers into matrix and the strong interaction between the amine and hydroxyl functionalities in the clay structure and the polar PVC/NBR matrix [30].

11.3.9 X-Ray Diffraction

The X-ray diffraction (XRD) patterns are generated by destructive intervention of the single-colored x-ray beam spread at special angles from each group of network surfaces in the specimen. The intensity of the peaks is determined by the distribution of atoms in the network. As a result, the pattern of XRD is a fingerprint of periodic atomic arrangements in a material [31]. XRD technique can be used for realizing the presence of nano additives in the polymer matrix and also the structural changes, creating pending intercalation of polymeric chains. The intercalation of the polymer chains into the silicate layers leads to a decrease in intensity and the broadening of peaks, and eventually, the stacks of layered silicates become more exfoliated [32].

11.3.10 SAXS and WAXS Analysis

Wide-angle XRD and small-angle X-ray scattering (SAXS) are x-ray-diffraction methods where X-rays are elastically scattered by polymeric composite containing nanomaterials at small angles and wide angles; the distance difference between sample and detector and structural information of polymers between 1 and 200 nm obtained by SAXS, when filler is added into the PVC/rubber blend—these techniques provide information about the shape, size, distribution, and surface-to-volume ratio of dispersed nano additives in the polymeric matrix and also determine the crystalline structure of polymers [33].

11.4 Conclusions

From the review studies on PVC-based blends, it is found that nanomaterials have significant interaction with polymer matrix, with increase of nanomaterials loading in the PVC/rubber nano-blend considerably improving strength and stability. From dynamical mechanical analysis, it was confirmed that the presence of a single peak in between the T_g of the pure components shows miscibility of the polymer blend. The uniform distribution of nanomaterials into the polymer matrix has been shown to reduce stress concentration, thereby affecting the results of mechanical properties. The study of anhydride additives' influence on rheological properties shows that the torque difference for NBR/PVC blends containing additive was higher compared with the virgin blends due to the strong interaction and interfacial adhesion between phases.

AI disclosure statement: The author acknowledges the use of AI-based tools for minor language editing or grammatical correction. However, all intellectual contributions, analyses, and conclusions presented in this chapter are entirely the author's own.

References

1. Liu, C., Luo, Y., Jia, Z., Particle configuration and properties of poly (vinyl chloride)/halloysite nanotubes nanocomposites *via in situ* suspension polymerization. *Polym. Compos.*, 35, 856–863, 2014.
2. Kaushik, P., Rajasekar, R., Kang, D.J., Effect of fillers and nitrile blended PVC on natural rubber/high styrene rubber with nanosilica blends: morphology and wear. *Mater. Des.*, 31, 25–34, 2010.
3. Yang, L., Toughening and reinforcement of rigid PVC with silicone rubber/nano-$CaCO_3$ shell-core structured fillers. *Appl. Polym. Sci.*, 102, 2560–2567, 2006.
4. Liu, Z. and Zhu, X., Effects of interfacial adhesion on the rubber toughening of poly (vinyl chloride). *Polymer*, 42, 737–746, 2001.
5. Yamak, H.B., Emulsion polymerization: effects of polymerization variables on the properties of vinyl acetate based emulsion polymers, in: *Polymer Science*, 2013.
6. Kotoulas, A Generalized Population Balance Model for the Prediction of Particle Size Distribution in Suspension Polymerization Reactors. *Chem. Eng. Sci.*, *61*, 332–346, 2006.
7. Arshady, R., Suspension, emulsion, and dispersion Polymerization: A Methodological Survey. *Colloid Polym. Sci.*, *270*, 717–732, 1992.
8. Wang, H., Xie, G., Fang, M., Ying, Z., Tong, Y., Mechanical reinforcement of graphene/poly(vinyl chloride) composites prepared by combining the *in-situ* suspension polymerization and melt-mixing methods. *Compos. Part B*, 113, 278–284, 2017.
9. Krug, P.A., Get your slurry in a hurry with high shear mixer. *Eng. Live*, 2009.
10. Parnell, S. and Min, K., Reactive blending of thermoplastic polyurethane *in situ* with poly (vinyl chloride). *Polym. Eng. Sci.*, 45, 876–887, 2005.
11. Rauwendaal, C., Improve mixing by generating elongational flow, Rauwendaal extrusion engineering, 2015.
12. Bouquey, M. and Loux, C., Morphological Study of Two-Phase Polymer Blends During Compounding in a Novel Compounder on the Basis of Elongational Flows. *Appl. Polym. Sci.*, 119, 482–490, 2011.

Also of Interest

Other Scrivener Publishing books by Visakh P. M

Polyvinyl Chloride (PVC)-Based Biocomposites and Bionanocomposites
Edited by Shiv Kumari Panda and Visakh P. M
Published 2026. ISBN 978-1-394-21495-2

Marine Biotechnology
Edited by Shaju S. S., Vipin. P. M. and Visakh P. M
Published 2026. ISBN 978-1-394-30129-4

Polyvinyl Alcohol-Based Biocomposites and Bionanocomposites
Edited by Visakh P. M and Olga B. Nazarenko
Published 2023. ISBN 978-1-119-5920-9

Metal Oxide Nanocomposites
Synthesis and Applications
Edited by B. Raneesh and Visakh P. M
Published 2021. ISBN 978-1-119-36357-6

High Performance Polymers and Their Nanocomposites
Edited by Visakh P. M and Semkin A. O.
Published 2019. ISBN 978-1-119-36365-1

Polypropylene-Based Biocomposites and Bionanocomposites
Edited by Visakh P. M and Matheus Poletto
Published 2018. ISBN 978-1-119-28356-0

Soy Protein-Based Blends, Composites and Nanocomposites
Edited by Visakh P. M and Olga Nazarenko
Published 2017. ISBN 978-1-119-41830-6

Polyethylene-Based Biocomposites and Bionancomposites
Edited by Visakh P. M and Sigrid Lüftl
Published 2016. ISBN 978-11903845-0

Nanostructured Polymer Membranes
Volume 1: Processing and Characterization
Edited by Visakh P. M and Olga Nazarenko
Published 2016. ISBN 978-1-118-83173-1

Nanostructured Polymer Membranes
Volume 2: Applications
Edited by Visakh P. M and Olga Nazarenko
Published 2016. ISBN 978-1-118-83178-6

Polyethylene-Based Blends, Composites and Nanocomposites
Edited by Visakh P. M and María José Martínez Morlanes
Published 2015. ISBN 978-1-118-83128-1

Polyoxymethylene Handbook
Structure, Properties, Applications and Their Nanocomposites
Edited by Sigrid Lüftl, Visakh P. M and Sarath Chandran
Published 2014. ISBN 978-1-118-38511-1

Advances in Food Science and Technology
In two volumes
Edited by Visakh P. M, Sabu Thomas, Laura B. Iturriaga and Pablo Daniel Ribotta
Published 2013. ISBN 978-1-118-12102-3

Handbook of Engineering and Specialty Thermoplastics 4
Volume 4: Nylons
Edited by Sabu Thomas and Visakh P. M
Published 2012. ISBN 978-0-470-63925-2

Handbook of Engineering and Specialty Thermoplastics 3
Volume 3: Polyethers and Polyesters
Edited by Sabu Thomas and Visakh P. M
Published 2012. ISBN 978-0-470-63926-9

www.scrivenerpublishing.com